Two Manuals Combined into One Book

Military Intelligence

And

Security Force Assistance

Department of Defense

Pentagon
Publishing

Change Number 1

Headquarters
Department of the Army
Washington, DC, 17 May 2004

Intelligence

Contents

*This publication supersedes FM 34-1, 27 September 1994.

Preface

FM 2-0 is the Army's keystone manual for military intelligence (MI) doctrine. It describes—

- The fundamentals of intelligence operations.
- The operational environment (OE).
- Intelligence in unified action.
- The Intelligence Battlefield Operating System (BOS).
- Intelligence considerations in strategic readiness.
- The intelligence process.
- MI roles and functions within the context of Army operations.

This manual conforms to the overarching doctrinal precepts presented in FM 3-0.

This manual provides doctrinal guidance for the Intelligence BOS actions in support of commanders and staffs. It also serves as a reference for personnel who are developing doctrine; tactics, techniques, and procedures (TTP); materiel and force structure; and institutional and unit training for intelligence operations.

This manual provides MI guidance for all commanders, staffs, trainers, and MI personnel at all echelons. It forms the foundation for MI and the Intelligence BOS doctrine development, and applies equally to the Active Component (AC), United States Army Reserve (USAR), and Army National Guard (ARNG). It is also intended for commanders and staffs of joint and combined commands, US Naval and Marine Forces, units of the US Air Force, and the military forces of multinational partners.

Headquarters, US Army Training and Doctrine Command is the proponent for this publication. The preparing agency is the US Army Intelligence Center and School. Send written comments and recommendations on DA Form 2028 (Recommended Changes to Publications and Blank Forms) directly to: Commander, ATZS-FDT-D (FM 2-0), 550 Cibeque Street, Fort Huachuca, AZ 85613-7017. Send comments and recommendations by e-mail to *ATZS-FDC-D@hua.army.mil*. Follow the DA Form 2028 format or submit an electronic DA Form 2028.

Unless otherwise stated, masculine nouns and pronouns do not refer exclusively to men.

This manual contains Army tactical task (ART) description taken verbatim from the *Army Universal Task List* (AUTL). ART task descriptions are followed by a reference to FM 7-15 in parentheses and the ART number; for example (See FM 7-15, ART 5.1.1.).

PART ONE

Intelligence in the Operational Environment

Part One discusses MI's role in peace, conflict, and war. Supporting the warfighter with effective intelligence is the primary focus of Military Intelligence. Intelligence provides commanders and decisionmakers with the requisite information facilitating their situational understanding so that they may successfully accomplish their missions in full spectrum operations.

Chapter 1 describes the operational environment and the roles of MI within the operational environment. It introduces the Intelligence warfighting function, the intelligence tasks, and the intelligence process, which are the mechanisms through which MI supports the warfighter. This chapter also introduces the intelligence disciplines, which are explained in detail in Part Three of this manual.

Chapter 2 describes the interaction of MI within the nation's intelligence community structure, providing an overview of the intelligence community at the national level and the unified action level—joint, multinational, and interagency aspects of full spectrum operations. This chapter also discusses the concepts and components of intelligence reach.

Chapter 1

Intelligence and the Operational Environment

ROLE OF INTELLIGENCE

1-1. The commander requires intelligence about the enemy and the area of operations (AO) prior to engaging in operations in order to effectively execute battles, engagements, and other missions across the full spectrum of operations. Intelligence assists the commander in visualizing the AO, organizing the forces, and controlling operations to achieve the desired tactical objectives or end-state. Intelligence supports force protection (FP) by alerting the commander to emerging threats and assisting in security operations.

1-2. The unit may need to deal with multiple threats. The commander must understand how current and potential enemies organize, equip, train, employ, and control their forces. Intelligence provides an understanding of the enemy, which assists in planning, preparing, and executing military operations. The commander must also understand the operational environment and its effects on both friendly and enemy operations. The commander receives mission-oriented intelligence on enemy forces and the

AO from the G2/S2. The G2/S2 depends upon the intelligence, surveillance, and reconnaissance (ISR) effort to collect and provide information on the enemy and AO.

1-3. One of the most significant contributions that intelligence personnel can accomplish is to accurately predict future enemy events. Although this is an extremely difficult task, predictive intelligence enables the commander and staff to anticipate key enemy events or reactions and develop corresponding plans or counteractions. Intelligence analysis conclusions, assessments, or products that attempt to define, describe, present, or portray the future situation or conditions of the enemy, terrain, weather, and civil considerations.

1-4. The most important purpose of intelligence is to influence decisionmaking. Commanders must receive the intelligence, understand it (because it is tailored to the commander's requirements), believe it, and act on it. Through this doctrinal concept, intelligence drives operations.

THE INTELLIGENCE WARFIGHTING FUNCTION

1-5. *The intelligence warfighting function is the related tasks and systems that facilitate understanding of the operational environment.* (FM 3-0) It includes tasks associated with intelligence, surveillance, and reconnaissance (ISR) operations, and is driven by the commander. Intelligence is more than just collection. It is a continuous process that involves analyzing information from all sources and conducting operations to develop the situation. The intelligence warfighting function includes the following tasks:

- Support to force generation.
- Support to situational understanding.
- Perform intelligence, surveillance, and reconnaissance.
- Support to targeting and information operations.

1-6. The intelligence warfighting function is one of six warfighting functions—movement and maneuver, intelligence, fires, sustainment, command and control, and protection. *A warfighting function is a group of tasks and systems (people, organizations, information, and processes) united by a common purpose that commanders use to accomplish missions and training objectives* (FM 3-0). (See FM 3-0, chapter 4, for a detailed discussion of the warfighting functions.) The intelligence warfighting function is a flexible force of personnel, organizations, and equipment that, individually or collectively, provide commanders with the timely, relevant, and accurate intelligence required to visualize the battlefield, assess the situation, and direct military actions. Additionally, the intelligence warfighting function is—

- A complex system that operates worldwide, from "mud-to-space," in support of an operation, to include the ability to leverage theater and national capabilities.
- Requires cooperation and division of labor internally, higher, lower, adjacent, and across components and the coalition.

1-7. The intelligence warfighting function not only includes assets within the MI branch but also includes the assets of all branches or warfighting functions that conduct intelligence warfighting function tasks. Every Soldier, as a part of a small unit, is a potential information collector and an essential component to help reach situational understanding. Each Soldier develops a special level of awareness simply due to exposure to events occurring in the AO and has the opportunity to collect and report information by observation and interaction with the population (see paragraph 3-3).

1-8. Planning and executing military operations will require intelligence regarding the threat (traditional, irregular, catastrophic, and disruptive) and the AO. The intelligence warfighting function generates intelligence and intelligence products that portray the enemy and aspects of the environment. These intelligence products enable the commander to identify potential courses of action (COAs), plan operations, employ forces effectively, employ effective tactics and techniques, and implement protection.

1-9. The intelligence warfighting function is always engaged in supporting the commander in offensive, defensive, stability, and, when directed, civil support operations. Intelligence provides products that are timely, relevant, accurate, and predictive. Hard training, thorough planning, meticulous preparation, and aggressive execution posture the Army for success. In the current environment we must maintain intelligence readiness to support operations on "no notice." This support is comprehensive and reaches across full spectrum operations and levels of war to produce the intelligence required to successfully accomplish the mission. A combination of space, aerial, seaborne, and ground-based systems provide the most comprehensive intelligence possible. During force projection operations, the intelligence warfighting function supports the commander with accurate and responsive intelligence from predeployment through redeployment.

1-10. The intelligence warfighting function architecture provides specific intelligence and communications structures at each echelon from the national level through the tactical level. These structures include intelligence organizations, systems, and procedures for collecting, processing, analyzing, and delivering intelligence and other critical information in a useable form to those who need it, when they need it. Effective communications connectivity and automation are essential components of this architecture.

INTELLIGENCE TASKS (METL)

1-11. The personnel and organizations within the Intelligence warfighting function conduct four primary intelligence tasks that facilitate the commander's visualization and understanding of the threat and the AO. These tasks are interactive and often take place simultaneously. (Refer to FM 7-15 for the complete subordinate task listing.) Figure 1-1 shows these tasks tailored to the commander's needs.

INTELLIGENCE TASKS	COMMANDER'S FOCUS	COMMANDER'S DECISIONS
Support to Situational Understanding - Perform IPB - Perform Situation Development - Provide Intelligence Support to Force Protection - Conduct Police Intelligence Operations	Plan a mission. Execute the operation. Secure the force.	Which COA will I implement? Which enemy actions are expected?
Support to Strategic Responsiveness - Perform I&W - Ensure Intelligence Readiness - Conduct Area Studies of Foreign Countries - Support Sensitive Site Exploitation	Orient on contingencies.	Should I increase the unit's level of readiness? Should I implement the OPLAN?
Conduct Intelligence, Surveillance, and Reconnaissance - Perform Intelligence Synchronization - Perform ISR Integration - Conduct Tactical Reconnaissance - Conduct Surveillance	Plan the mission. Prepare. Execute. Assess.	Which DPs, HPTs, etc., are linked to the enemy actions? Are the assets available and in position to collect on the DPs, HPTs, etc.? Have the assets been repositioned for contingency mission?
Provide Intelligence Support to Effects - Provide Intelligence Support to Targeting - Provide Intelligence Support to IO - Provide Intelligence Support to Combat Assessment	Destroy/suppress/ neutralize targets. Reposition intelligence or attack assets.	Is my fire (lethal or non-lethal) and maneuver effective? Should I refire the same targets?

Figure 1-1. Intelligence Tailored to the Commander's Needs.

SUPPORT TO SITUATIONAL UNDERSTANDING

1-12. This task centers on providing information and intelligence to the commander, which facilitates the commander's understanding of the enemy and the environment. It supports the command's ability to make sound decisions. Support to situational understanding comprises four subtasks: perform intelligence preparation of the battlefield (IPB), perform situation development, provide intelligence support to FP, and conduct police intelligence operations.

Perform Intelligence Preparation of the Battlefield

1-13. The G2/S2 is the staff proponent for IPB. IPB is the staff planning activity undertaken by the entire staff to define and understand the battlefield and the options it presents to friendly and threat forces. IPB includes input from the whole staff. There is only one IPB in each headquarters with inputs from all affected staff cells; they are not separate warfighting function or staff section IPBs throughout the headquarters. It is a systematic process of analyzing and visualizing the operational environment in a specific geographic area for a specific mission or in anticipation of a specific mission. By applying IPB, the commander and staff gain the information necessary to selectively apply and maximize combat power at critical points in time and space. IPB is most effective when it integrates each staff element's expertise into the final products. To conduct effective IPB, the G2/S2 must—

- Produce IPB products that support the staff's preparation of estimates and the military decision-making process (MDMP).

- Identify characteristics of the AO, including the information environment, that will influence friendly and threat operations.
- Establish the area of interest (AOI) in accordance with the commander's guidance.
- Identify gaps in current intelligence holdings.
- Determine multiple enemy COAs (ECOAs) by employing predictive analysis techniques to anticipate future enemy actions, capabilities, or situations.
- Establish a database that encompasses all relevant data sets within and related to the operational environment.
- Determine the enemy order of battle (OB), doctrine, and TTP. Identify any patterns in enemy behavior or activities.
- Accurately identify and report hazards within the AO, including the medical threat and toxic industrial material (TIM).
- Accurately identify threat capabilities, high-value targets (HVTs), and threat models.
- Integrate IPB information into COA analysis and the MDMP.
- Update IPB products as information becomes available.

Perform Situation Development

1-14. Situation development is a process for analyzing information and producing current intelligence about the enemy and environment during operations. The process helps the intelligence officer recognize and interpret the indicators of enemy intentions, objectives, combat effectiveness, and potential ECOAs. Situation development—

- Confirms or denies threat COAs.
- Provides threat locations.
- Explains what the threat is doing in relation to the friendly force operations.
- Provides an estimate of threat combat effectiveness.

1-15. Through situation development, the intelligence officer is able to quickly identify information gaps, and explain enemy activities in relation to the unit's own operations, thereby assisting the commander in gaining situational understanding. Situation development helps the commander make decisions and execute branches and sequels. This reduces risk and uncertainty in the execution of the plan. The intelligence officer maintains, presents, and disseminates the results of situation development through intelligence input to the common operational picture (COP) and other intelligence products.

Provide Intelligence Support to Force Protection

1-16. Provide intelligence in support of protecting the tactical force's fighting potential so that it can be applied at the appropriate time and place. This task includes the measures the force takes to remain viable and functional by protecting itself from the effects of or recovery from enemy activities. FP consists of those actions taken to prevent or mitigate hostile actions against Department of Defense (DOD) personnel (to include Department of Army

[DA] civilians, contractors, uniformed personnel, and family members), resources, facilities, and critical information. These actions—

- Conserve the force's fighting potential for application at the decisive time and place.
- Incorporate coordinated and synchronized offensive and defensive measures.
- Facilitate the effective employment of the joint force while degrading the capabilities of and opportunities for the threat.

1-17. Intelligence to FP consists of monitoring and reporting the activities, intentions, and capabilities of adversarial groups and determining their possible COAs. Detecting the adversary's methods in today's OE requires a higher level of situational understanding, informed by current and precise intelligence. This type of threat drives the need for predictive intelligence based on analysis of focused information from intelligence, law enforcement, and security activities.

1-18. Intelligence analysis in support of FP employs analytical methodologies and tools to provide situational understanding and to predict the adversary's actions. Modified or standard time-event charts, association matrixes, activity matrixes, link diagrams, and overlays are beneficial in monitoring the actions of the adversary. Overlays may include (but are not limited to) threat training camps, organizations, finances, personalities, industrial sites, information systems, decisionmaking infrastructures, specific activities, and locations of previous attacks.

Conduct Police Intelligence Operations

1-19. Police intelligence operations (PIO) are a military police (MP) function that supports, enhances, and contributes to the commander's force protection program, COP, and situational understanding. The PIO function ensures that information collected during the conduct of other MP functions—maneuver and mobility support, area security, law and order, and internment and resettlement—is provided as input to the intelligence collection effort and turned into action or reports. PIO has three components: (See FM 7-15, ART 1.1.4)

- Collect police information.
- Conduct Police Information Assessment Program (PIAP).
- Develop police intelligence products.

Notes:

US Code, Executive Orders, DOD Directives, and Army Regulations contain specific guidance regarding prohibition on the collection of intelligence information on US citizens, US corporations, and resident aliens. These laws and regulations include criminal penalties for their violation. Any PIO directed against US citizens should undergo competent legal review prior to their initiation.

The inclusion of the PIO task branch in the Intelligence warfighting function does not change the intelligence process described in this manual.

SUPPORT TO STRATEGIC RESPONSIVENESS

1-20. Intelligence support to strategic responsiveness supports staff planning and preparation by defining the full spectrum of threats, forecasting future threats, and forewarning the commander of enemy actions and intentions. Support to strategic responsiveness consists of four subtasks: Perform I&W, ensure intelligence readiness, conduct area studies of foreign countries, and support sensitive site exploitation. (See FM 7-15, ART 1.2)

Perform Indications and Warnings

1-21. This activity provides the commander with forewarning of enemy actions or intentions; the imminence of threat actions. The intelligence officer develops I&W in order to rapidly alert the commander of events or activities that would change the basic nature of the operations. It enables the commander to quickly reorient the force to unexpected contingencies and shape the battlefield. (See FM 7-15, ART 1.2.1)

1-22. The G2/S2 at the operational and strategic levels develops I&W in order to rapidly alert the commander of events or activities that would change the basic nature of the operations so the commander can initiate the appropriate action in a timely manner. I&W reduce the risk of enemy actions that are counter to planning assumptions. I&W enable the commander to quickly reorient the force to unexpected events and to shape the battlefield by manipulating enemy activities.

Ensure Intelligence Readiness

1-23. Intelligence readiness operations support contingency planning and preparation by developing baseline knowledge of multiple potential threats and operational environments. These operations and related intelligence training activities engage the Intelligence warfighting function to respond effectively to the commander's contingency planning intelligence requirements. (See FM 7-15, ART 1.2.2) While still in garrison, intelligence defines the full spectrum of threats and forecasts future threats and dangers. Intelligence readiness operations accomplish the following:

- Provide intelligence to support contingency-based training and staff planning.
- Identify, consider, and evaluate all potential threats to the entire unit.
- Provide a broad understanding of the operational environment of the contingency area, which is developed through continuous exchange of information and intelligence with higher echelon and joint intelligence organizations.

Conduct Area Studies of Foreign Countries

1-24. Study and understand the cultural, social, political, religious, and moral beliefs and attitudes of allied, host nation (HN), or indigenous forces to assist in accomplishing goals and objectives. (See FM 7-15, ART 1.2.3)

Note: The inclusion of this task does not change the support to strategic responsiveness provided by MI organizations described in this manual.

Support Sensitive Site Exploitation

1-25. Sensitive site exploitation consists of a related series of activities inside a sensitive site captured from an adversary. A sensitive site is a designated, geographically limited area with special military, diplomatic, economic, or information sensitivity for the United States. This includes factories with technical data on enemy weapon systems, war crimes sites, critical hostile government facilities, areas suspected of containing persons of high rank in a hostile government or organization, terrorist money laundering, and document storage areas for secret police forces. These activities exploit personnel, documents, electronic data, and material captured at the site, while neutralizing any threat posed by the site or its contents. While the physical process of exploiting the sensitive site begins at the site itself, full exploitation may involve teams of experts located around the world. (See FM 7-15, ART 1.2.4)

Note: The inclusion of this task does not change the support to strategic responsiveness provided by MI organizations described in this manual.

CONDUCT INTELLIGENCE, SURVEILLANCE, AND RECONNAISSANCE

1-26. With staff participation, the intelligence officer synchronizes intelligence support to the ISR effort by focusing the collection, processing, analysis, and intelligence products on the critical needs of the commander. The operations officer, in coordination with the intelligence officer, tasks and directs the available ISR assets to answer the commander's critical information requirements (CCIRs). Through various detection methods and systematic observation, reconnaissance and surveillance obtains the required information. A continuous process, this task has four subtasks: perform intelligence synchronization, perform ISR integration, conduct tactical reconnaissance, and conduct surveillance. (See FM 7-15, ART 1.3, and FM 3-90).

Perform Intelligence Synchronization

1-27. The intelligence officer, with staff participation, synchronizes the entire collection effort to include all assets the commander controls, assets of lateral units and higher echelon units and organizations, and intelligence reach to answer the commander's priority intelligence requirements (PIRs) and information requirements (IRs). (See FM 7-15, ART 1.3.1)

1-28. The intelligence officer, with staff participation, supports the G3/S3 in orchestrating the entire ISR effort to include all assets the commander controls, assets of lateral units and higher echelon units and organizations, and intelligence reach to answer the CCIRs (PIRs and friendly force information requirements [FFIRs]) and other intelligence requirements. Intelligence synchronization activities include the following:

- Conducting requirements management (RM): anticipate, develop, analyze, validate, and prioritize intelligence requirements. Recommend PIRs to the commander. Manage the commander's intelligence requirements, requests for information (RFIs) from subordinate and lateral organizations, and tasks from higher headquarters. Eliminate satisfied requirements and add new requirements as necessary.

- Developing indicators.
- Developing specific IRs (specific information requirements [SIRs]).
- Converting the SIRs into intelligence tasks or ISR tasks. (See Figure 1-2 for the ISR task development process.) The S2/G2 assigns intelligence production and reach tasks to subordinate intelligence elements or personnel, submits RFIs to higher and lateral echelons, and coordinates with (or assists) the G3/S3 to develop and assign ISR tasks.
- Comparing the ISR tasks to the capabilities and limitations of the available ISR assets (in coordination with the operations officer).
- Forwarding SIRs that cannot be answered by available assets to higher or lateral organizations as RFIs.
- Assessing collection asset reporting and intelligence production to evaluate the effectiveness of the ISR effort.
- Maintaining situational understanding to identify gaps in coverage and to identify the need to cue or redirect ISR assets.
- Updating the intelligence synchronization plan. The G2/S2 manages and updates the intelligence synchronization plan as PIRs are answered and new requirements arise.

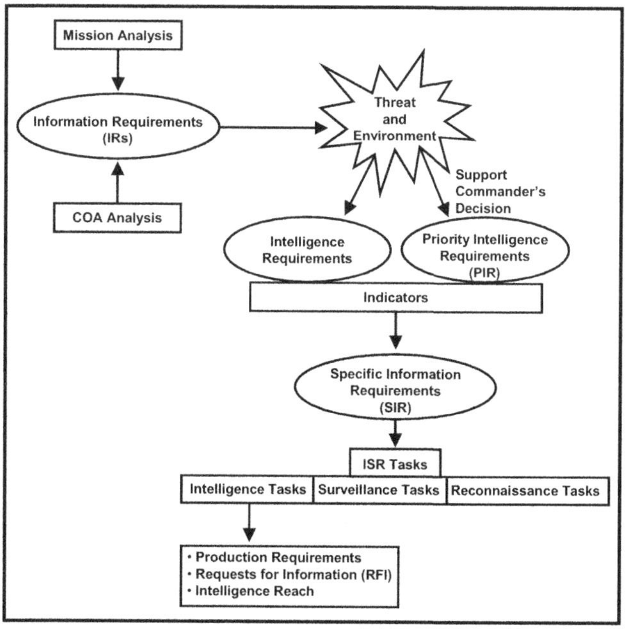

Figure 1-2. ISR Task Development Process.

1-29. **Intelligence Synchronization Considerations.** The G2/S2 generally follows six considerations in planning intelligence synchronization and ISR activities: anticipate, integrate, prioritize, balance, control, and reach. Refer to FM 2-01 for more information regarding intelligence synchronization.

- **Anticipate.** The intelligence staff must recognize when and where to shift collection or identify new intelligence requirements. The overall intent of this principle is to identify a new or adjust an existing requirement and present it to the commander for approval before waiting for the commander or staff to identify it.
- **Integrate.** The intelligence staff must be fully and continuously integrated into the unit's orders production and planning activities to ensure early identification of intelligence requirements. Early and continuous consideration of collection factors enhances the unit's ability to direct collection assets in a timely manner, ensures thorough planning, and increases flexibility in selecting assets.
- **Prioritize.** Prioritize each intelligence requirement based on its importance in supporting the commander's intent and decisions. Prioritization, based on the commander's guidance and the current situation, ensures that limited ISR assets and resources are directed against the most critical requirements.
- **Balance.** ISR capabilities complement each other. The intelligence staff should resist favoring or becoming too reliant on a particular unit, discipline, or system. Balance is simply planning redundancy, when required, eliminating redundancy when not desired, and ensuring an appropriate mix of ISR assets or types. The intelligence synchronization matrix (ISM) is useful in determining or evaluating balance.
- **Control.** To ensure timely and effective responses to intelligence requirements, a unit should first use ISR assets it controls. These assets usually are more responsive to their respective commander and also serve to lessen the burden on the ISR assets of other units, agencies, and organizations.
- **Reach.** Although usually not as responsive as a unit's own assets, intelligence reach may be the only way to satisfy an intelligence requirement. If at all possible, one should not depend solely on intelligence reach to answer a PIR.

1-30. **Develop IRs**. The intelligence staff develops a prioritized list of what information needs to be collected and produced into intelligence. Additionally, the intelligence staff dynamically updates and adjusts those requirements in response to mission adjustments and changes. This list is placed against a latest time intelligence is of value to ensure intelligence and information are reported to meet operational requirements. (See FM 7-15, ART 1.3.1) Figure 1-3 shows a comparison of IR, PIR, and intelligence requirements.

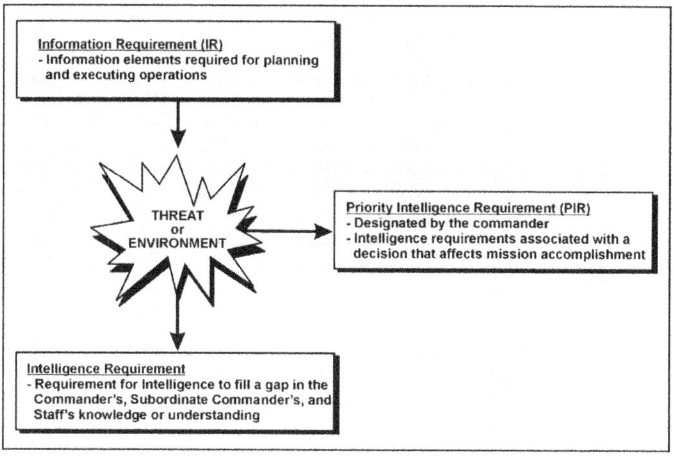

Figure 1-3. IR, PIR, and Intelligence Requirements Comparison.

1-31. An effective discussion of ISR has to include an understanding of the CCIRs. The CCIRs are elements of information required by commanders that directly affect decisionmaking and dictate the successful execution of military operations. The commander decides what information is critical based on the commander's experience, the mission, the higher commander's intent, and the staff's input (initial IPB, information, intelligence, and recommendations). Refer to FM 3-0 for more information regarding CCIRs.

1-32. Based on the CCIRs, two types of supporting IRs are generated: PIRs and FFIRs. However, commanders may determine that they need to know whether one or more essential elements of friendly information (EEFI) have been compromised or that the enemy is collecting against a designated EEFI. In those cases, commanders may designate that question as one of their CCIRs. Figure 1-4 shows the CCIR composition.

1-33. IRs are all of the information elements required by the commander and staff for the successful planning and execution of operations; that is, all elements necessary to address the factors of mission, enemy, terrain and weather, troops and support available, time available, civil considerations (METT-TC). Vetting by the commander or a designated representative turns an IR into either a PIR or an intelligence requirement. IRs are developed during COA analysis based on the factors of METT-TC.

1-34. PIRs are those intelligence requirements for which a commander has an anticipated and stated priority during the task of planning and decisionmaking. PIRs are associated with a decision based on an enemy action or inaction or the AO that will affect the overall success of the commander's mission. **The commander designates intelligence requirements tied directly to decisions as CCIR (PIR and FFIR).**

Answers to the PIRs help produce intelligence essential to the commander's situational understanding and decisionmaking. For information on PIR development, see FM 2-01.

1-35. The G2/S2 recommends to the commander those IRs produced during the MDMP that meet the criteria for PIR. They do not become CCIR (PIR and FFIR) until approved by the commander. Additionally, The commander may unilaterally designate PIRs. The IRs that are not designated by the commander as PIRs become intelligence requirements. The intelligence requirement is a gap in the command's knowledge or understanding of the AO or threat that the Intelligence warfighting function must fill.

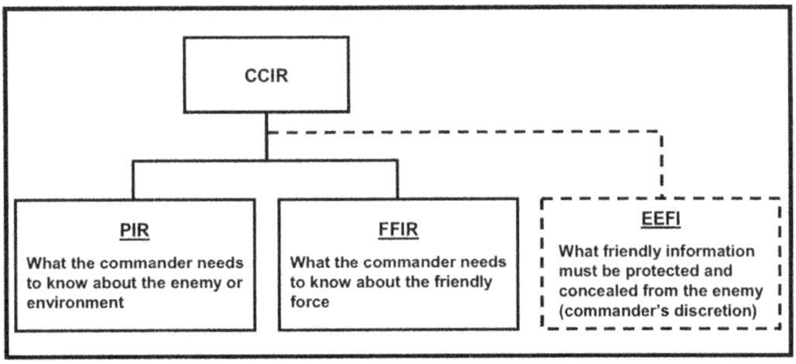

Figure 1-4. Commander's Critical Information Requirements Composition.

1-36. The G3/S3 then tasks the unit's assets to answer both the PIR and intelligence requirements through the ISR plan. PIR should—

- Ask only one question.
- Support a decision.
- Identify a specific fact, event, activity (or absence thereof) which can be collected.
- If linked to an ECOA, indicate an ECOA prior to, or as early as possible in, its implementation.
- Indicate the latest time the information is of value (LTIOV). The LTIOV is the absolute latest time the information can be used by the commander in making the decision the PIR supports. The LTIOV can be linked to time, an event, or a point in the battle or operation.

1-37. **Friendly Force IRs.** The staff also develops FFIRs which, when answered, provide friendly force information that the commander and staff need to achieve situational understanding and to make decisions.

1-38. **Essential Elements of Friendly Information**. EEFI establish information to protect, not information to obtain. However, commanders may determine that they need to know whether one or more EEFI have been compromised or that the enemy is collecting against a designated EEFI. In those cases, commanders may designate that question as one of their CCIRs, which generates PIRs and/or FFIRs. For example, a commander may determine that if the enemy discovers the location and movement of the friendly reserve, the operation is at risk. In this case, the location and movement of the friendly reserve are EEFI. The commander designates determining whether the enemy has discovered the location and movement of the friendly reserve as one of the CCIR. That CCIR, in turn, generates PIR and FFIR to support staff actions in determining whether the EEFI has been compromised.

1-39. **Develop the Intelligence Synchronization Plan.** The entire unit staff develops their IRs and determines how best to satisfy them. The staff uses reconnaissance and surveillance assets to collect information. The intelligence synchronization plan includes all assets that the operations officer can task or request and coordination mechanisms to ensure adequate coverage of the AOIs. (See FM 7-15, ART 1.3.1.2)

1-40. The intelligence synchronization plan, often presented in a matrix format as an ISM, aids in synchronizing the entire ISR effort with the overall operation and the commander's decisions and/or decision points (DPs). The intelligence synchronization plan is often produced in conjunction with the ISR plan. However, before performing intelligence synchronization and finalizing the intelligence synchronization plan, the G2/S2 must have—

- The CCIR (PIR and FFIR).
- A prioritized list of the remaining intelligence requirements.
- Evaluated ISR assets and resources.
- All of the assigned ISR tasks.

Perform ISR Integration

1-41. The operations officer, in coordination with the intelligence officer and other staff members, orchestrates the tasking and directing of available ISR assets to answer the CCIR. The operations officer, with input from the intelligence officer, develops tasks from the SIRs which coincide with the capabilities and limitations of the available ISR assets and the latest time information is of value (LTIOV). Intelligence requirements are identified, prioritized, and validated and an ISR plan is developed and synchronized with the scheme of maneuver. (See FM 7-15, ART 1.3.2, and FM 3-90)

1-42. The G3/S3, in coordination with the G2/S2 and other staff members, orchestrates the tasking and directing of available ISR assets to answer the CCIRs (PIRs and FFIRs) and IRs. The result of this process is the forming of the ISR plan. The ISR plan provides a list of all the ISR tasks to be accomplished. The G2/S2 and the G3/S3 develop tasks from the SIRs. These tasks are then assigned based on the capabilities and limitations of the available ISR assets and the LTIOV.

1-43. **Develop the ISR Plan.** The operations officer is responsible for developing the ISR plan. The entire unit staff analyzes each requirement to determine how best to satisfy it. The staff will receive orders and RFIs from both subordinate and adjacent units and higher headquarters. The ISR plan includes all assets that the operations officer can task or request and coordination mechanisms to ensure adequate coverage of the area of interest. (See FM 7-15, ART 1.3.2.1, and FM 3-90)

1-44. **The ISR Plan**. The ISR plan is usually produced as the ISR Annex to an OPORD (Annex L, Intelligence Surveillance, and Reconnaissance). Refer to FM 5-0 for specific and authoritative information on the ISR Annex. ISR is a continuous combined arms effort led by the operations and intelligence staffs in coordination with the entire staff that sets reconnaissance and surveillance in motion. The PIRs and other intelligence requirements drive the ISR effort. The commander takes every opportunity to improve situational understanding and the fidelity of the COP about the enemy and terrain through the deployment of ISR assets. Commanders integrate reconnaissance and surveillance to form an integrated ISR plan that capitalizes on their different capabilities. The ISR plan is often the most important part of providing information and intelligence that contributes to answering the CCIRs. For the G2/S2, an effective ISR plan is critical in answering the PIR. Also see FM 3-55 for more information on the ISR plan.

1-45. The ISR plan is not an MI-specific product—the G3/S3 is the staff proponent of the ISR plan— it is an integrated staff product executed by the unit at the direction of the commander. The G2/S2, however, must maintain situational understanding in order to recommend to the commander and G3/S3 changes or further development of the ISR plan. Based on the initial IPB and CCIRs, the staff—primarily the G2/S2—identifies gaps in the intelligence effort and develops an initial ISR plan based on available ISR assets. The G3/S3 turns this into an initial ISR Annex that tasks ISR assets as soon as possible to begin the collection effort.

1-46. The G3/S3, assisted by the G2/S2, uses the ISR plan to task and direct the available ISR assets to answer the CCIRs (PIRs and FFIRs) and intelligence requirements. Conversely, the staff revises the plan as other intelligence gaps are identified if the information is required to fulfill the CCIRs or in anticipation of future intelligence requirements. With staff participation, the G2/S2 intelligence officer synchronizes the ISR effort through a complementary product to the ISR plan—the intelligence synchronization plan.

1-47. **Execute and Update the ISR Plan**. The operations officer updates the ISR plan based on information received from the intelligence officer. The operations officer is the integrator and manager of the ISR effort through an integrated staff process and procedures. As PIRs are answered and new information requirements arise, the intelligence officer updates intelligence synchronization requirements and provides the new input to the operations officer who updates the ISR plan. He works closely with all staff elements to ensure the unit's organic collectors receive appropriate taskings. This ISR reflects an integrated collection strategy and employment, production and dissemination scheme that will effectively answer the commander's PIR. (See FM 7-15, ART 1.3.2.2, and FM 3-90)

Conduct Tactical Reconnaissance

1-48. To obtain, by visual observation or other detection methods, such as signals, imagery, measurement of signature or other technical characteristics, human interaction and other detection methods about the activities and resources of an enemy or potential enemy, or to secure data concerning the meteorological, hydrographic, or geographic characteristics and the indigenous population of a particular area. This task includes the conduct of NBC reconnaissance and the tactical aspects of SOF special reconnaissance. The five subtasks are—

- Conduct a Zone Reconnaissance.
- Conduct an Area Reconnaissance.
- Conduct a Reconnaissance in Force.
- Conduct a Route Reconnaissance.
- Conduct a Reconnaissance Patrol.

Note: This task branch includes techniques by which ART 1.1.4.1 (Collect Police Information) may be performed. (FM 7-15)

1-49. Reconnaissance is a mission undertaken to obtain by visual observation or other detection methods, information about the activities and resources of an enemy or potential enemy, and about the meteorological, hydrographic, or geographic characteristics of an AO. MI personnel and organizations can conduct reconnaissance through obtaining information derived from signals, imagery, measurement of signatures, technical characteristics, human interaction, and other detection methods. When performing reconnaissance, it is important to—

- Orient the reconnaissance asset on the named area of interest (NAI) and/or reconnaissance objective in a timely manner.
- Report all information rapidly and accurately.
- Complete the reconnaissance mission not later than (NLT) the time specified in the order.
- Answer the requirement that prompted the reconnaissance task.

Conduct Surveillance

1-50. To systematically observe the airspace, surface, or subsurface areas, places, persons, or things in the AO by visual, aural (audio), electronic, photographic, or other means. Other means may include but are not limited to space-based systems, and using special NBC, artillery, engineer, SOF, and air defense equipment. (See FM 7-15, ART 1.3.4, and FM 3-90)

Note: This task is a technique by which ART 1.1.4.1 (Collect Police Information) may be performed.

1-51. Conducting surveillance is systematically observing the airspace, surface, or subsurface areas, places, persons, or things in the AO by visual, aural, electronic, photographic, or other means. Surveillance activities include—

- Orienting the surveillance asset on the NAI and/or the surveillance objective in a timely manner.
- Reporting all information rapidly and accurately.
- Completing the surveillance mission NLT the time specified in the order.
- Answering the requirement that prompted the surveillance task.

PROVIDE INTELLIGENCE SUPPORT TO EFFECTS

1-52. The task of providing the commander information and intelligence support for targeting of the threat's forces, threat organizations, units and systems through lethal and non-lethal fires to include electronic attack and information operations. This task includes three subtasks: provide intelligence support to targeting, provide intelligence support to information operations, and provide intelligence support to combat assessment. (See FM 7-15, ART 1.4)

Provide Intelligence Support to Targeting

1-53. The intelligence officer, supported by the entire staff, provides the commander information and intelligence support for targeting the threat's forces and systems through employment of direct, indirect lethal and non-lethal fires. It includes identification of threat capabilities and limitations. The Intelligence warfighting function plays a crucial role by providing intelligence throughout the steps of the targeting process: decide, detect, deliver, assess. Support to targeting includes identifying threat capabilities and limitations. The G2/S2 supports the unit's lethal and non-lethal targeting effort through target development, target acquisition (TA), electronic warfare (EW), and combat assessment. This task has two subtasks:

- *Provide Intelligence Support to Target Development* is the systematic analysis of the enemy forces and operations to determine HVTs, systems, and system components for potential attack through maneuver, fires, or information.
- *Provide Intelligence Support to Target Detection* establishes procedures for dissemination of targeting information. The targeting team develops the sensor/attack system matrix to determine the sensor required to detect and locate targets. The intelligence officer places the following requirements into the integrated ISR plan.
  - Requires reconnaissance and surveillance operations to identify, locate, and track high-payoff targets (HPTs) for delivery of lethal or non-lethal effects.
  - Includes employing fires, offensive IO, and other attack capabilities against enemy C2 systems as part of the unit's FS plan and IO objectives.

Provide Intelligence Support to Information Operations

1-54. IO are actions taken to affect adversary information, influence other's decisionmaking processes and information systems while protecting one's own information and information systems. Overall operational continuity and mission success requires close, mutual coordination and synchronization of intelligence plans and operations with IO elements and related activities. Refer to Appendix A, this manual, for more information. This task has three subordinate tasks: (See FM 7-15, ART 1.4.2)

- Provide Intelligence Support to Offensive IO.
- Provide Intelligence Support to Defensive IO.
- Provide Intelligence Support to Activities Related to IO.

Provide Intelligence Support to Combat Assessment

1-55. Combat assessment is the determination of the overall effectiveness of force employment during military operations. The objective of combat assessment is to identify recommendations for the course of military operations. It answers the question, "Were the objectives met by force employment?" Although the assessment is primarily an intelligence responsibility, it requires input from and coordination with operations and fire support staffs. Combat assessment consists of conducting physical damage, functional damage, and target system assessments. (See FM 7-15, ART 1.4.3)

1-56. The staff determines how combat assessment relates to a specific target by conducting physical damage, functional damage, and target system assessments.

- **Conduct Physical Damage Assessment (PDA).** A PDA is an estimate of the extent of physical damage to a target based upon observed or interpreted damage. This post-attack target analysis is a coordinated effort among all units.
- **Conduct Functional Damage Assessment (FDA).** The FDA estimates the remaining functional or operational capability of a targeted facility or object. The staff bases FDA on the assessed physical damage and estimates of the threat's ability to recuperate (to include the time required to resume normal operations). Multiple echelons typically conduct this all-source analysis. The targeting or combat assessment cell integrates the initial target analyses with other sources, including intelligence, and then compares the original objective with the current status of the target to determine if the objective has been met.
- **Conduct Target System Assessment (TSA).** The TSA is an estimate of the overall impact of force employment against an adversary's target system. The unit, supported by higher echelon assets, normally conducts this assessment. The analyst combines all combat assessment reporting on functional damage to targets within a target system and assesses the overall impact on that system's capabilities. This process lays the groundwork for future recommendations for military operations in support of operational objectives.

1-57. **Munitions Effects Assessment (MEA).** MEA takes place concurrently and interactively with combat assessment since the same signatures used to determine the level of physical damage also give clues to munitions effectiveness. MEA is primarily the responsibility of operations and FS personnel, with input from the G2/S2. After the same weapon is used to attack several targets of a specific type, MEA should be accomplished to evaluate weapon performance. MEA analysts seek to identify through systematic trend analysis any deficiencies in weapon system and munitions performance or combat tactics by answering the question, "Did the weapons employed perform as expected?" Using a variety of input (targeting analysts, imagery analysts, structural engineers, and mission planners) analysts prepare a report assessing munitions performance. If combat troops capture attacked targets, it is then possible to collect detailed information on the target. Reports should detail weapon performance against specified target types. This information could have a crucial impact on future operations.

1-58. **Re-attack Recommendation**. Re-attack recommendations follow directly from both battle damage assessment (BDA) and MEA efforts. Basically re-attack recommendations answer the question, "Have we achieved the desired effects against our targeted objectives?" Evolving objectives, target selection, timing, tactics, weapons, vulnerabilities, and munitions are all factors in the new recommendations, combining both operations and intelligence functions.

1-59. **Targeting Meeting**. The role of the G2/S2 in targeting meetings is critical to ensuring intelligence supports the targeting process. The G2/S2 must come to the targeting meeting prepared. Additionally, the G2/S2 must understand the steps of the targeting process and tailor the intelligence products and G2/S2 participation according to the targeting process steps of: decide, detect, deliver, and assess.

1-60. There are two major areas in which the G2/S2 must prepare in order to support the targeting meeting.

- **Analysis:** Determine what the enemy is doing (current situation) and anticipated threat models; ECOA situation templates, ECOA sketches and statements, HVTs, and other associated products.
- **IPB Products:** Updated from the initial IPB effort, including analytical results (see above bullet), and tailored to the targeting requirements.

1-61. **Decide**. At the targeting meeting, the G2/S2 should be prepared to do the following:

- Brief the current and projected future enemy situation with an event template.
- Brief any combat assessments.
- Brief HVTs for potential selection of HPTs.
- Brief the current PIR portion of CCIRs, the PIRs of the CCIRs for the next phase or in accordance with the same timeframe of the targeting meeting.
- Brief the current ISM or intelligence synchronization plan.
- Participate in wargaming.
- Refine the initial ISR plan (in conjunction with the FS and G3/S3 representatives).

1-62. **Detect**. Based on information from the decide portion of the targeting meeting, the G2/S2 should be prepared to do the following:

- Identify specific ISR assets.
- Develop IRs and propose pertinent IRs to the commander for designation as PIRs.
- Complete the initial ISR plan; ensure it addresses the HPTs.

1-63. **Electronic Warfare**. EW provides targeting information to the commander based on transmitter location data. There are three major subdivisions within EW: electronic support (ES), electronic protection (EP), and electronic attack (EA).

- **ES** – Also referred to as electronic warfare support (EWS) in JP 1-02, involves actions to search for, intercept, identify, and locate or localize sources of intentional and unintentional radiated electromagnetic energy. In addition to supporting the overall ISR effort, ES is needed to produce intelligence required to support EA missions.
- **EP** – Involves passive and active means taken to protect personnel, facilities, and equipment from any effects of friendly or enemy employment of EW that degrades, neutralizes, or destroys friendly combat capabilities.
- **EA** – Involves the use of electromagnetic energy, directed energy, or anti-radiation weapons to attack personnel, facilities, or equipment with the intent of degrading, neutralizing, or destroying enemy combat capability and is considered a form of fires. EA includes actions taken to prevent or reduce an enemy's effective use of the electromagnetic spectrum (EMS), such as jamming and electromagnetic deception.

1-64. **Deliver**. The G2/S2 should be prepared to do the following:

- Brief how the Intelligence warfighting function will work to detect and track the target through the entire targeting process.
- Brief the implementation of the intelligence synchronization and ISR plans, including the LTIOV and communications structure.

1-65. **Assess**. Depending on the factors of METT-TC and the PIRs, the same ISR assets that detect and track the targets may be required to support combat assessments in order to determine if re-attack is required. Conversely, if re-attack is a decision that the commander will make based upon the effects on the enemy, combat assessment should be reflected in the PIR, and supported in the corresponding ISR plan.

THE OPERATIONAL ENVIRONMENT

1-66. The operational environment can be defined as a composite of the conditions, circumstances, and influences that affect the employment of capabilities and bear on the decision of the commander. The operational environment encompasses physical areas and factors of the air, land, maritime, and space domains. It also includes the information environment and enemy, adversary, friendly, and neutral systems.

OPERATIONAL VARIABLES (PMESII-PT)

1-67. Analysis of the operational environment—in terms of political, military, economic, social, infrastructure, and information with the addition of physical environment and time (PMESII-PT) variables—provides relevant information that senior commanders use to understand, visualize, and describe the operational environment. As a set, these operational variables are often abbreviated as PMESII-PT. A description of each of the PMESII-PT variables follows:

- **Political**. Describes the distribution of responsibility and power at all levels of governance or cooperation.
- **Military**. Explores the military capabilities of all relevant actors in a given operational environment.
- **Economic**. Encompasses individual behaviors and aggregate phenomena related to the production, distribution, and consumption of resources.
- **Social**. Describes the cultural, religious, and ethnic makeup within an operational environment.
- **Information**. Describes the nature, scope, characteristics, and effects of individuals, organizations, and systems that collect, process, disseminate, or act on information.
- **Infrastructure**. Is composed of the basic facilities, services, and installations needed for the functioning of a community or society.
- **Physical Environment**. Defines the physical circumstances and conditions that influence the execution of operations throughout the domains of air, land, sea, and space.
- **Time**. Influences military operations within an operational environment in terms of the decision cycles, operational tempo, and planning horizons. (FM 3-0)

1-68. For the intelligence portion, when identifying threats based on the systems approach to PMESII-PT, there are four primary components of the operational environment. These are the threat, terrain, weather, and civil considerations.

1-69. While the operational variables are directly relevant to campaign planning, they are too broad to be applied directly to tactical planning. That does not mean they are not of value at all levels; they are fundamental to developing the understanding of the operational environment necessary to conduct planning at any level, in any situation. The degree to which each operational variable provides useful information depends on the situation and echelon. For example, social and economic variables often receive close analysis as part of enemy and civil considerations at brigade and higher levels. They may affect the training and preparation of small units. However, they are not relevant to a small-unit leader's mission analysis. That leader may only be concerned with such questions as "Who is the tribal leader for this village?" "Is the electrical generator working?" "Does the enemy have antitank missiles?"

MISSION VARIABLES

1-70. Upon receipt of a warning order (WARNO) or mission, Army leaders narrow their focus to six mission variables. Mission variables are those aspects of the operational environment that directly affect a mission. They outline the situation as it applies to a specific Army unit. The mission variables are mission, enemy, terrain and weather, troops and support available, time available, and civil considerations (METT-TC). These are the categories of relevant information used for mission analysis. Army leaders use the mission variables to synthesize operational variables and tactical-level information with local knowledge about conditions relevant to their mission. The METT-TC categories are listed below:

- **Mission**. The mission is the task, together with the purpose, that clearly indicates the action to be taken and the reason.
- **Enemy**. Relevant information regarding the enemy may include the following:
  - Threat characteristics (previously order of battle [OB] factors).
    - Doctrine (or known execution patterns).
    - Personal habits and idiosyncrasies.
    - Equipment, capabilities, and vulnerabilities.
  - Threat COAs.
- **Terrain and Weather**. Terrain and weather are natural conditions that profoundly influence operations. Terrain and weather are neutral; they favor neither side unless one is more familiar with—or better prepared to operate in—the environment.
- **Troops and Support Available**. The number, type, capabilities, and condition of available friendly troops and support available. These include the resources available from joint, interagency, intergovernmental, multinational, host nation (HN), commercial (via contracting), and private organizations.
- **Time Available**. Time is critical to all operations. Controlling and exploiting time is central to initiative, tempo, and momentum. By exploiting time, commanders can exert constant pressure, control the relative speed of decisions and actions, and force exhaustion on enemy forces.
- **Civil Considerations**. Civil considerations comprise six characteristics expressed in the memory aid ASCOPE:
  - Areas.
  - Structures.
  - Capabilities.
  - Organizations.
  - People.
  - Events.

Note. For additional information on ASCOPE and the IPB process, see FM 2-01.3. Understanding the operational environment requires understanding the civil aspects of the AO. Civil considerations reflect how the manmade infrastructure, civilian institutions, and the attitudes and activities of the civilian leaders, populations, and organizations within an AO influence the conduct of military operations.

1-71. METT-TC enables leaders to synthesize operational level information with local knowledge relevant to their missions and tasks in a specified AO. Tactical and operational leaders can then anticipate the consequences of their operations before and during execution. See FM 3-0 for a detailed discussion of PMESII-PT and METT-TC.

THE INTELLIGENCE PROCESS

1-72. Intelligence operations generally include the five functions that constitute the intelligence process: **plan, prepare, collect, process,** and

produce. Additionally, there are three common tasks that occur across the five functions of the intelligence process: **analyze, disseminate,** and **assess**. The three common tasks are discussed after the last function, produce. The intelligence process functions are not necessarily sequential; this is what differentiates the Army's intelligence process from the Joint intelligence cycle. The intelligence process provides a common model with which to guide one's thinking, discussing, planning, and assessing about the threat or AOI environment. The intelligence process generates information about the threat, the AOI, and the situation, which allows the commander and staff to develop a plan, seize and retain the initiative, build and maintain momentum, and exploit success.

PLAN

1-73. The plan step consists of the activities that identify pertinent IRs and develop the means for satisfying those requirements. The CCIRs (PIRs and FFIRs) drive the planning of the ISR effort. The intelligence officer supports the G3/S3 in arranging the ISR effort, based on staff planning, to achieve the desired collection effects. Planning activities include, but are not limited to—

- Conducting IPB.
- Managing requirements.
- Submitting RFIs and using intelligence reach to fill information gaps.
- Evaluating reported information.
- Establishing the intelligence communications and dissemination architecture.
- Developing, managing, and revising the intelligence synchronization plan and the ISR plan as mission requirements change.

PREPARE

1-74. The failure of MI units or personnel to accomplish their tasks or missions can often be attributed to their failure to prepare. The prepare step includes those staff and leader activities which take place upon receiving the operation order (OPORD), operation plan (OPLAN), warning order (WARNO), or commander's intent in order to improve the unit's ability to execute tasks or missions.

1-75. The most habitual and egregious preparation failures committed by leaders (as evidenced by performance at the combat training centers) is not the conduct of specific tasks, but the failure to adequately coordinate for more generic combat requirements such as—

- Friendly forward unit liaison.
- Departing or reentering friendly lines or AOs.
- Fire support: enroute, at mission location, and return.
- Casualty evacuation procedures.
- Passwords (running, forward of friendly lines), recognition signals, call signs, and frequencies.
- Resupply.
- Movement through friendly AOs.

COLLECT

1-76. The collect step involves collecting and reporting information in response to ISR tasking. ISR assets collect information and data about threat forces, activities, facilities, and resources as well as information concerning the environmental and geographical characteristics of a particular AO. A successful ISR effort results in the timely collection and reporting of relevant and accurate information. This collected information forms the foundation of intelligence databases, intelligence production, and the G2's/S2's situational understanding. The requirements manager evaluates the reported information for its responsiveness to the CCIRs (PIRs and FFIRs).

PROCESS

1-77. Processing involves converting collected data and information into a form that is suitable for analyzing and producing intelligence. Examples of processing include enhancing imagery, translating a document from a foreign language, converting electronic data into a standardized report that can be analyzed by a system operator, and correlating information.

1-78. Processing data and information is performed unilaterally and cooperatively by both humans and automated systems.

PRODUCE

1-79. The produce step involves evaluating, analyzing, interpreting, synthesizing, and combining information and intelligence from single or multiple sources into intelligence or intelligence products in support of known or anticipated requirements. Production also involves combining new information and intelligence with existing intelligence in order to produce intelligence in a form that the commander and staff can apply to the MDMP and that supports and helps facilitate situational understanding. During the produce step, the intelligence staff manipulates information by—

- Analyzing the information to isolate significant elements.
- Evaluating the information to determine accuracy, timeliness, usability, completeness, precision, and reliability.
- Combining the information with other relevant information and previously developed intelligence.
- Applying the information to estimate possible outcomes.
- Presenting the information in a format that will be most useful to its eventual user.

1-80. The intelligence staff deals with numerous and varied production requirements based on PIRs and intelligence requirements; diverse missions, environments, and situations; and presentation requirements. Through analysis, collaboration, and intelligence reach, the G2/S2 and the staff use the collective intelligence production capability of higher, lateral, and subordinate echelons to meet the production requirements. Proficiency in these techniques and procedures facilitates the intelligence staff's ability to answer command and staff requirements regardless of the factors of METT-TC.

1-81. The three common tasks, discussed below, can occur throughout the intelligence process.

Analyze

1-82. The intelligence staff analyzes intelligence and information about the enemy's capabilities, friendly vulnerabilities, and the battlefield environment as well as issues and problems that arise within the intelligence process itself to determine their nature, origin, and interrelationships. This analysis enables commanders, staffs, and leaders to determine the appropriate action or reaction and to focus or redirect assets and resources to fill information gaps or alleviate pitfalls. It is also within the analyze function that intelligence analysts sort through large amounts of collected information and intelligence to obtain only that information which pertains to the CCIRs (PIRs and FFIRs), maintenance of the COP, and facilitates the commander's situational understanding.

Disseminate

1-83. Disseminating is communicating relevant information of any kind from one person or place to another in a usable form by any means to improve understanding or to initiate or govern action. Disseminating intelligence entails using information management techniques and procedures to deliver timely, relevant, accurate, predictive, and usable intelligence to the commander. Determining the product format and selecting the means to deliver it are key aspects of dissemination. Information presentation may be in a verbal, written, interactive, or graphic format. The type of information, the time allocated, and the individual preference of the commander all influence the information format.

Assess

1-84. Assessment plays an integral role in all aspects of the intelligence process. Assessing includes evaluating the effectiveness of intelligence in supporting the operation. Assessing the situation and available information begins upon receipt of the mission and continues throughout the intelligence process. The continual assessment of intelligence operations and ISR efforts, available information and intelligence, and various aspects of the factors of METT-TC are critical to ensure—

- The G2/S2 answers all CCIRs (PIRs and FFIRs).
- The G2/S2 and G3/S3 redirect ISR assets to support changing requirements.
- Using information and intelligence properly.

INTELLIGENCE DISCIPLINES

1-85. Intelligence disciplines are categories of intelligence functions. The Army's intelligence disciplines are All-Source Intelligence, Human Intelligence (HUMINT), Geospatial Intelligence (GEOINT), Imagery Intelligence (IMINT), Signals Intelligence (SIGINT), Measurement and Signatures Intelligence (MASINT), Open-Source Intelligence (OSINT), Technical Intelligence (TECHINT), and Counterintelligence (CI). For more information regarding the intelligence disciplines, see Part Three of this manual, as well as the respective manuals, which covers each individual intelligence discipline.

ALL-SOURCE INTELLIGENCE

1-86. All-source intelligence is defined as the intelligence products, organizations, and activities that incorporate all sources of information and intelligence, including open-source information, in the production of intelligence. All-source intelligence is a separate intelligence discipline, as well as the name of the task used to produce intelligence from multiple intelligence or information sources.

HUMAN INTELLIGENCE

1-87. HUMINT is the collection of foreign information—by a trained HUMINT Collector—from people and multimedia to identify elements, intentions, composition, strength, dispositions, tactics, equipment, personnel, and capabilities. It uses human sources as a tool, and a variety of collection methods, both passively and actively, to collect information.

GEOSPATIAL INTELLIGENCE

1-88. Title 10 U.S. Code § 467 establishes GEOINT. The definition of geospatial intelligence is—

> *The exploitation and analysis of imagery and geospatial information to describe, assess, and visually depict physical features and geographically referenced activities on the Earth. GEOINT consists of imagery, imagery intelligence, and geospatial information.*

1-89. The Army implementation of GEOINT is a result of the Army's organization, manning, and training. There are multiple types of data and information that various Army units and organizations collect, provide, and analyze in order to support the GEOINT enterprise. The two primary GEOINT service providers in the Army are MI units and organizations and Engineer (topographic) units and organizations. MI units and organizations provide imagery and IMINT to the enterprise. Engineer (topographic) units and organizations provide geospatial data and information to the enterprise. Therefore, while some of the collection, analysis, and exploitation of imagery and geospatial information occur within the intelligence warfighting function; some of the collection, analysis, and exploitation of imagery and geospatial information occur outside intelligence.

IMAGERY INTELLIGENCE

1-90. IMINT is intelligence derived from the exploitation of imagery collected by visual photography, infrared, lasers, multi-spectral sensors, and radar. These sensors produce images of objects optically, electronically, or digitally on film, electronic display devices, or other media.

SIGNALS INTELLIGENCE

1-91. SIGINT is a category of intelligence comprising either individually or in combination all communications intelligence (COMINT), electronic intelligence (ELINT), and foreign instrumentation signals intelligence (FISINT), however transmitted. SIGINT is derived from communications, electronics, and foreign instrumentation signals.

MEASUREMENT AND SIGNATURES INTELLIGENCE

1-92. MASINT is technically derived intelligence that detects, locates, tracks, identifies, and/or describes the specific characteristics of fixed and dynamic target objects and sources. It also includes the additional advanced processing and exploitation of data derived from IMINT and SIGINT collection. MASINT collection systems include but are not limited to radar, spectroradiometric, electro-optical (E-O), acoustic, radio frequency (RF), nuclear detection, and seismic sensors, as well as techniques for gathering chemical, biological, radiological, and nuclear (CBRN), and other material samples.

OPEN-SOURCE INTELLIGENCE

1-93. The *National Defense Authorization Act for Fiscal Year 2006* states:

> *Open source intelligence is produced from publicly available information that is collected, exploited, and disseminated in a timely manner to an appropriate audience for the purpose of addressing a specific intelligence requirement.*

1-94. Expressed in terms of the Army intelligence process, OSINT is relevant information derived from the systematic collection, processing, and analysis of publicly available information in response to intelligence requirements.

1-95. The Army does not have a specific military occupational specialty (MOS), additional skill identifier (ASI), or special qualification identifier (SQI) for OSINT. With the exception of the Asian Studies Detachment, the Army does not have base tables of organization and equipment (TOE) for OSINT units or staff elements. OSINT missions and tasks are imbedded within existing missions and force structure or accomplished through task organization.

TECHNICAL INTELLIGENCE

1-96. TECHINT is intelligence derived from the collection and analysis of threat and foreign military equipment and associated materiel for the purposes of preventing technological surprise, assessing foreign scientific and technical (S&T) capabilities, and developing countermeasures designed to neutralize an adversary's technological advantages.

COUNTERINTELLIGENCE

1-97. CI counters or neutralizes intelligence collection efforts through collection, counterintelligence investigations, operations, analysis, and production, and functional and technical services. CI includes all actions taken to detect, identify, track, exploit, and neutralize the multidiscipline intelligence activities of friends, competitors, opponents, adversaries, and enemies; and is the key intelligence community contributor to protect US interests and equities. CI assists in identifying EEFI, identifying vulnerabilities to threat collection, and actions taken to counter collection and operations against US forces.

UNIFIED ACTION

1-98. Unified action is the synchronization, coordination, and/or integration of the activities of governmental and nongovernmental entities with military operations to achieve unity of effort. (JP 1) Joint operations focus and maximize the complementary and reinforcing effects and capabilities of each service. Joint force commanders (JFCs) synchronize the complementary capabilities of the service components that comprise the joint force.

1-99. The employment of MI in campaigns and major operations must be viewed from a joint perspective, and the "mud-to-space" intelligence concept must establish a fully interoperable and integrated joint intelligence capability. ARFOR intelligence assets work with multinational and interagency partners to accomplish their missions. Ideally, multinational and interagency intelligence partners provide cultures, perspectives, and capabilities that reinforce and complement Army MI strengths and capabilities. Close intelligence coordination is the foundation of successful unified action. (See chapter 2.)

1-100. A critical part of current operations is the execution of the joint doctrinal concept of persistent surveillance. Joint doctrine defines persistent surveillance as a collection strategy that emphasizes the ability of some collection systems to linger on demand in an area to detect, locate, characterize, identify, track, target, and possibly provide battle damage assessment and re-targeting in real or near-real time. Persistent surveillance facilitates the formulation and execution of preemptive activities to deter or forestall anticipated adversary courses of action (JP 2-0).

THE NATURE OF LAND OPERATIONS

1-101. Landpower is the ability—by threat, force, or occupation—to gain, sustain, and exploit control over land, resources, and people (FM 3-0). Army operations reflect expeditionary and campaign capabilities that constantly adapt to each campaign's unique circumstances. Expeditionary capabilities require forces organized to be modular, versatile, and rapidly deployable. Rapidly deployed expeditionary force packages provide immediate options for seizing or retaining the operational initiative and also allow the conduct of sustained operations for as long as necessary. Army forces are organized, trained, and equipped for endurance. The Army's preeminent challenge is to balance expeditionary agility and responsiveness with the endurance and adaptability needed to complete a campaign successfully.

1-102. The capability to prevail in close combat is indispensable and unique to land operations. The outcome of battles and engagements depends on Army forces' ability to prevail in close combat. Many factors inherent in land combat combine to complicate the situation. These factors include chaos, complexity, insufficient intelligence, errors in understanding or planning, difficult terrain, the civilian population, and an adaptive and lethal enemy. The axiom "intelligence drives operations" continues to be true especially for land operations; operations and intelligence are inextricably linked.

1-103. Four considerations are preeminent in generating expeditionary capabilities and Army force packages.

- **Scope.** Considers and strives to understand an enemy throughout the depth of an operational area. Commanders rely on intelligence in order to use maneuver, fires, and other elements of combat power to defeat or destroy enemy forces.
- **Duration.** Forces routinely conduct missions prior to, during, and after the commitment of land combat forces. Intelligence is always engaged.
- **Terrain.** Missions occur among a complex variety of natural and manmade features. Employing forces in the complexity of the ground environment requires thorough planning.
- **Permanence.** Forces are integrated with, or assigned to, land combat forces as they seize or secure ground.

1-104. Several attributes of the land environment affect the application of landpower. These attributes include—

- The requirement to deploy and employ Army forces rapidly.
- The requirement for Army forces to operate for protracted periods.
- The nature of close combat.
- Uncertainty, chance, friction, and complexity.

1-105. Reconnaissance, Surveillance, and Target Acquisition (RSTA)/ISR is the means the Army uses to implement the joint doctrinal concept of persistent surveillance in support of tactical operations. Dependable technology and responsive intelligence lessen the effects of uncertainty, chance, friction, and complexity. Complex and dynamic Army tactical operations require extensive ISR capabilities to satisfy the commander's information requirements to detect, locate, characterize, identify, track, and target HPTs, and to provide combat assessment in real time within a very fluid operational environment.

1-106. *Reconnaissance, Surveillance, and Target Acquisition/ Intelligence, Surveillance, and Reconnaissance* is a full spectrum combined arms mission that integrates ground and air capabilities to provide effective, dynamic, timely, accurate, and assured combat information and multidiscipline actionable intelligence for lethal and non-lethal effects and decisions in direct support of the ground tactical commander.

FORCE PROJECTION OPERATIONS

1-107. Force projection is the military component of power projection. It is a central element of the national military strategy. Army organizations and installations, linked with joint forces and industry, form a strategic platform to maintain, project, and sustain ARFOR, wherever they deploy. Force projection operations are inherently joint and require detailed planning and synchronization. Force projection encompasses a range of processes—mobilization, deployment, employment, sustainment, and redeployment—discussed below.

1-108. The Army must change its mindset from depending on an "intelligence buildup" to performing intelligence readiness on a daily basis in order to meet the requirements for strategic responsiveness. MI personnel, even in garrison at the lowest tactical echelons, must use their analytic and other systems and prepare for possible operations on a daily basis.

1-109. Built on a foundation of intelligence readiness, the Intelligence warfighting function provides the commander with the intelligence needed to plan, prepare, and execute force projection operations. Successful intelligence during force projection operations relies on continuous collection and intelligence production before and during the operation. In a force projection operation, higher echelons will provide intelligence to lower echelons until the tactical ground force completes entry and secures the lodgment area. The joint force J2 must exercise judgment when providing information to subordinate G2s/S2s so not to overwhelm them. Figure 1-5 is an example of force projection intelligence.

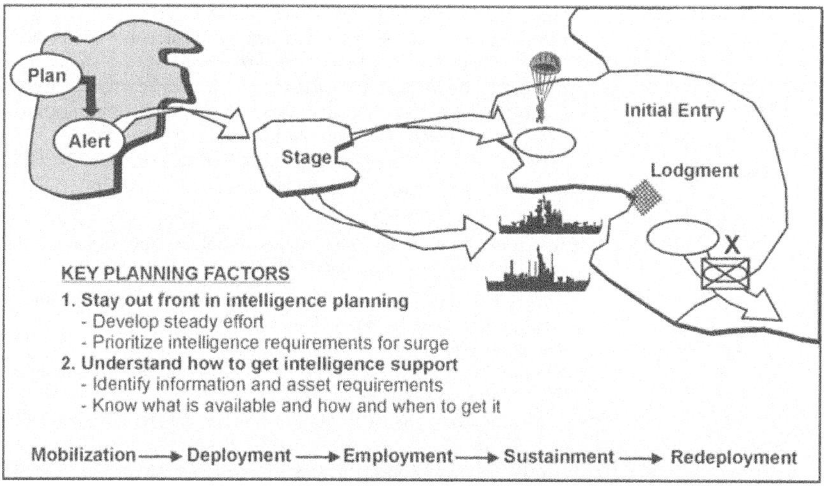

Figure 1-5. Force Projection Intelligence.

1-110. The G2/S2 must anticipate, identify, consider, and evaluate all threats to the entire unit throughout force projection operations. This is critical during the deployment and entry operations stages of force projection. During these stages, the unit is particularly vulnerable to enemy actions because of its limited combat power and knowledge of the AO. Intelligence personnel must, therefore, emphasize the delivery of combat information and intelligence products that indicate changes to the threat or AO developed during predeployment IPB.

1-111. The G2/S2 should—

- Review available databases on assigned contingency AOIs, conduct IPB on these AOIs, and develop appropriate IPB products.
- Comply with higher headquarters standing operating procedures (SOPs) and manuals for specific intelligence operations guidance.
- Coordinate for and rehearse electronic message transfers (for example, Internet Protocol addresses, routing indicators) using the same communications protocols with theater, higher headquarters, subordinate, and lateral units that the unit would use when deployed.
- Plan, train, and practice surging intelligence functions on likely or developing contingency crises.
- Prepare and practice coordination from predeployment through redeployment with other elements and organizations (for example, HUMINT, IMINT, SIGINT, MASINT, IO, staff weather officer [SWO], civil affairs [CA], psychological operations [PSYOP], and special operations forces [SOF] units, to include databases and connectivity).
- Include the following as a part of daily (sustainment) operations:
  - US Army Reserve and other augmentation.
  - A linguist plan with proficiency requirements. (Alert linguists through early entry phases of deployment.)
  - Training (individual and collective).
- Establish formal or informal intelligence links, relationships, and networks to meet developing contingencies.
- Forward all RFIs to higher headquarters in accordance with SOPs.
- Establish statements of intelligence interests (SIIs), other production, and I&W requirements.

1-112. To draw intelligence from higher echelons and focus intelligence downward, based on the commander's needs, the G2/S2 must—

- Understand the J2's multiple echelon and broadcast dissemination capability to ensure near-real time (NRT) reporting to all deployed, in transit, or preparing to deploy forces.
- Maintain or build intelligence databases on the environment and threats for each probable contingency.
- State and record the CCIR (as a minimum, list the PIRs and ISR tasks or requests).

1-113. Until the unit's collection assets become operational in the AO, the G2/S2 will depend upon intelligence from the ARFOR or JTF to answer the unit's intelligence needs. The following paragraphs describe the intelligence and ISR considerations during force projection.

MOBILIZATION

1-114. Mobilization is the process by which the armed forces or part of them are brought to a state of readiness for war or other national emergency. It assembles and organizes resources to support national objectives. Mobilization includes activating all or part of the US Army Reserve, and assembling and organizing personnel, supplies, and materiel. A unit may be brought to a state of readiness for a specific mission or other national emergency. This process, called mobilization, is where specific US Army

Active or Reserve units, capabilities, and personnel are identified and integrated into the unit. During mobilization, the G2/S2 must—

- Monitor intelligence reporting on threat activity and I&W data.
- Manage IRs and RFIs from their unit and subordinate units to include updating ISR planning.
- Establish habitual training relationships with their US Army Active and Reserve augmentation units and personnel as well as higher echelon intelligence organizations as identified in the existing OPLAN. Support the US Army Reserve units and augmentation personnel by preparing and conducting intelligence training and threat update briefings and by disseminating intelligence.
- Identify ISR force requirements for the different types of operations and contingency plans (CONPLANs).
- Identify individual military, civilian, and contractor augmentation requirements for intelligence operations. The Army, and the Intelligence warfighting function in particular, cannot perform its missions without the support of its Department of the Army Civilians (DACs) and contractors. The force increasingly relies on the experience, expertise, and performance of non-uniformed personnel and has fully integrated these non-uniformed personnel into the warfighting team.

1-115. During mobilization the G2/S2, in conjunction with the rest of the staff, must ensure the adequacy of training and equipping of US Army Active and Reserve MI organizations and individual augmentees to conduct intelligence operations.

1-116. The G2/S2 supports peacetime contingency planning with IPB products and databases on likely contingency areas. The G2/S2 establishes an intelligence synchronization plan that will activate upon alert notification. For smooth transition from predeployment to entry, the G2/S2 must coordinate intelligence synchronization and communications plans before the crisis occurs. The intelligence synchronization plan identifies the intelligence requirements supporting those plans, to include—

- ISR assets providing support throughout the AOI.
- Command and support relationships of ISR assets at each echelon.
- Report and request procedures not covered in unit SOPs.
- Sequence of deployment of ISR personnel and equipment. Early deployment of key ISR personnel and equipment is essential for force protection and combat readiness. Composition of initial and follow-on deploying assets is influenced by the factors of METT-TC, availability of communications, and availability of lift.
- Communications architecture supporting both intelligence staffs and ISR assets.
- Friendly vulnerabilities to hostile intelligence threats and plans for conducting FP measures. The staff must begin this type of planning as early as possible to ensure adequate support to FP of deploying and initial entry forces.
- Monitor time-phased force and deployment data (TPFDD) and recommend changes in priority of movement, unit, or capability to enable ISR operations.

1-117. The G2/S2 must continually monitor and update the OPLANs to reflect the evolving situation, especially during crisis situations. National intelligence activities monitor regional threats throughout the world and can answer some intelligence requirements supporting the development of OPLANs.

1-118. Upon alert notification, the G2/S2 updates estimates, databases, IPB products, and other intelligence products needed to support command decisions on force composition, deployment priorities and sequence, and the AOI. Units reassess their collection requirements immediately after alert notification. The G2/S2 begins verifying planning assumptions within the OPLANs. CI and ISR personnel provide FP support and antiterrorism measures.

1-119. Throughout mobilization, unit intelligence activities will provide the deploying forces with the most recent intelligence on the AO. The intelligence staff will also update databases and situation graphics. The G2/S2 must—

- Fully understand the unit, ARFOR, and joint force intelligence organizations.
- Revise intelligence and intelligence-related communications architecture and delete or integrate any new systems and software with the current architecture.
- Support 24-hour operations and provide continuous intelligence.
- Plan all required intelligence reach procedures.
- Determine transportation availability for deployment and availability when deployed.
- Determine all sustainability requirements.
- Determine intelligence release requirements and restrictions; releasability to multinational and HN sources.
- Review status of forces agreements (SOFAs), rules of engagement (ROE), international laws, and other agreements, emphasizing the effect that they have on intelligence collection operations. (Coordinate with the staff judge advocate [SJA] on these issues.)
- Ensure ISR force deployment priorities are reflected in the TPFDD to support ISR operations based upon the factors of METT-TC.
- Ensure intelligence links provide the early entry commander vital access to multi-source army and joint intelligence collection assets, processing systems, and databases.
- Review the supporting unit commanders' specified tasks, implied tasks, task organization, scheme of support, and coordination requirements with forward maneuver units. Address issues or shortfalls and direct or coordinate changes.
- Establish access to national HUMINT, IMINT, SIGINT, MASINT, and CI databases, as well as automated links to joint service, multinational, and HN sources.

DEPLOYMENT

1-120. Deployment is the movement of forces and materiel from their point of origin to the AO. This process has four supporting components:

predeployment activities, fort to port, port to port, and port to destination; these components are known collectively as reception, staging, onward movement, and integration (RSO&I) activities. Success in force projection operations hinges on timely deployment. The size and composition of forces requiring lift are based on the factors of METT-TC, availability of pre-positioned assets, the capabilities of HN support, and the forward presence of US forces. Force or tactical tailoring is the process used to determine the correct mix and sequence of deploying units.

1-121. During deployment, intelligence organizations at home station or in the rear area take advantage of modern SATCOM, broadcast technology, and automated data processing (ADP) systems to provide graphic and textual intelligence updates to the forces enroute. Enroute updates help eliminate information voids and, if appropriate, allow the commander to adjust the plan prior to arrival in theater in response to changes in the OE or enemy actions.

1-122. Intelligence units extend established networks to connect intelligence staffs and collection assets at various stages of the deployment flow. Where necessary, units establish new communications paths to meet unique demands of the mission. The theater and corps analysis and control elements (ACEs) play a critical role in making communications paths, networks, and intelligence databases available to deploying forces.

1-123. Space-based systems are key to supporting intelligence during the deployment and the subsequent stages of force projection operations by—

- Monitoring terrestrial AOIs through ISR assets to help reveal enemy location and disposition, attempting to identify the enemy's intent.
- Providing communications links between forces enroute and in the continental United States (CONUS).
- Permitting MI collection assets to accurately determine their position through the Global Positioning System (GPS).
- Providing timely and accurate data on meteorological, oceanographic, and space environmental factors that might affect operations.
- Providing warning of theater ballistic missile launches.
- Providing timely and accurate weather information to all commanders through the Integrated Meteorological System (IMETS).

1-124. Situation development dominates intelligence operations activities during initial entry operations. The G2/S2 attempts to identify all threats to arriving forces and assists the commander in developing FP measures. During entry operations, echelons above corps (EAC) organizations provide intelligence. This support includes providing access to departmental and joint intelligence and deploying scalable EAC intelligence assets. The entire effort focuses downwardly to provide tailored support to deploying and deployed echelons in response to their CCIRs (PIRs and FFIR).

1-125. Collection and processing capabilities are enhanced, as collection assets build up in the deployment area, with emphasis on the build-up of the in-theater capability required to conduct sustained ISR operations. As the build-up continues, the G2/S2 strives to reduce total dependence on extended split-based intelligence from outside the AO. As assigned collection assets arrive into the theater, the G2/S2 begins to rely on them for tactical intelligence although higher organizations remain a source of intelligence.

1-126. As the ARFOR enter the theater of operations, the joint force J2 implements and, where necessary, modifies the theater intelligence architecture. Deploying intelligence assets establishes liaison with staffs and units already present in the AO. Liaison personnel and basic communications should be in place prior to the scheduled arrival of parent commands. ISR units establish intelligence communications networks.

1-127. CONUS and other relatively secure intelligence bases outside the AO continue to support deployed units. Systems capable of rapid receipt and processing of intelligence from national systems and high capacity, long-haul communications systems are critical to the success of split-based support of a force projection operation. These systems provide a continuous flow of intelligence to satisfy many operational needs.

1-128. The G2/S2, in coordination with the G3/S3, participates in planning to create conditions for decisive operations. The G2/S2 also adjusts collection activities as combat strength builds. During entry operations the G2/S2—

- Monitors FP indicators.
- Monitors the ISR capability required to conduct sustained intelligence operations.
- Monitors intelligence reporting on threat activity and I&W data.
- Develops measurable criteria to evaluate the results of the intelligence synchronization plan.
- Assesses—
  - Push versus pull requirements of intelligence reach.
  - Effectiveness of the intelligence communications architecture.
  - Reporting procedures and timelines.
  - Intelligence to OPLANs and OPORDs, branches, and sequels (to include planning follow-on forces).

EMPLOYMENT

1-129. Employment is the conduct of operations to support a JFC commander. Employment encompasses an array of operations, including but not limited to—

- Entry operations (opposed or unopposed).
- Shaping operations (lethal and non-lethal).
- Decisive operations (combat or support).
- Postconflict operations (prepare for follow-on missions or redeployment).

Entry Operations

1-130. Enemies often possess the motives and means to interrupt the deployment flow of ARFOR. Threats to deploying forces may include advanced conventional weaponry (air defense, mines, etc.) and WMD. Sea and air ports of debarkation (PODs) should be regarded as enemy HPTs because they are the entry points for forces and equipment. PODs are vulnerable because they are fixed targets with significant machinery and equipment that is vulnerable to attack; in addition to military forces and materiel, HN support personnel, contractors, and civilians may all be working there. An enemy attack, or even the threat of an enemy attack, on a POD can have a major impact on force projection momentum. Commanders at all levels require predictive intelligence so that they may focus attention on security actions that reduce vulnerabilities. To avoid, neutralize, or counter threats to entry operations, the commanders rely on the ability of the G2/S2 to support future operations by accurately identifying enemy reactions to US actions, anticipating their response to our counteractions and predicting additional ECOAs.

1-131. Predictive intelligence also supports the decisions the commander and staff must make about the size, composition, structure, and deployment sequence of the force in order to create the conditions for success. Commanders rely on predictive intelligence to identify potential friendly decisions before the actual event. While thorough planning develops friendly COAs to meet possible situations, the nature of an operation can change significantly before execution. The G2/S2 must provide timely, accurate, and predictive intelligence to ensure the commander can retain the initiative to implement the plan or make decisions before losing the opportunity to do so.

Shaping Operations

1-132. Shaping operations create and preserve conditions for the success of the decisive operation. Shaping operations include lethal and non-lethal activities conducted throughout the AO. They support the decisive operation by affecting enemy capabilities and forces, or by influencing enemy decisions. G2/S2 intelligence analysis and ISR activities support the development and execution of shaping operations by identifying threat centers of gravity and decisive points on the battlefield. The G2/S2 also ensures the intelligence process focuses on the CCIRs (PIRs and FFIRs). It is critical that the G2/S2 provide timely, accurate, and predictive intelligence to the commander to facilitate the execution of shaping operations. Predictive intelligence should provide sufficient time for the commander to understand how the enemy will react to US COAs so that appropriate shaping operations can be implemented.

Decisive Operations

1-133. Decisive operations are those that directly accomplish the task assigned by the higher headquarters. Decisive operations conclusively determine the outcome of major operations, battles, and engagements. Continuously synchronized ISR activities, coupled with predictive intelligence results and products during all stages of force projection, combine to ensure the commander is prepared to employ the right forces with the right support to conduct decisive operations at the most appropriate place and time. In addition to coordinating with the G3/S3 to synchronize all ISR activities, the G2/S2 must answer the CCIRs (PIRs and FFIRs) and provide continuous assessments of the enemy's current situation and predict the enemy's subsequent COA, branches, and/or sequels.

Postconflict Operations

1-134. Upon cessation of hostilities or truce, deployed forces enter a new stage of force projection operations. Postconflict operations focus on restoring order, reestablishing HN infrastructure, preparing for redeployment of forces, and planning residual presence of US forces. While postconflict operations strive to transition from conflict to peace, there remains a possibility of resurgent hostilities by individuals and forces. The ISR effort, particularly predictive intelligence analysis, remains just as critical in postconflict operations as in the other employment operations. ISR operations support the postconflict emphasis on restoration operations exemplified by commanders redirecting their CCIRs (PIRs and FFIRs) and IRs to support units conducting these missions. These operations might include—

- Engineer units conducting mine clearing or infrastructure reconstruction operations.
- Medical and logistics units providing humanitarian relief.
- MP units providing law and order assistance.
- CA units reestablishing local control and preparing for the orderly transition of local governments and to support civil-military operations (CMO).

SUSTAINMENT

1-135. Sustainment involves providing and maintaining levels of personnel and materiel required to sustain the operation throughout its duration. It is essential to generating combat power. CSS may be split-based between locations within and outside CONUS. These operations include ensuring units have the MI assets required to accomplish the mission, such as personnel (including linguists), communications systems, ISR systems, and appropriate maintenance support.

REDEPLOYMENT

1-136. Redeployment is the process by which units and materiel reposture themselves in the same theater; transfer forces and materiel to support another JFC's operational requirements; or return personnel and materiel to the home or demobilization station upon completion of the mission. Redeployment operations encompass four phases:

- Recovery, reconstitution, and pre-deployment activities.
- Movement to and activities at the port of embarkation (POE).
- Movement to the POD.
- Movement to home station.

1-137. As combat power and resources decrease in the AO, FP and I&W become the focus of the commander's intelligence requirements. This in turn drives the selection of those assets that must remain deployed until the end of the operation and those that may redeploy earlier. The S2—

- Monitors intelligence reporting on threat activity and I&W data.
- Continues to conduct intelligence to FP.
- Requests ISR support (theater and national systems) and intelligence in support of redeployment.

1-138. After redeployment, MI personnel and units recover and return to predeployment activities. ISR units resume contingency-oriented peacetime intelligence operations. US Army Reserve ISR units demobilize and return to peacetime activities. G2/S2s must—

- Monitor intelligence reporting on threat activity and I&W data.
- Update or consolidate databases.
- Maintain intelligence readiness.
- Provide their input into the Force Design Update (FDU) process to refine modified table of organizations and equipment (MTOE) and evaluate the need for Individual Mobilization Augmentee (IMA) personnel.

Chapter 2

Intelligence and Unified Action

UNIFIED ACTION

2-1. As discussed in JP 1-02 and Chapter 1, unified action describes the wide scope of actions (including the synchronization of activities with governmental and NGO agencies) taking place within unified commands, subordinate unified commands, or JTFs under the overall direction of the commanders of those commands. Under unified action, commanders integrate joint, single-service, special, and supporting intelligence operations with interagency, nongovernmental, and multinational operations. The ARFOR often brings unique ISR capabilities to unified action operations.

2-2. In a unified action, the intelligence staff takes on additional responsibilities, relationships, and infrastructure to execute multi-service and multinational intelligence operations. This chapter discusses the synchronization of Army intelligence efforts with joint and other national and international partners to achieve unity of effort and to accomplish the commander's intent.

THE LEVELS OF WAR

2-3. The levels of war are doctrinal perspectives that clarify the links between strategic objectives and tactical actions. Although there are no finite limits or boundaries between them, the three levels of war are strategic, operational, and tactical (see Figure 2-1).

- The strategic level is that level at which a nation, often as one of a group of nations, determines national and multinational security objectives and guidance and develops and uses national resources to accomplish them.
- The operational level is the level at which campaigns and major operations are conducted and sustained to accomplish strategic objectives within theaters or AOs. It links the tactical employment of forces to strategic objectives.
- The tactical level is the employment of units in combat. It includes the ordered arrangement and maneuver of units in relation to each other, the terrain, and the enemy to translate potential combat power into victorious battles and engagements.

2-4. Understanding the interdependent relationship of all three levels of war helps commanders visualize a logical flow of operations, allocate resources, and assign tasks. Actions within the three levels are not associated with a particular command level, unit size, equipment type, or force or component type. The concept of strategic, operational, and tactical intelligence operations aids JFCs and their J2s in visualizing the flow of intelligence from one level to the next. The concept facilitates allocating required collection, analysis, production, and dissemination resources; and facilitates the assignment of appropriate intelligence tasks to national, theater, component, and supporting intelligence elements.

STRATEGIC

2-5. The President and the Secretary of Defense use strategic intelligence to develop national strategy and policy, monitor the international situation, prepare military plans, determine major weapon systems and force structure requirements, and conduct strategic operations.

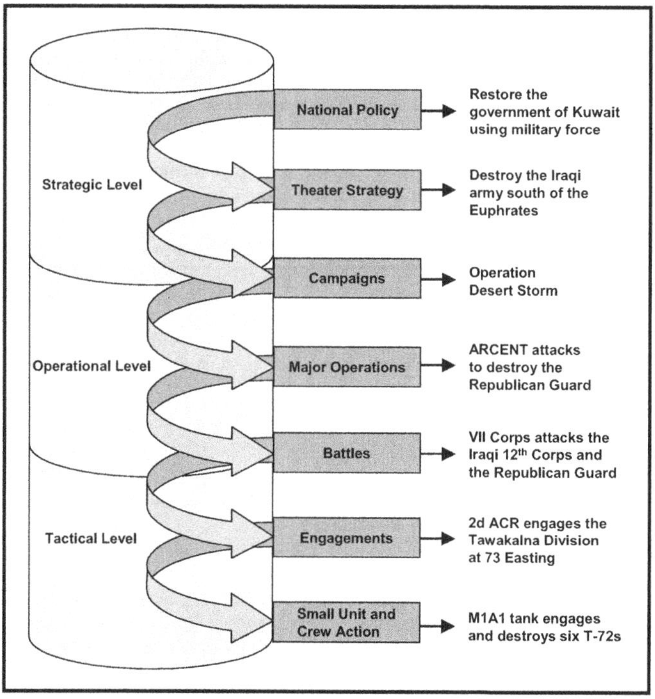

Figure 2-1. The Levels of War.

2-6. Intelligence supports joint operations across the full spectrum of military operations. It determines the current capabilities, and forecasts future developments, of adversaries or potential adversaries that could affect the national security and interests. A strategic operation also produces the intelligence required by combatant commanders to prepare strategic estimates, strategies, and plans to accomplish missions assigned by higher authorities. Theater intelligence includes determining when, where, and in what strength the adversary will stage and conduct theater level campaigns and strategic unified operations. The intelligence staff should also focus predictive analysis efforts on identifying strategic threat events and how these events will impact US actions at the strategic, operational, and tactical levels. Intelligence operations support IO as well. Intelligence operations support strategic planning by—

- Developing strategic intelligence policy.
- Preparing the strategic collection plan.
- Allocating national intelligence resources.

2-7. Intelligence requirements are those requirements generated from the staff's IRs regarding the enemy and environment that are not a part of the CCIRs (PIRs and FFIRs). Intelligence requirements require collection and can provide answers in order to identify indicators of enemy actions or intent, which reduce the uncertainties associated with an operation. Significant changes (that is, branches and sequels) with an operation usually lead to changes in intelligence requirements. Of particular importance is information relating to enemy or threat strategic vulnerabilities, strategic forces, strategic centers of gravity, and any capabilities relating to the development and employment of CBRNE.

2-8. Global and regional issues and threats are identified and reported to the President and the Secretary of Defense, as well as to the senior military leadership and the combatant commanders. Intelligence requirements include any foreign developments that could threaten the US, its citizens abroad, or multinational military, political, or economic interests. Intelligence also includes identifying hostile reactions to US reconnaissance activities and indications of impending terrorist attacks (I&W). For a complete listing of strategic tasks, refer to CJCSM 3500.04C.

OPERATIONAL

2-9. Combatant commanders and subordinate JFCs and their component commanders are the primary users of intelligence. At the operational echelons, intelligence—

- Focuses on the military capabilities and intentions of enemies and threats.
- Provides analysis of events within the AOI and helps commanders determine when, where, and in what strength the adversary might stage and conduct campaigns and major operations.
- Supports all phases of military operations, from mobilization all the way through redeployment of US forces, and continues during sustainment.
- Supports all aspects of the joint campaign.
- Identifies adversary centers of gravity and HVTs.
- Provides critical support to friendly IO.

2-10. The JFC and staff allocate intelligence resources and request support from national agencies, other theaters, and multinational partners. During stability operations and support operations, operational intelligence includes training and assisting multinational partners in conducting intelligence operations.

2-11. Combatant commanders use intelligence concerning the nature and characteristics of the battlespace to determine the type and scale of operations. Intelligence also aids in determining the impact of significant regional features and hazards on the conduct of both friendly and adversary operations. Significant regional factors include the natural environment, political, informational, economic, industrial, infrastructure, geographic, demographic, topographic, hydrographic, climatic, populace, cultural, medical, lingual, historical, and psychological aspects of the AOI. Intelligence analysis also assists in determining the ROE and other restrictions which will affect operations in the JFC's AO.

2-12. Intelligence relating to the adversary's military and nonmilitary capabilities assists in determining the adversary's ability to conduct military operations. Factors that operational intelligence addresses include mobilization potential, force structure (including alliance forces), force dispositions, equipment, military doctrine, C2 structure, and their MDMP. Intelligence includes the continuous refinement of the OBs for the entire array of the enemy's forces in the AOI. For a complete listing of operational tasks, refer to CJCSM 3500.04C.

US Army Intelligence and Security Command (INSCOM)

2-13. The U.S. Army has vested its intelligence at the operational level with INSCOM, the major Army command (MACOM) responsible for the Army's intelligence forces above Corps. INSCOM's mission is to conduct and support dominant intelligence, security, and information operations for military commanders and national decisionmakers. The INSCOM strategy is to provide superior information and information capabilities to Army commanders, while denying the same to adversaries. Headquarters (HQ), INSCOM, in coordination with its major subordinate commands (MSCs), provides a myriad of general intelligence support operations. INSCOM is providing a globally focused, rapidly deployable, knowledge based, adaptively force packaged capability, supporting commanders and leaders with actionable intelligence at the point of decision. INSCOM serves as the national to tactical Intelligence bridge.

Army Space Program Office (ASPO)

2-14. ASPO executes the Army's Tactical Exploitation of National Capabilities Program (TENCAP). The program focuses on exploiting current and future tactical potential of national systems and integrating the capabilities into the Army's tactical decisionmaking process. Army TENCAP systems enable the tactical commander maximum flexibility to satisfy intelligence needs under a wide range of operational scenarios. ASPO is the point of contact (POC) for all tactical activities between MACOMs or users and the National Reconnaissance Office (NRO).

TACTICAL

2-15. Tactical commanders use intelligence for planning and conducting battles and engagements. Relevant, accurate, predictive, and timely intelligence allows tactical units to achieve an advantage over their adversaries. Precise and predictive intelligence, on the threat and targets, is essential for mission success. Predictive intelligence also enables the staff to better identify or develop ECOAs. Tactical intelligence—

- Identifies and assesses the enemy's capabilities, COAs, and vulnerabilities, as well as describes the battlespace.
- Seeks to identify when, where, and in what strength the enemy will conduct tactical level operations.
- Provides the commander with information on imminent threats to the force including those from terrorists, saboteurs, insurgents, and foreign intelligence collection.
- Provides critical support to friendly IO.
- Develops and disseminates targeting information and intelligence.

2-16. Intelligence provides the tactical commander with the information and intelligence he requires to successfully employ combat forces against enemy forces. Thus, intelligence tasks support the execution of battles and engagements. These intelligence tasks are different from those at other levels due to their ability to immediately influence the outcome of the tactical commander's mission. They include information gathered from tactical sources, such as combat information, interrogations, debriefings, and eliciting information from captured or misplaced personnel. For a complete listing of tactical tasks, refer to FM 7-15.

CATEGORIES OF INTELLIGENCE

2-17. Directly or indirectly, the unit receives intelligence from throughout the US intelligence community. The intelligence operations and products of national, joint, and service organizations that make up the intelligence community fall into one of six categories:

- I&W.
- Current intelligence.
- GMI.
- Target intelligence.
- S&T intelligence.
- CI.

2-18. The categories of intelligence are distinguishable from each other primarily by the purpose of the intelligence product. The categories can overlap and some of the same intelligence is useful in more than one category. Depending upon the echelon, intelligence organizations use specialized procedures to develop each category of intelligence. The following information describes each category.

INDICATIONS AND WARNINGS

2-19. Indications and warnings (I&W) are those intelligence activities intended to detect and report time-sensitive intelligence information on foreign developments that could involve a threat to the US or allied and/or coalition military, political, or economic interests or to US citizens abroad. I&W includes forewarning of enemy actions or intentions; the imminence of hostilities; insurgency; nuclear or non-nuclear attack on the US, its overseas forces, or alliled and/or coalition nations; hostile reactions to US reconnaissance activities; terrorist attacks; and other similar events. (JP 3-13). I&W comes from time-sensitive information and analysis of developments that could involve a threat to the US and multinational military forces, US political or economic interests, or to US citizens. While the G2/S2 is primarily responsible for producing I&W intelligence, each element, such as the MPs conducting PIO, within every unit contributes to I&W through awareness of the CCIRs and reporting related information.

CURRENT INTELLIGENCE

2-20. Current intelligence involves the integration of time-sensitive, all-source intelligence and information into concise, accurate, and objective reporting on the battlespace and current enemy situation. One of the most important forms of current intelligence is the enemy situation portion of the COP. The G2/S2 is responsible for producing current intelligence for the unit. Current intelligence supports ongoing operations across full spectrum operations. In addition to the current situation, current intelligence should provide projections of the enemy's anticipated situations (estimates) and their implications on the friendly operation.

GENERAL MILITARY INTELLIGENCE

2-21. GMI is intelligence concerning military capabilities of foreign countries or organizations or topics affecting potential US or multinational military operations relating to armed forces capabilities, including OB, organization, training, tactics, doctrine, strategy, and other factors bearing on military strength and effectiveness and area and terrain intelligence. This broad category of intelligence is normally associated with long-term planning at the national level. However, GMI is also an essential tool for the intelligence staff and should be in place long before the start of preparations for a particular military operation. An up-to-date, comprehensive intelligence database is critical to the unit's ability to plan and prepare rapidly for the range of operations and global environments in which it may operate. GMI supports the requirement to quickly respond to differing crisis situations with corresponding intelligence spanning the globe. One of the many places to get information for GMI is the medical intelligence database. The G2/S2 planner develops his initial IPB from GMI products.

2-22. The G2/S2 develops and maintains the unit's GMI database on potential threat forces and environments based on the commander's guidance. As an essential component of intelligence readiness, this database supports the unit's planning, preparation, execution, and assessment of operations. The G2/S2 applies and updates the database as it executes its intelligence production tasks.

TARGET INTELLIGENCE

2-23. Target intelligence is the analysis of enemy units, dispositions, facilities, and systems to identify and nominate specific assets or vulnerabilities for attack, re-attack, or exploitation (for intelligence). It consists of two mutually supporting production tasks: target development and combat assessment.

- **Target development** is the systematic evaluation and analysis of target systems, system components, and component elements to determine HVTs for potential attack through maneuver, fires, or non-lethal means.
- Once attacked, **combat assessment** provides a timely and accurate estimate of the affects of the application of military force (lethal or non-lethal) an IO on targets and target systems based on predetermined objectives.

SCIENTIFIC AND TECHNICAL INTELLIGENCE

2-24. Scientific and technical intelligence (S&TI) is the product resulting from the collection, evaluation, analysis, and interpretation of foreign S&T information which covers foreign developments in basic and applied research and in applied engineering techniques and S&T characteristics, capabilities, and limitations of all foreign military systems, weapons, weapon systems, and materiel, the research and development (R&D) related thereto, and the production methods employed for their manufacture.

2-25. S&T intelligence concerns foreign developments in basic and applied sciences and technologies with warfare potential. It includes characteristics, capabilities, vulnerabilities, and limitations of all weapon systems, subsystems, and associated materiel, as well as related R&D. S&T intelligence also addresses overall weapon systems and equipment effectiveness. Specialized organizations—such as the Defense Intelligence Agency (DIA) Missile and Space Intelligence Center (MSIC), INSCOM, Air Missile Defense (AMD), Area Air and Missile Defense Command (AAMDC), and National Ground Intelligence Center (NGIC)—produce this category of intelligence. The G2/S2 establishes instructions within SOPs, orders, and plans for handling and evacuating captured enemy materiel (CEM) for S&T intelligence (S&TI) exploitation.

COUNTERINTELLIGENCE

2-26. As previously defined at the end of Chapter 1, CI attempts to identify and recommend countermeasures to the threat posed by foreign intelligence security services (FISS) and the ISR activities of non-state entities such as organized crime, terrorist groups, and drug traffickers. CI analysis incorporates information from all sources as well as the results of CI collection, investigations, and operations to analyze the multidiscipline threat posed by foreign intelligence services and activities.

INTELLIGENCE COMMUNITY

2-27. There are many organizations in the intelligence community that support military operations by providing specific intelligence products and services. The J2/G2/S2 and his staff must be familiar with these organizations and the methods of obtaining information from them as necessary. Figure 2-2 shows organizations that compose the intelligence community.

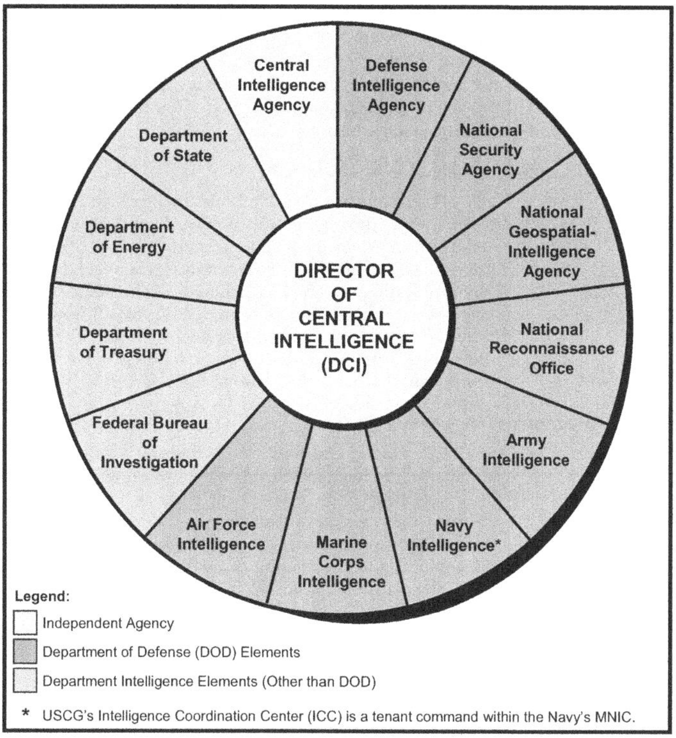

Figure 2-2. Intelligence Community Membership.

DOD AGENCIES

Defense Intelligence Agency (DIA)

2-28. DIA is a combat support agency and a major collector and producer in the defense intelligence community. The DIA supports the full spectrum of operations to include basic MI, counterterrorism, counterdrugs, medical intelligence, WMD and proliferation, United Nations (UN) peacekeeping and multinational support, missile and space intelligence, noncombatant evacuation operations (NEOs), targeting, combat assessment, and battle damage assessment (BDA).

National Security Agency (NSA)

2-29. NSA ensures cryptologic planning and support for joint operations. Working with the tactical cryptologic units of a command, the NSA provides SIGINT and information security (INFOSEC), encompassing communications security (COMSEC) and computer security, as well as telecommunications support and operations security (OPSEC). The people and equipment providing SIGINT, INFOSEC, and OPSEC constitute the United States Cryptologic System (USCS). The NSA, through the USCS, fulfills cryptologic command and/or management, readiness, and operational responsibilities in support of military operations according to the Secretary of Defense tasking, priorities, and standards of timeliness.

National Geospatial-Intelligence Agency (NGA)

2-30. The NGA (formerly National Imagery and Mapping Agency [(NIMA)] mission is to provide timely, relevant, and accurate intelligence and geospatial information in support of national security objectives of the United States. The Director of NGA advises the Secretary of Defense, Director Central Intelligence (DCI), Chairman of the Joint Chiefs of Staff (JCS), and the combatant commanders on imagery, IMINT, and geospatial information. The Operations Directorate, Customer Support Office, is the focal point for interface with external customers, including the JCS, combatant commands, services, and national and defense agencies.

National Reconnaissance Office (NRO)

2-31. The NRO develops and operates unique and innovative space reconnaissance systems and conducts intelligence-related activities essential for US security. The role of the NRO is to enhance US government and military information superiority across full spectrum operations. NRO responsibilities include supporting I&W, monitoring arms control agreements, and performing crisis support to the planning and conduct of military operations. The NRO accomplishes its mission by building and operating IMINT and SIGINT reconnaissance satellites and associated communications systems

US Navy (USN)

2-32. Naval intelligence products and services support the operating naval forces, the Department of the Navy, and the maritime intelligence requirements of national level agencies. Naval intelligence operates the National Maritime Intelligence Center (NMIC). Naval intelligence responsibilities include maritime intelligence on global merchant affairs, counter-narcotics, fishing issues, ocean dumping of radioactive waste, technology transfer, counter-proliferation, cryptologic related functions, CI, I&W support, management of Coast Guard collection, and development of new weapons systems and countermeasures.

US Marine Corps (USMC)

2-33. USMC intelligence provides pre-deployment training and force contingency planning for requirements that are not satisfied by theater, other service, or national capabilities. The Marine Corps Intelligence Agency (MCIA) handles the integration, development, and application of GMI, technical information, all-source production, and open-source information.

US Air Force (USAF)

2-34. USAF ISR assets fill a variety of roles to meet US national security requirements. The USAF operates worldwide ground sites and an array of airborne ISR platforms to meet national level intelligence requirements. To support day-to-day USAF operations and to meet specific USAF requirements, intelligence professionals at the wing and squadron levels use suites of interoperable analysis tools and dissemination systems to tailor information received from all levels and agencies in the intelligence community. USAF responsibilities include all-source information on aerospace systems and potential adversaries' capabilities and intentions, cryptologic operations, I&W, IO, and criminal investigative and CI services.

NON-DOD AGENCIES

2-35. Although the primary focus of non-DOD members of the intelligence community is strategic intelligence and support to the President and the Secretary of Defense, these agencies also produce intelligence and intelligence products that support operational and tactical ARFOR. This responsibility includes assessing potential issues and situations that could impact US national security interests and objectives. These agencies identify global and regional issues and threats. Some of the intelligence products and services these agencies provide are essential to accurate assessment of the threat and battlespace, particularly during stability operations and support operations.

Central Intelligence Agency (CIA)

2-36. The CIA's primary areas of expertise are in HUMINT collection, imagery, all-source analysis, and the production of political and economic intelligence. CIA and military personnel staff the CIA's Office of Military Affairs (OMA). As the CIA's single POC for military support, OMA negotiates, coordinates, manages, and monitors all aspects of agency support for military operations. This support is a continuous process, which the agency enhances or modifies to respond to a crisis or developing operation. Interaction between OMA and the DCI representatives to the Office of the Secretary of Defense (OSD), the Joint Staff, and the combatant commands facilitates providing national level intelligence in support of joint operations, contingency and operations planning, and exercises.

Department of State, Bureau of Intelligence and Research

2-37. The Bureau of Intelligence and Research coordinates programs for intelligence, analysis, and research; it produces intelligence studies and current intelligence analyses essential to foreign policy determination and execution. Its subordinate Bureau of International Narcotics Matters develops, coordinates, and implements international narcotics control assistance activities. It is the principal POC and provides policy advice on international narcotics control matters for the Office of Management and Budget (OMB), the NSC, and the White House Office of National Drug Control Policy (ONDCP). The Bureau also oversees and coordinates the international narcotics control policies, programs, and activities of US agencies.

Department of Energy, Office of Nonproliferation and National Security

2-38. The Department of Energy's Office of Nonproliferation and National Security assists in the development of the State Department's policy, plans, and procedures relating to arms control, nonproliferation, export controls, and safeguard activities. Additionally, this office is responsible for—

- Managing the department's R&D program.
- Verifying and monitoring arms implementation and compliance activities.
- Providing threat assessments and support to headquarters and field offices.

Department of the Treasury

2-39. The US Treasury Department's intelligence-related missions include producing and disseminating foreign intelligence relating to US economic policy and participating with the Department of State in the overt collection of general foreign economic information.

Federal Bureau of Investigation (FBI)

2-40. The FBI is the principal investigative arm of the Department of Justice (DOJ) and has primary responsibility for CI and counterterrorism operations conducted in the US. CI operations contemplated by any other organizations in the US must be coordinated with the FBI. Any overseas CI operation conducted by the FBI must be coordinated with the CIA.

Department of Homeland Security

2-41. The mission of the Department of Homeland Security is to develop and coordinate the implementation of a comprehensive national strategy to secure the US from terrorist threats or attacks. The organizational construct of homeland security mission is a framework of prepare, deter, preempt, defend, and respond. Component agencies will analyze threats and intelligence, guard our borders and airports, protect our critical infrastructure, and coordinate the response of our nation for future emergencies. The component agencies of the department include—

- US Coast Guard (USCG).
- US Customs Service.
- US Border Patrol.
- Federal Emergency Management Agency (FEMA).
- Immigration and Naturalization Service (INS).
- Transportation Security Administration.
- Federal Protective Service.
- Office of Domestic Preparedness.

United States Coast Guard (USCG)

2-42. The USCG, subordinate to the Department of Homeland Security, has unique missions and responsibilities as both an armed force and a law enforcement agency (LEA), which makes it a significant player in several national security issues. The USCG intelligence program supports counter-drug operations, mass seaborne migration operations, alien migration interdiction operations, living marine resource enforcement, maritime intercept operations, port status and/or safety, counterterrorism, coastal and harbor defense operations, and marine safety and/or environmental protection.

Other Agencies

2-43. There are a number of US Government agencies and organizations, not members of the intelligence community, that are responsible for collecting and maintaining information and statistics related to foreign governments and international affairs. Organizations such as the Library of Congress, the Departments of Agriculture and Commerce, the National Technical Information Center, US Information Agency, US Information Service, and the US Patent Office are potential sources of specialized information on political, economic, and military-related topics. The intelligence community may draw on these organizations to support and enhance research and analysis and to provide relevant information and intelligence for planners and decisionmakers. Many other US Government agencies may become directly involved in supporting DOD especially during stability operations and support operations. (See JP 2-02 for a description of agency support to joint operations and intelligence.) These organizations include—

- Department of Transportation.
- Disaster Assistance Response Team within the Office of Foreign Disaster.
- US Agency for International Development.
- NGA.

UNIFIED ACTION INTELLIGENCE OPERATIONS

2-44. In unified action, ARFOR synchronize their actions with those of other participants to achieve unity of effort and to accomplish the combatant commander's objectives. Unified action links subordinates to the combatant commander under combatant command (command authority) (COCOM). Multinational, interagency, and nonmilitary forces work with the combatant commander through cooperation and coordination. Combatant commanders form theater strategies and campaigns, organize joint forces, designate operational areas, and provide strategic guidance and operational focus to subordinates. The aim is to achieve unity of effort among many diverse agencies in a complex environment. Subordinate JFCs synchronize joint operations in time and space, direct the action of foreign military forces (multinational operations), and coordinate with governmental and NGOs (interagency coordination) to achieve the same goal.

2-45. The J2 staff provides intelligence promptly, in an appropriate form, and by any suitable means to those who need it. Intelligence personnel ensure that the consumers understand the intelligence and assist them as they apply the intelligence to their operations.

2-46. Dissemination requires establishing appropriate communications systems and procedures. The J2 and other intelligence personnel must fully participate in all operation planning and execution, and develop close working relationships with the JFC and other staff elements.

2-47. The commander and staff assess intelligence operations to determine their effectiveness and to make any necessary improvements. The intelligence process functions presented in Chapter 4 provide the criteria for evaluating intelligence operations.

JOINT INTELLIGENCE OPERATIONS

2-48. The JTF commanders and his intelligence staff must—

- Understand the intelligence requirements of superior, subordinate, and component commands.
- Identify organic intelligence capabilities and shortfalls.
- Access theater and national systems to ensure appropriate intelligence and CI products are available to the JTF.

2-49. The JTF's intelligence effort focuses on integrating multi-source information and multi-echelon intelligence into all-source intelligence products that provide clear, relevant, and timely knowledge of the enemy and battlespace. These products must be in formats that are readily understood and directly usable by the recipient in a timely manner. They must neither overload the user nor the communications architecture.

2-50. The J2 is the JTF commander's focal point for intelligence. The J2 directly supports the JFC's responsibilities for determining objectives, directing operations, and evaluating the effects of those operations. The J2 supports the execution of the plan with the intelligence needed to sustain the operation, attain joint force objectives, provide support to subordinate commands, and continually support FP efforts. The J2 analyzes the potential threat situation and provides assessments to support friendly opportunities. The J2 then supports the execution of the plan with the operational intelligence needed to sustain the operations, attain joint force objectives, and support FP. To maintain the initiative, the JFC will seek to get inside the adversary's decisionmaking cycle; that is, the JFC will seek to develop procedures and an organization in order to receive new and accurate intelligence and respond to the new situation faster than the adversary. The J2 must help in identifying the adversary's decisionmaking cycle and identifying vulnerabilities that may be exploited. The J2 also—

- Ensures the provision of the required ISR support to the JTF and its subordinate functional and service components.
- Assists the JTF commander in defining intelligence responsibilities and PIRs.
- Actively participates in joint staff planning and in planning, preparing, executing, and assessing ISR efforts. This includes leading the JTF's joint collection board and providing representation at the joint targeting board.

2-51. Figure 2-3 shows a typical JTF J2 organization. The overall organization of the JTF and operations will dictate actual composition of the J2. At a minimum, a core element of analytical, ISR management, and administrative capabilities is required.

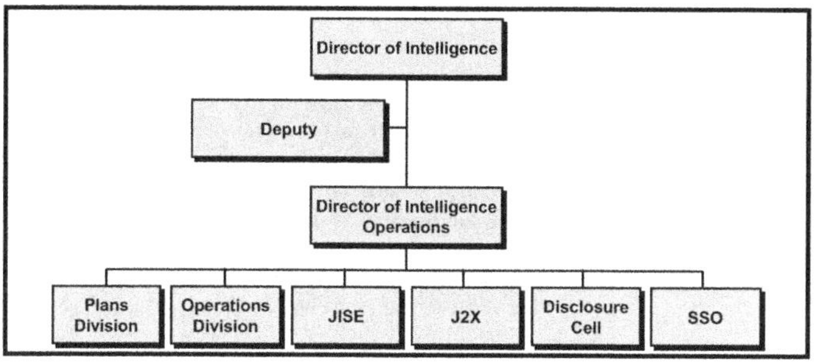

Figure 2-3. Typical Joint Task Force J2 Organization.

Considerations In Joint Intelligence Operations

2-52. The Army, Air Force, Navy, Marine Corps, and SOF must provide globally responsive assets to support the combatant commanders' theater strategies and the national security strategy. The capabilities of the other armed forces complement those of ARFOR. During joint operations, they provide support consistent with JFC-directed missions. When conducting joint intelligence operations, there are a number of unique problems that can arise due to the complexity of integrating the efforts of the different services and commands. Elements affecting joint intelligence operations among the different services include the following:

- Intelligence liaison is critical to the success of intelligence operations and requires early establishment, particularly between units that have not routinely trained together and possess differing capabilities. As a minimum, organizations exchange liaison teams with the higher echelon organization. Additional liaison may be necessary to facilitate joint force collection, production, or dissemination requirements. Liaison teams—
  - Support planning and C2 of intelligence operations.
  - Ensure timely two-way flow of intelligence between commands.
  - Manage intelligence and resource requirements of the subordinate command.
  - Advise the commander on service ISR capabilities, limitations, and employment.
- Commanders and staffs use IPB to understand the battlespace and develop or refine plans and orders. IPB products exchanged between echelons ensure a common picture of the battlespace and estimate of the situation.
- Communications considerations for joint operations include—
  - Planning for intelligence communications transition to facilitate execution of branches or sequels to the plan or to accommodate shifting of the main effort from one force to another.

  - Identifying the initial communications architecture to include establishing procedures and protocols for information exchanges (databases, text, imagery, voice, and video).
  - Balancing the availability of service-unique intelligence systems between echelons or services. This may require each service providing additional resources. The senior commander is responsible for allocating resources.
  - Disseminating intelligence between commands and services. Additional communications equipment, intelligence terminals, and personnel may be required to balance capabilities between services.
- Identifying the databases each service possesses or has access to; determining which databases will support the operation and, if necessary, merge them into a single database; and ensure access by the entire force.
- Providing a focal point for subordinate command access to national or joint intelligence is essential. The senior commander will request and allocate resources required to support this access.

2-53. The JFC's intelligence requirements, concept of operation, and intent drive the ISR effort. The different organizations and services participating in joint intelligence operations must continuously share information, intelligence, and products to satisfy requirements. See FM 2-01 for details on intelligence requirements and RM. Activities required to facilitate an effecttive joint collection effort include—

- Coordinating intelligence and ISR operations to optimize capabilities of collection assets and reduce duplication of effort.
- Integrating supporting national and theater intelligence collection assets into the intelligence synchronization plan.
- Establishing procedures for tracking and handing off HPTs between services and echelons.
- Establishing procedures for cueing Army and other service collection assets.
- Coordinating airspace for ISR assets.
- Maximizing available linguist capabilities. Shortages of military linguists trained in target languages may require cross-leveling Army and other service linguists.

2-54. Reporting and intelligence production considerations in joint operations include—

- Establishing production criteria and thresholds that produce timely and relevant intelligence keyed to the commander's intelligence and targeting requirements. The ISM, attack guidance matrix (AGM), and HPT list are examples of tools used to support joint intelligence and targeting efforts. The sharing of these products to all echelons and services is crucial.
- Establishing common methodology and criteria for producing the BDA and supporting the combat assessment function.
- Establishing set standards for the number and frequency of periodic reports such as intelligence summaries (INTSUMs).

Joint Intelligence Architecture

2-55. In addition to the J2 staffs at every joint level of command, the key organizations in the joint intelligence architecture are the National Military Joint Intelligence Center (NMJIC) (see Figure 2-4), the Joint Intelligence Centers (JICs) or Joint Analysis Centers (JACs) of the unified commands, and, when formed, the JTF's joint intelligence support element (JISE). Working together, these organizations play the primary role in managing and controlling joint intelligence operations. The formal relationships which link these organizations facilitate information management and optimize complementary intelligence functions by echelon without obstructing the timely flow of intelligence up, down, or laterally.

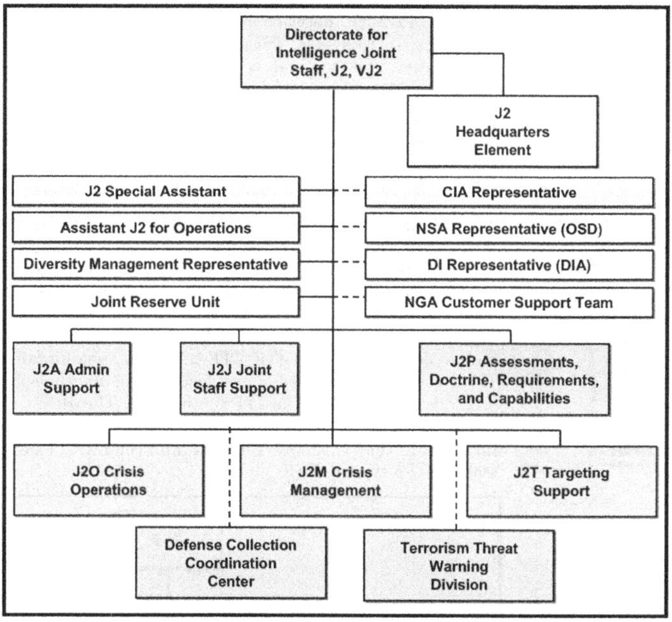

Figure 2-4. National Military Joint Intelligence Center.

2-56. The NMJIC is the focal point for intelligence activities in support of joint operations. The NMJIC is the CJCS J2 intelligence watch within the National Military Command Center (NMCC) in the Pentagon. When the assigned or attached assets cannot satisfy the combatant commander and the JTF commander's crisis-related and time-sensitive intelligence requirements, the NMJIC, as part of the J2's staff, sends tasks to appropriate agencies to fill the requirements. During crises, the NMJIC expands as necessary to establish a working group, intelligence task force, or an expanded intelligence task force. The NMJIC participates in targeting by developing national level target lists. When requested, the NMJIC supports the theater in performing BDA. Besides supporting the combatant commands and JTFs, the NMJIC supports any multinational partners and prescribed international organizations.

2-57. The JIC is organized in accordance with the combatant commander's prerogatives, but normally performs the general functions described in JP 2-0 and specific unified command intelligence TTP. The JIC is responsible for providing intelligence to JTF and its subordinates during military operations. If the JIC cannot meet the combatant commander's requirements, the JIC forwards requests to the NMJIC or to subordinate command levels through established channels, using standard command procedures. In some cases, the JIC may also ensure timely support by approving a direct communication path, in advance, between requesters such as the JTF JISE and outside producers. In this case, the parties must inform the JIC of all requests as they occur. This method is most appropriate when the parties require products and services, which the JIC does not routinely produce.

2-58. At the discretion of the JTF commander, the J2 can establish a JISE during the crisis or preparation stage for operations to augment the J2 staff. The JISE is a tailored subset of the JIC or the intelligence organization of the service component designated as the JTF headquarters.

2-59. Under the direction of the J2, the JISE normally manages the intelligence collection, production, and dissemination for the JTF. The JISE provides intelligence to JTF operational forces and performs common intelligence functions. By design, the JISE is scaleable and can expand to meet the needs of the JTF and the operating environment. It is composed of analytical experts, analysis teams, and ISR managers that provide services and products, which the JTF, JTF staff, and subordinate components require. These experts, mentioned above, focus on solving the operational intelligence problems of concern to the JTF commander. The JISE's capability to perform all-source analysis and ISR synchronization is key to producing operational intelligence that is timely, relevant, and complete. Figure 2-5 illustrates the features of a typical JISE.

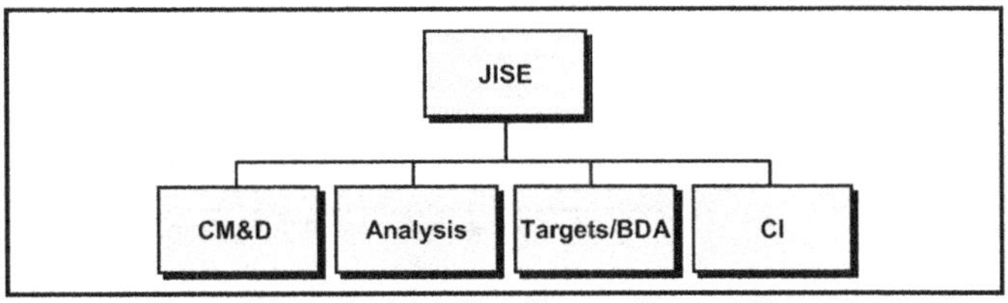

Figure 2-5. Typical Joint Intelligence Support Element.

Joint Task Force Intelligence Organizations

2-60. In addition to the JISE, the JTF commander and J2 may require other supporting JICs or teams based on projected operations. The JTF commander may make a request to the NMJIC for specific national intelligence agency capabilities. The NMJIC evaluates and coordinates the JTF commander's requirements with the J3, J5, and national intelligence agencies and tailors the composition of the deployment packages to meet those needs. The deployment packages, such as the National Intelligence Support Team (NIST), provide access to the entire range of capabilities resident in the national intelligence agencies and can focus those capabilities on the JTF commander's intelligence requirements. The J2X manages and coordinates the HUMINT and CI activities of national, theater, and service components operating within the JTF's JOA. The Joint Captured Materiel Exploitation Center (JCMEC) assists in management of recovery, exploitation, and disposal of captured enemy equipment (CEE). The JTF commander's requirements dictate the composition and tailoring of such deployment packages.

Augmentation Considerations

2-61. Depending on the scale of the operations, the intelligence organizations described above and those of the JTF's subordinate command may require personnel augmentation. Optimum use of available intelligence assets is essential to ensure quality support in meeting the JTF commander's requirements.

2-62. The demand for additional intelligence increases significantly during crisis and wartime operations. As the need arises for more intelligence personnel, the intelligence presence also increases at all command levels. Locating additional intelligence personnel and knowing how to integrate those personnel into the operation is vital. The JTF J2 should identify intelligence personnel augmentation requirements in accordance with the Chairman, Joint Chiefs of Staff (CJCS) Instruction 1301.01. The combatant commander's joint table of mobilization and distribution (JTMD) also contains the standing augmentation requirements. The JTMD should reflect the need for either IMA or Individual Ready Reserve (IRR) personnel. The combatant commander and the JTF refine personnel requirements and initiate requests when they anticipate or start an operation.

2-63. A consideration for the JTF when requesting support or augmentation is that these national level teams and individual augmentees are not totally self-contained elements; rather they require logistic, information, and other support from the supported command. Each deployment is unique based on mission, duration, team composition, and capabilities required. A full NIST, for example, requires a private, access-controlled area within a sensitive compartmented information facility (SCIF) work environment and dedicated secure communications.

2-64. For more information on intelligence operations as they apply to other armed services, see the individual service intelligence doctrine. See also JP 2-0 series, JP 3-0, JP 3-55, and JP 5-0 for more details on joint intelligence operations and considerations.

Multinational Intelligence

2-65. Multinational intelligence operations take place within the structure of an alliance or coalition. Some multinational military organizations, such as the North Atlantic Treaty Organization (NATO) and the UN Command in the Republic of Korea (ROK), are highly structured and enduring. Others, such as the coalition formed during the Gulf War, are less formal and temporary.

2-66. In multinational operations, the multinational force commander (MNFC) exercises command authority over a military force composed of elements from two or more nations. Therefore, in most multinational operations, the JTF must share intelligence, as necessary, for mission accomplishment with foreign military forces and coordinate exchange of intelligence with those forces.

2-67. In some circumstances, the JTF may need to seek authority to go outside the usual political-military channels to provide information to NGOs. The JTF must tailor intelligence policy and dissemination criteria to each multinational operation.

INTELLIGENCE REACH

2-68. Military forces use intelligence reach to rapidly access information, receive support, and conduct collaboration and information sharing with other units and organizations unconstrained by geographic proximity, echelon, or command.

2-69. Intelligence reach supports minimizing the deployed footprint of ISR assets. By providing enhanced information and tailored intelligence products, intelligence reach can greatly enhance the intelligence capabilities of the unit and play a significant role in improving the commander's ability to make decisions on the battlefield. It allows the commander to harness national, joint, foreign, and other military organization resources. Table 2-1 shows examples of partners and sources for intelligence reach.

2-70. Detailed planning and training are critical to the success of intelligence reach operations. The following are steps that the staff can take to ensure optimal use, operability, and effectiveness of intelligence reach:

- Establish data exchange methods and procedures.
- Establish electronic message transfer procedures.
- Establish homepages for identified forces.
- Establish POCs for I&W centers, production centers, Theater JICs, DIA, INSCOM, and their major subordinate commands such as NGIC and the theater intelligence brigades and groups.
- Ensure the intelligence staff has the necessary personnel, training, automated systems, bandwidth, and resources to conduct intelligence reach.
- Determine IRs through staff planning. Develop production requirements for identified intelligence gaps.
- Order geospatial products for the projected joint AOI.
- Establish and maintain a comprehensive directory of intelligence reach resources before deployment and throughout operations. The value of intelligence reach will greatly increase as the staff develops and maintains ready access to rich information resources. These resources are numerous and may include, for example, Army, Joint, DOD, non-DOD, national, commercial, foreign, and university research programs. Know what types of information the resources can provide. Continuously expand the resource directory through identification of new resources.

- Use intelligence reach first to fill intelligence gaps and requirements and answer RFIs. This technique can preclude unnecessary tasking or risk to limited ISR assets.
- Maintain continuous situational understanding and anticipate intelligence requirements. Use intelligence reach to fulfill these requirements and provide the results to the commander and staff for planning and decisionmaking.
- Exchange intelligence reach strategies with other units.
- Present the information retrieved through intelligence reach in a usable form. Share the information derived from intelligence reach with subordinate, lateral, and higher echelons. Ensure follow-on forces have all information as well.

Table 2-1. Examples of Intelligence and Reach Partners and Sources.

ARMY	SERVICES	JOINT	DOD
ACE ISE TSE (TIB/TIG) 902d NGIC ATCAE ARISCs	ONI NMIC AIA NAIC MCIA	USEUCOM JAC USSOUTHCOM JIC USSOCOM JIC USSPACECOM CIC USJFCOM AIC USCENTCOM JIC USTRANSCOM JIC USSTRATCOM USPACOM JIC USNORTHCOM JIC	DIA CMO MSIC AFMIC DHS NGA NSA RSOCs
NON-DOD	**NATIONAL**	**COMMERICAL**	**FOREIGN**
DOE FBI DOS DEA FEMA INS	DCI NIC CIA NRO INR DOT, Office of Intel Support	RAND Jane's Defence Weekly Economic Intelligence Unit CNN Reuters Associated Press United Press International	DIS NFHQ DIO

INTELLIGENCE REACH COMPONENTS

2-71. Intelligence reach requires the G2/S2 to develop a strategy on how best to support the unit's mission with intelligence reach capabilities. There are six basic components of the strategy:

- Push
- Database Access.
- Pull.
- Broadcast Services.
- Collaborative Tools.
- Requirements Management.

Push

2-72. Push occurs when the producers of intelligence or information are knowledgeable of the customer's requirements and are able to send the desired intelligence to the customer without further requests. Push is accomplished through the most appropriate or efficient dissemination means. Unless coordinated as an acceptable dissemination method, push is not accomplished solely by posting intelligence products on a web page. The entity that posts the document must ensure the intended recipient is aware of the product's location or has received the product.

2-73. Push begins with the statement of intelligence interest (SII). The SII establishes the unit's intelligence and IRs. The SII is prepared by intelligence staff organizations to register their interest in receiving recurring hardcopy and softcopy reports, studies, and publications covering a wide variety of intelligence subjects. The J2/G2/S2 works with the Department of the Army Production and Dissemination Management (DAPDM) to establish a profile for information transfer. Intelligence reports or information meeting the unit's requirements are then automatically sent directly to the unit's classified or unclassified networks or communication systems.

2-74. Organizations that push information and data down to subordinate units must be careful of the amount, detail, and focus of the information they are sending. Too much information will overwhelm the subordinate unit. Crucial information may be lost in the midst of an overabundant flow of information, much of which is of little use. Intelligence reach often runs more efficiently if the user pulls the information or at least focuses the intelligence producers to send the appropriate information with the correct level of detail at the appropriate time.

Database Access

2-75. Access to local, theater, DOD, non-DOD, and commercial databases allows analysts to leverage stored knowledge on topics ranging from basic demographics to OB information. A validated DIA Customer Number (acquired by the J2/G2/S2) in combination with Secret Internet Protocol Network (SIPRNET) and Joint Worldwide Intelligence Communications System (JWICS) connectivity establishes access to most of the databases online. The challenge for an analyst is to gain an understanding of the structure, contents, strengths, and weaknesses of the database regardless of database type. Additionally, the procedures are often difficult for extracting portions or downloading and transferring the database to unit level automated information systems.

2-76. Each intelligence discipline has unique databases established and maintained by a variety of agencies. Database access is accomplished through unit or agency homepages via SIPRNET (Intelink-S) and JWICS (Intelink). The DAPDM office, upon approving the unit's SII, validates the requirement for access to the majority of these databases. Units coordinate with the DA dissemination and program manager (DPM) and/or the agency for access to those databases requiring passwords and permissions beyond normal Intelink-S and Intelink access.

Pull

2-77. Pull occurs when the requestor is familiar enough with existing databases and products to be able to anticipate the location of the desired information. Knowledge of both the types and locations of intelligence databases can greatly increase the efficiency of the Intelligence BOS by saving time and effort on the part of analysts at every echelon.

2-78. The G2/S2 must also ensure that intelligence reach capability extends to multinational forces, augmentees, and other services or organizations working as part of the JTF. The G2/S2 must forward relevant intelligence to units which do not possess the necessary automation to conduct intelligence reach.

Broadcast Services

2-79. Broadcast services are an integrated, interactive dissemination system, focusing on tactical user's IRs using joint message formats. The theater broadcast's data prioritization and processing occur through information management. Selected tactical, theater, DOD, and national collectors broadcast their messages using prescribed message formats. The analyst decides which types of messages are required. Broadcast service data is responsive to the operational commander's information needs and allows tactical users to construct successively detailed pictures of the battlespace.

2-80. Tactical users require intelligence and information to be available at the lowest classification level possible (for example, collateral SECRET) and releasable, as necessary, to multinational partners. Broadcast service information management will adhere to national policy as it relates to the provision of data to NATO or multinational partners.

Collaborative Tools

2-81. Collaboration is the sharing of knowledge, expertise, and information, normally online, and may take numerous forms. Collaborative tools are computer-based tools (groupware) that help individuals work together and share information. They allow for virtual online meetings and data sharing. As an example of the use of collaborative tools, the President or the Secretary of Defense and DOD during a crisis situation will establish a number of crisis action teams (CATs) or joint interagency working groups (JIAWGs). These groups or teams are formed to address specified subjects or topics in support of the warfighter or decisionmaker. The groups will focus on the crisis and normally publish their products on a homepage. Analysts with online access can participate in a number of ways. This includes passively accessing the homepages to study the products, sending queries to the identified POCs, or by having their organization join the CAT or JIAWG. Table 2-2 shows some examples of collaborative tools.

Requirements Management

2-82. The RM system provides a mechanism for submitting RFIs, tasking, and managing ISR assets. Analysts who are trained and familiar with the RM process and the various tasking procedures can leverage its systems for refined information. Each intelligence discipline has established procedures for requesting specific information. It is therefore advantageous to have someone familiar with each discipline's procedures to participate in synchronizing the ISR effort. For example, an analyst receives a HUMINT-based intelligence information report, which provides information on an event or subject. If he requires more refined data or clarification, he should submit a time-sensitive requirement or a source-directed requirement. This creates a two-way communication from the field to the collectors without creating additional standing requirements. The more familiar the analyst is with the RM process, the better he can leverage it.

2-83. The intelligence reach component of RM includes the ability of an intelligence officer at any level to request information, which is beyond what is available at his location. The normal procedure for obtaining intelligence or information not obtainable through the use of available ISR assets is to submit an RFI to the next higher echelon. Users enter RFIs into the RM system. See ST 2-33.5 for more information on intelligence reach operations and FM 2-01 for more information on RM.

Table 2-2. Examples of Collaborative Tools.

Tool	Description
Chat (Audio and Text)	Used to conduct conversations online.
Whiteboard	Permits real-time display of drawings, pictures, or documents for group discussion and comment. Participants can annotate in real time as well.
Bulletin Board	Used to post notices and facilitate discussions on any topic.
Video	Used at a desktop computer or a video teleconferencing (VTC) center to allow the person or group to see with whom they are communicating.
Discussion Groups (Newsgroups)	Topics are posted to a website for discussion and comment where participants can follow a line of discussion on a topic.
File-Sharing Tools	Virtual file cabinets allow information to be stored on web servers, and are available to anyone having access to the site and electronic permission to use the files.
Presentation Tools	Used in a virtual auditorium to allow lectures and briefings to be given to an audience.
Application Sharing	An entire team can use an application running on one computer to revise documents.
Text Tools	Allows live text input and editing by group members. Once complete, the text document can be copied into word processing software.
E-mail	Electronic mail.
Persistent Capability	The ability to preserve files, briefings, or other team or project material for future reference. Properly organized, it becomes an information management device and is invaluable to a long-term effort.
Instant Messaging	This allows real-time exchange of notes and messages.

PART TWO

Intelligence in Full Spectrum Operations

Part Two provides a primer of what and who constitutes both MI and the Intelligence BOS and the process that these entities use in order to provide warfighters and decisionmakers with the intelligence they require in order to accomplish their respective missions.

Chapter 3 discusses the role of MI and the Intelligence BOS within full spectrum operations. It provides an overview of intelligence readiness, particularly the intelligence requirements associated with force projection. The concept of MI asset technical control is discussed as a complement to, not a replacement of, the Army's command and support relationships.

Chapter 4 presents the Intelligence Process—the Intelligence BOS methodology that accomplishes the primary focus of intelligence in full spectrum operations, which provides the warfighter with effective intelligence.

Chapter 3

Fundamentals in Full Spectrum Operations

FULL SPECTRUM OPERATIONS

3-1. Army commanders at all echelons may combine different types of operations simultaneously and sequentially to accomplish missions. Full spectrum operations include offensive, defensive, stability, and support operations. Missions in any environment require ARFOR that are prepared to conduct any combination of these operations:

- **Offensive operations** aim to destroy or defeat an enemy. Their purpose is to impose US will on the enemy and achieve decisive victory.
- **Defensive operations** defeat an enemy attack, buy time, economize forces, or develop conditions favorable for offensive operations. Defensive operations alone normally cannot achieve a decision. Their purpose is to create conditions for a counteroffensive that allow ARFOR to regain the initiative.
- **Stability operations** promote and protect US national interests by influencing the threat, political, and information dimensions of the OE through a combination of peacetime developmental, cooperative activities, and coercive actions in response to crises. Regional security is supported by a balanced approach that enhances regional stability

and economic prosperity simultaneously. ARFOR presence promotes a stable environment.

- **Support operations** employ ARFOR to assist civil authorities, foreign or domestic, as they prepare for or respond to crises and relieve suffering. Domestically, ARFOR respond only when the President or Secretary of Defense direct. ARFOR operate under the lead federal agency and comply with provisions of US law, to include the Posse Comitatus and Stafford Acts. (See FM 3-0.)

3-2. Intelligence supports the commander across full spectrum operations. It helps the commander decide when and where to concentrate sufficient combat power to overwhelm the enemy. ISR is essential for the commander to preclude surprise from the enemy, maintain the initiative on the battlefield, and win battles. Commanders at all levels synchronize intelligence with the other BOSs to maximize their ability to see and strike the enemy simultaneously throughout the AO.

3-3. Every soldier in the command is responsible for detecting and reporting enemy activities, dispositions, and capabilities. This task is critical because the environment we operate in is characterized by violence, uncertainty, complexity, and asymmetric methods by the threat. The increased situational awareness that soldiers develop through personal contact and observation is a critical element of that unit's ability to more fully understand the OE. However, soldiers collect information, they are not intelligence collectors. While medical personnel cannot be assigned ISR tasks due to their Geneva Convention category status, medical personnel who gain information through casual observation of activities in plain view while discharging their humanitarian duties will report the information to their supporting intelligence element.

OFFENSIVE OPERATIONS

3-4. Offensive operations at all levels require effective intelligence to help the commander avoid the enemy's main strength and to deceive and surprise the enemy. During offensive operations, intelligence must provide the commander with the composition, disposition, limitations, employment characteristics, and anticipated enemy actions in a timely enough manner for the commander to significantly affect the enemy commander's decision cycle. It ensures commanders have the intelligence they need to conduct offensive operations with minimum risk of surprise.

DEFENSIVE OPERATIONS

3-5. The immediate purpose of defensive operations is to defeat an enemy attack. Commanders defend to buy time, hold key terrain, hold the enemy in one place while attacking in another, or destroy enemy combat power while reinforcing friendly forces. Intelligence should determine the enemy's strength, COAs, and location of enemy follow-on forces. The defending commander can then decide where to arrange his forces in an economy-of-force role to defend and shape the battlefield. Intelligence support affords him the time necessary to commit the striking force precisely.

3-6. Intelligence in area defensive operations identifies, locates, and tracks the enemy's main attack and provides the commander time to allocate sufficient combat power to strengthen the defense at the point of the enemy's main effort. Intelligence should also identify where and when the commander can most decisively counterattack the enemy's main effort or exploit enemy vulnerabilities.

3-7. Although the battlefield is normally organized as decisive, shaping, and sustaining operations, commanders conducting defensive operations may delineate intelligence in battlefield organizational terms of deep area, close area, and rear area.

STABILITY OPERATIONS

3-8. The environment is often much more complex during stability operations and as a result intelligence is often more complex. In fact, intelligence is even more important a factor (or operational multiplier) during stability operations. As a result, the commander must be even more involved in and knowledgeable of ISR (to include ISR operations the commander controls and other higher level ISR operations that may be occurring within his AO) during stability operations. For example, the commanders must understand the complex details of HUMINT and special access program (SAP) operations.

3-9. The commander requires the appropriate intelligence and IPB products in order to determine how best to influence the threat, political and information dimensions of the operational environment, and enhance regional stability. The identification and analysis of characteristics of the terrain and weather, politics, infrastructure, health status, civilian press, attitudes, and culture of the local populace and all possible threats are important in conducting stability operations. A lack of knowledge concerning local politics, customs, and culture could lead to US actions which attack inappropriate targets or which may offend or cause mistrust among the local population.

SUPPORT OPERATIONS

3-10. Support operations are usually nonlinear and noncontiguous. Commanders designate the decisive, shaping, and sustaining operations necessary for mission success. However, determining the intelligence requirements that drive the ISR effort intending to identify any potential threat's centers of gravity and decisive points may require a more complex and unorthodox thought process than that used in offensive and defensive operations. The G2/S2 may have to define the enemy differently. In support operations, the adversary is often the effects of disease, hunger, or disaster on a civilian population. US forces conducting support operations must also fully understand the organization and identity of key figures or groups within the region where they are operating, as these figures may influence greatly the actions of the population—both civilian and military.

ELEMENTS OF COMBAT POWER

3-11. Combat power is the ability to fight. It is the total means of destructive or disruptive force, or both, that a military unit or formation can apply against the adversary at a given time. Commanders combine the elements of combat power to meet constantly changing requirements and defeat an enemy. The elements of combat power are—

- **Manuever** is the employment of forces, through movement combined with fire or fire potential, to achieve a position of advantage with respect to the enemy to accomplish the mission. Maneuver is the means by which commanders concentrate combat power to achieve surprise, shock, momentum, and dominance.
- **Firepower** provides the destructive force essential to overcoming the enemy's ability and will to fight. Firepower magnifies the effects of maneuver by destroying enemy forces and restricting his ability to counter friendly actions; maneuver creates the conditions for the effective use of firepower.
- **Leadership** is the most dynamic element of combat power. Confident, audacious, and competent leadership focuses on the other elements of combat power and serves as the catalyst that creates conditions for success. Leadership provides purpose, direction, and motivation in all operations.
- **Protection** is the preservation of fighting potential of a force so the commander can apply maximum force at the decisive time and place. Protection is neither timidity nor risk avoidance.
- **Information** enhances leadership and magnifies the effects of maneuver, firepower, and protection. Today, ARFOR use information collected to increase their situational understanding before engaging the enemy. Information from the COP and running estimates, transformed into situational understanding, allows commanders to combine the elements of combat power in new ways.

THE FOUNDATIONS OF ARMY OPERATIONS

3-12. Understanding the principles of war and tenets of Army operations is fundamental to operating successfully across the range of military operations. The principles of war and tenets of Army operations form the foundation of Army operational doctrine. (Refer to FM 3-0 for a full description of the Principles of War and Tenets of Army Operations.)

- **The Principles of War** provide general guidance for conducting war and military operations other than war at the strategic, operational, and tactical levels. The principles are the enduring bedrock of Army doctrine.
- **The Tenets of Army Operations** build upon the principles of war. They further describe the characteristics of successful operations. These tenets are essential to victory.

THE OPERATIONAL FRAMEWORK

3-13. The operational framework consists of the arrangement of friendly forces and resources in time, space, and purpose with respect to each other and the enemy or situation. It consists of the **AO, battlespace**, and the **battlefield organization**. The framework establishes an area of geographic and operational responsibility and provides a way for commanders to visualize how to employ forces against the enemy. Commanders design an operational framework to accomplish their mission by defining and arranging its three components. They use the framework to focus combat power. The operational framework provides a mechanism through which the commander can focus the Intelligence BOS effort. Understanding the operational framework, AOIR, and intelligence coordination line (ICL) and their relationship to each other is key to planning and executing ISR operations.

- **AO** is an operational area defined by the JFC for land and naval forces. AOs do not typically encompass the entire operational area of the JFC but should be large enough for component commanders to accomplish their missions and protect their forces. AOs should also allow component commanders to employ their available systems to the limits of their capabilities. The AO is the basic control measure for all operations that defines the geographical area for which a particular unit is responsible. The commander—
  - Assumes responsibility for intelligence, maneuver, fires, terrain management, security, and movement within his AO.
  - Establishes control measures within his AO to assign responsibilities, coordinate intelligence, fires, and maneuver, and to control other activities.
- **Battlespace** is the environment, factors, and conditions that must be understood to successfully apply combat power, protect the force, or complete the mission. This includes air, land, sea, space, and the included enemy and friendly forces; facilities; weather; terrain; the EMS; and the information environment within the operational areas and AOIs. The G2/S2 performs IPB and synchronizes ISR activities throughout the battlespace as determined by the commander's METT-TC considerations.
  - Area of influence is the geographical area in which a commander can directly influence operations by maneuver of FS systems normally under the commander's command or control. Areas of influence surround and include the associated AO.
  - AOI is that area of concern to the commander, including the area of influence, areas adjacent thereto, and extending into enemy territory to the objectives of current or planned operations. This area also includes areas occupied by enemy forces who could jeopardize the accomplishment of this mission. (See JP 1-02.)
  - Information environment refers to a commander's battlespace that encompasses information activity affecting the operation. To envision that part of the information environment that is within their battlespace, commanders determine the information activities that affect their own operational capabilities and opposing C2 and information systems.
  - Force Projection Bases are the intermediate staging bases and power projection platforms.
  - Home Stations are the permanent locations of active and RC units. To a significant degree, events occurring at home station affect the morale and performance of deployed forces. Thus, the commander's battlespace encompasses all home station functions.

- **Battlefield Organization** is the allocation of forces in the AO by purpose. It consists of three all-encompassing categories of operations:
  - Decisive Operations are those that directly accomplish the task assigned by higher headquarters. Decisive operations conclusively determine the outcome of major operations, battles, and engagements.
  - Shaping Operations are operations at any echelon that create and preserve conditions for the success of decisive operations.
  - Sustaining Operations are operations at any echelon that enable shaping and decisive operations by providing CSS, rear area and base security, movement control, terrain management, and infrastructure development.

3-14. AOIs extend into enemy territory, to the objectives of current or planned operations. They include areas occupied by enemy forces that could jeopardize the accomplishment of the mission. AOIs also serve to focus intelligence development and IO directed at factors outside the AO that may affect the operation. (FM 3-0, para 4-79) The scheme of support includes the coordination of reconnaissance and surveillance missions and AOIs with a joint force or higher headquarters and lateral units to answer the intelligence requirements within the AOI. The G2/S2 monitors enemy activities within the AOI and provides intelligence on enemy activities that may affect the unit.

3-15. The AOIR is an area allocated to a commander in which the commander is responsible for the provision of intelligence within the means at the commander's disposal. It is an area allocated to the commander in which the commander is responsible for the collection of information concerning the threat and the environment and the analysis of that information in order to produce intelligence. Higher headquarters also ensure through intelligence handovers, collection management, and deconfliction that problems with duplication, confliction, and command and control do not occur in the AOIR. (FM 2-19.402/FM 34-80-2) They include the available ISR assets, capability of the G2/S2 section, available intelligence architecture, and METT-TC considerations. The AOIR cannot extend beyond a unit's AOI; however, it can be smaller than its AO as well as vary (expand or contract) during an operation. An example of when a unit's AOIR is smaller than its AO is when a higher headquarters ISR effort covers an area within the unit's AO.

3-16. ICLs designate the boundary between AOIRs. The G2/S2 establishes ICLs to facilitate coordination between higher, lateral, and subordinate units; coordinates with the G3/S3 to direct subordinates to track enemy units and HPTs in their areas; and hands over intelligence responsibility for areas of the battlefield. The establishment of ICLs ensures that there are no gaps in the collection effort; that all echelons are aware of the location, mission, and capabilities of other assets. The G2/S2 keeps abreast of collection activities in progress (all echelons) and battlefield developments through the ICLs. ICLs help—

- Facilitate coordination between higher, lateral, and subordinate units.

- Ensure higher, lateral, and subordinate units share information and intelligence as enemy entities and HPTs move into, within, or transition between AOIRs.
- Specify intelligence responsibility for areas of the battlefield.

ARMY CAPABILITIES

3-17. Commanders combine AC and RC ARFOR—consisting of different types of units with varying degrees of modernization—with multinational forces and civilian agencies to achieve effective and efficient unified action. A broad range of organizations makes up the institutional Army that supports the field Army. Institutional Army organizations design, staff, train, and equip the force. The institutional Army assists in effectively integrating Army capabilities. It does this through leadership and guidance regarding force structure, doctrine, modernization, and budget. (See FM 100-11.)

TASK ORGANIZATION

3-18. The Army supports JFCs by providing tailored force packages to accomplish joint missions and dominate enemies and situations on land. Trained and equipped AC and RC units comprise these force packages. Within these force packages, Army commanders organize groups of units for specific missions. They reorganize for subsequent missions when necessary. This process of allocating available assets to subordinate commanders and establishing their command and support relationships is called task organizing. A temporary grouping of forces designed to accomplish a particular mission is a task organization. The ability of ARFOR to tailor (select forces based upon a mission) and task organize (temporarily organize units to accomplish a tactical mission) gives them extraordinary agility. It allows operational and tactical level commanders to organize their units to make the best use of available resources. The ability to task organize means ARFOR can shift rapidly among offensive, defensive, stability, and support operations.

COMBINED ARMS

3-19. The fundamental basis for the organization and operations of ARFOR is combined arms. Combined arms is the synchronized or simultaneous application of several arms—such as infantry, armor, field artillery, engineers, air defense, and aviation—to achieve an effect on the enemy that is greater than if each arm was used against the enemy separately or sequentially. The ultimate goal of Army organization for operations remains success in joint and combined arms warfare. Its combined arms capability allows commanders to form Army combat, combat support (CS), and CSS forces into cohesive teams focused on common goals.

COMMAND AND SUPPORT RELATIONSHIPS

3-20. Establishing clean command and support relationships is fundamental in organizing for all operations. These relationships can achieve clear responsibilities and authorities among subordinate and supporting units. The commander designates command and support relationships within his authority to weigh the decisive operation and support his scheme of maneuver. Some forces available to a commander are given command or support relationships that limit his authority to prescribe additional relationships. Command and support relationships carry with them varying responsibilities to the subordinate unit by parent and gaining units. By knowing the inherent responsibilities of each command and support relationship, a commander may organize his force to establish clear relationships.

3-21. Command relationships establish the degree of control and responsibility commanders have for forces operating under their tactical control (TACON). When commanders establish command relationships, they determine if the command relationship includes administrative control (ADCON). Figure 3-1 shows command and support relationships.

3-22. Support relationships define the purpose, scope, and effect desired when one capability supports another. Support relationships establish specific responsibilities between supporting and supported unit. Army support relationships are direct support (DS), general support (GS), general support-reinforcing (GSR), and reinforcing.

3-23. While not an actual C2 function, technical control often affects certain intelligence operations. Technical control ensures adherence to existing policies or regulations and provides technical guidance for MI activities, particularly HUMINT, SIGINT, and CI operations. Commanders direct operations but often rely on technical expertise to plan, prepare, execute, and assess portions of the unit's collection effort. Technical control also involves translating ISR tasks into the specific parameters used to focus highly technical or legally sensitive aspects of the ISR effort. Technical control includes, but is not limited to—

- Defining, managing, or guiding the employment of specific ISR assets.
- Identifying critical technical collection criteria such as technical indicators.
- Recommending collection techniques, procedures, or assets.
- Conducting operational reviews.
- Conducting operational coordination.
- Conducting specialized training for specific MI personnel or units.

3-24. An example of technical control is the Prophet control team converting the PIR and SOR sets from the MDMP process and assigning times and anticipated enemy frequencies for subordinate Prophets to collect.

	IF RELATIONSHIP IS:	INHERENT RESPONSIBILITIES ARE:							
		Has Command Relationship with:	May Be Task Organized by:	Receives CSS from:	Assigned Position or AO By:	Provides Liaison To:	Establishes/ Maintains Communications with:	Has Priorities Established by:	Gaining Unit Can Impose Further Command or Support Relationship of:
COMMAND	Attached	Gaining unit	Gaining unit	Gaining unit	Gaining unit	As required by gaining unit	Unit to which attached	Gaining unit	Attached; OPCON; TACON; GS; GSR; R; DS
COMMAND	OPCON	Gaining unit	Parent unit and gaining unit; gaining unit may pass OPCON to lower HQ. Note 1	Parent unit	Gaining unit	As required by gaining unit	As required by gaining unit and parent unit	Gaining unit	OPCON; TACON; GS; GSR; R; DS
COMMAND	TACON	Gaining unit	Parent unit	Parent unit	Gaining unit	As required by gaining unit	As required by gaining unit and parent unit	Gaining unit	GS; GSR; R; DS
COMMAND	Assigned	Parent unit	Parent unit	Parent unit	Gaining unit	As required by parent unit	As required by parent unit	Parent unit	Not Applicable
SUPPORT	Direct Support (DS)	Parent unit	Parent unit	Parent unit	Supported unit	Supported unit	Parent unit; Supported unit	Supported unit	Note 2
SUPPORT	Reinforcing (R)	Parent unit	Parent unit	Parent unit	Reinforced unit	Reinforced unit	Parent unit; reinforced unit	Reinforced unit: then parent unit	Not Applicable
SUPPORT	General Support Reinforcing (GSR)	Parent unit	Parent unit	Parent unit	Parent unit	Reinforced unit and as required by parent unit	Reinforced unit and as required by parent unit	Parent unit; then reinforced unit	Not Applicable
SUPPORT	General Support (GS)	Parent unit	Parent unit	Parent unit	Parent unit	As required by parent unit	As required by parent unit	Parent unit	Not Applicable

NOTE 1. In NATO, the gaining unit may not task organize a multinational unit (see TACON).

NOTE 2. Commanders of units in DS may further assign support relationships between their subordinate units and elements of the supported unit after coordination with the supported commander.

Figure 3-1. Army Command and Support Relationships and Inherent Responsibilities.

Chapter 4

Intelligence Process in Full Spectrum Operations

THE INTELLIGENCE PROCESS

4-1. Commanders use the operations process of plan, prepare, execute, and assess to continuously design and conduct operations (see Figure 4-1). The commander cannot successfully accomplish the activities involved in the operations process without information and intelligence. The design and structure of intelligence operations support the commander's operations process by providing him with intelligence regarding the enemy, the battlefield environment, and the situation.

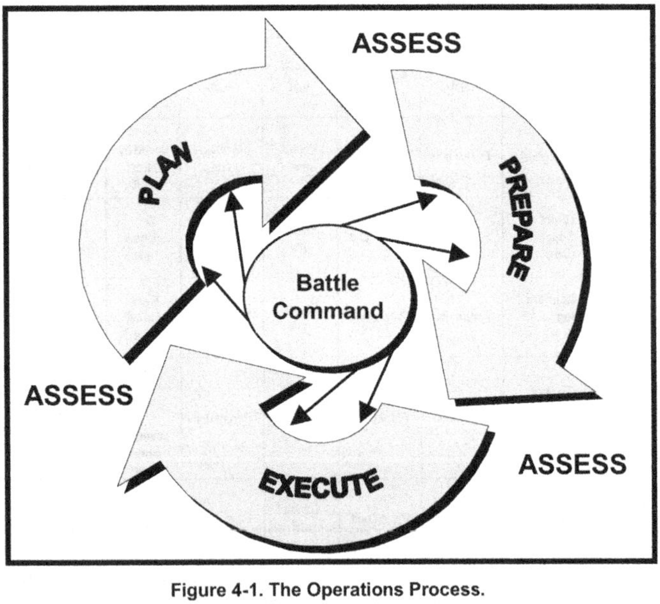

Figure 4-1. The Operations Process.

4-2. Intelligence operations consist of the functions that constitute the intelligence process: **plan, prepare, collect, process, produce**, and the three common tasks of **analyze, disseminate,** and **assess**. Just as the activities of the operations process overlap and recur as circumstances demand, so do the functions of the intelligence process. Additionally, the analyze, disseminate, and assess functions of the intelligence process occur continuously throughout the intelligence process.

4-3. The operations process and the intelligence process are mutually dependent. The commander, through the operations process, provides the guidance and focus through CCIRs and PIRs that drives the intelligence process; the intelligence process provides the continuous intelligence essential to the operations process. Intelligence about the enemy, the battlefield environment, and the situation allows the commander and staff to develop a plan, seize and retain the initiative, build and maintain momentum, and exploit success (see Figure 4-2). The intelligence process is just one of the mechanisms that provides input to build the COP and facilitate the commander's situational understanding.

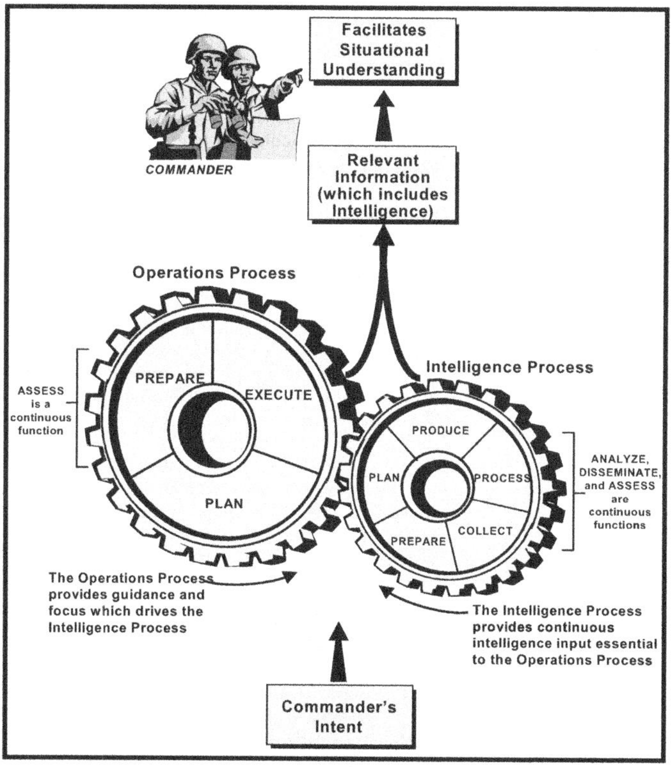

Figure 4-2. The Relationship Between the Operations and Intelligence Processes.

PLAN

4-4. The planning step of the intelligence process consists of activities that include assessing the situation, envisioning a desired outcome (also known as setting the vision), identifying pertinent information and intelligence requirements, developing a strategy for ISR operations to satisfy those requirements, directing intelligence operations, and synchronizing the ISR effort. The commander's intent, planning guidance, and CCIRs (PIRs and FIRs) drive the planning of intelligence operations. Planning, managing, and coordinating these operations are continuous activities necessary to obtain information and produce intelligence essential to decisionmaking.

COORDINATE

4-5. Staff and leaders coordinate with various elements, units, and organizations to ensure the necessary resources, linguist support (see Appendix B), information, intelligence, training, and procedures are in place to facilitate effective intelligence operations.

- Coordination for Movement of ISR Assets. All ISR assets at one time or another will move through or near another unit's AO. To avoid fratricide, ISR elements must coordinate with units, G3/S3, G2/S2, and each other, as well as coordinate with the fire support officer (FSO) to establish no-fire areas and/or other control measures around ISR assets and the air defense officer (ADO) in reference to aerial ISR assets in order to establish the appropriate weapons control status.
- Coordination for Information and Intelligence. The intelligence staff must prepare and practice coordination with personnel from all MI units, non-MI units, other service components, and multinational organizations that may contribute to or facilitate the collection effort. This coordination enables the G2/S2 to share and update databases, information, and intelligence and ensures connectivity with those organizations. All units are sources of relevant information regarding the enemy and the operational environment.
- Liaison. In order to accomplish the mission, exchange information and intelligence, move through certain areas and ensure FP, it may be necessary to coordinate with many different elements, organizations, and local nationals of the country in which friendly forces are conducting operations. Local nationals include police, town officials, foreign military forces, and political and other key figures within the AO. Operations may also necessitate coordination with other US and multinational forces; for example, the International Police Task Force (IPTF), Joint Commission Observers (JCO), Organization for Security and Cooperation in Europe (OSCE), Allied Military Intelligence Battalion (AMIB), and Defense HUMINT Service (DHS).
- Movement. Coordination with the G3/S3 ensures ease of movement and safe passage of friendly forces through an area. Coordinating movement also helps avoid fratricide.

INTELLIGENCE BOS CONSIDERATIONS

4-6. The Intelligence BOS is a unified system that anticipates and satisfies intelligence needs. Commanders ensure its proper employment by clearly articulating intent, designating CCIRs (PIRs and FFIRs), and prioritizing targets. Commanders must, however, understand the limitations of the Intelligence BOS to preclude unrealistic expectations of the system. The following are Intelligence BOS limitations:

- Intelligence only reduces uncertainty on the battlefield; it does not eliminate it entirely. The commander will always have to determine the presence and degree of risk involved in conducting a particular mission.
- The Intelligence BOS is composed of finite resources and capabilities. Intelligence systems and soldiers trained in specific ISR skills are limited in any unit. Once lost to action or accident, these soldiers and systems are not easily replaceable; for some, it may not be possible to replace them during the course of the current operation. The loss of soldiers and equipment can result in the inability to detect or analyze enemy actions. The loss of qualified language-trained soldiers, especially soldiers trained in low-density languages or skills, could adversely affect intelligence operations as well.
- The Intelligence BOS cannot effectively and efficiently provide intelligence without adequate communications equipment, capacity, and connectivity. Commanders and G2/S2s must ensure communications support to intelligence has the appropriate priority.
- Commanders and G2/S2s cannot expect that higher echelons will automatically send them everything they need. While intelligence reach is a valuable tool, the push of intelligence products from higher echelons does not relieve subordinate staffs from conducting detailed analysis and focusing the efforts of higher headquarters. Nor can they expect products pushed to them to be always at the level of detail they require. Commanders and G2/S2s must focus higher echelons by clearly articulating and actively pursuing intelligence requirements. By providing higher echelons with a clear picture of the required intelligence products, commanders can also narrow the flow of intelligence and information and preclude being overwhelmed by too much information.

4-7. Commanders should be aware that intelligence collection is enabled by, and subject to, laws, regulations, and policies to ensure proper conduct of intelligence operations. While there are too many to list here specifically, categories of these legal considerations include United States Codes (USCs), Executive Orders, National Security Council Intelligence Directives (NCSIDs), Army Regulations, United States Signal Intelligence Directives (USSIDs), SOFAs, ROE, and other international laws and directives.

PREPARE

4-8. The prepare step includes those staff and leader activities which take place upon receiving the OPORD, OPLAN, WARNO, or commander's intent to improve the unit's ability to execute tasks or missions and survive on the battlefield. These activities include—

- Effecting necessary coordination in accordance with the OPORD, METT-TC, unit SOP.
- Establishing and testing the intelligence architecture. This activity includes complex and technical issues like hardware, software, communications, COMSEC materials, network classification, technicians, database access, liaison officers (LNOs), training, funding, and TTP.
- Establishing an intelligence team attitude. This activity includes knowing different unit's and organization's capabilities, training the necessary collective skills, establishing effective relationships with different units and organizations, developing mutual battle rhythms and TTP, and leveraging the right architectures and collaboration tools.
- Coordinating effective analytic collaboration. Effective analytic collaboration is necessary to maximize the complementary analytic capabilities of different units and organizations that produce intelligence within the same theater of operations. Coordinating this collaboration is an effort-intensive activity that requires careful mutual planning, division of labor, defined responsibilities, and procedures for adapting to changing circumstances as they develop.
- Establishing reporting procedures.
- Conducting IPB.
- Producing Intelligence Estimates.
- Presenting briefings.
- Ensuring staff and personnel are trained. If personnel are not adequately trained at this point, they must be trained or the leader must evaluate the risk they bring to the operation.
- Planning refinement, brief-backs, SOP reviews, rehearsals, and coordinating with various elements and organizations.
- Establishing other troop-leading procedures (TLPs) or coordination, as necessary, in accordance with METT-TC factors.

G2/S2 PREPARATION ACTIVITIES

4-9. The G2/S2 takes numerous steps before mission execution to ensure intelligence operations run smoothly and effectively. These steps include, but are not limited to, the following:

- Conduct rehearsals.
- Conduct communication rehearsals.
- Review and update available databases and IPB products.
- Review applicable SOPs, Army Regulations, DA Pamphlets, Field Manuals, and ROE for guidance in conducting intelligence operations.
- Plan and practice actions supporting likely contingencies, or the branches or sequels to an operation.
- Ensure coordination measures are still in effect.
- Ensure training (individual and collective).

- Verify communications protocols with theater and higher headquarters and subordinate and lateral units.
- Update intelligence databases.
- Update the forces with the most recent intelligence on the AO immediately before mission execution.

VERIFICATION

4-10. Coordination for or requesting provisions or services is only the first step in acquiring them. It is crucial that staff and leaders check to verify that procedures, personnel, equipment, and services are in place and ready for mission execution.

REHEARSALS

4-11. Rehearsals help units prepare for operations by either verifying that provisions and procedures are in place and functioning or identifying inadequacies, which staff and leaders must remedy. They allow participants in an operation to become familiar with and to translate the plan into specific actions that orient them to their environment and other units when executing the mission. They also imprint a mental picture of the sequence of key actions within the operation and provide a forum for subordinate and supporting leaders and units to coordinate. (FM 6-0)

REPORTING PROCEDURES

4-12. The timely and accurate reporting of CCIRs (PIRs and FFIRs) and IRs is key to successful operations. All assets should know when, how often, and what format to use when reporting. The G2/S2 must verify the frequencies, alternate frequencies, and reactions during jamming, as well as the LTIOV for specific information to be reported. Unit SOPs provide the proper reporting procedures.

4-13. The G2/S2 coordinates with the unit staff, subordinate and lateral commands, and higher echelon units to ensure that specific reporting assets, personnel, equipment (especially communications), and procedures are in place. The G2/S2 requests or establishes the appropriate message addresses, routing indicators, mailing addresses, and special security office (SSO) security accreditation for units.

COMMUNICATIONS

4-14. Staff and leaders must work closely with the G6/S6 or signal officer (SIGO) to coordinate for the required communication links. The unit may require classified and unclassified network connections for their equipment. If elements of the unit will be working outside the range of the unit's communications systems, then it is necessary to coordinate for global or extended range communications. Leaders must obtain the required type and amount of communications equipment and related components as well as the latest fills and frequencies. They must possess and be familiar with all the instructions, passwords, policies, regulations, and directives conducive to OPSEC. They must also ensure soldiers are trained in the use and procedures involved in operating communications equipment.

SITUATION UPDATES

4-15. Each staff section and element conducts activities to maximize the operational effectiveness of the force. Coordination and preparation are just as important, if not more important, as developing the plan. Staff preparation includes assembling and continuously updating estimates. For example, continuous IPB provides accurate situational updates for commanders.

INTELLIGENCE HANDOFF

4-16. A well-executed intelligence hand-off will ensure a smooth and seamless transition between units. It is important that the incoming unit becomes familiar with the operation as soon as possible to avoid compromising the intelligence production and flow of the mission. The following are points to consider during a mission hand-off:

- Briefings and reports (learn what briefings are required and when as well as report formats and requirements).
- Past, present, and planned activities within the AOI.
- Established SOPs (know procedures for reporting; intelligence contingency funds [ICFs] and incentive use if applicable; emplacement and use of ISR equipment).
- Key personalities (introductions are required; establish rapport and a good working relationship with all key personalities).
  - Key personnel on the base or camp (their responsibilities; how to contact them).
  - Key personnel in other US and multinational service components (coordinate for exchange of information and intelligence).
  - Key personalities from surrounding towns (local figures).
  - Key national level political and military figures.
- Supporting units (know where to go for provisions, information, or assistance and POCs within those organizations).
- Current attitudes (understand current attitudes and perspectives of the local populace).
- Equipment operation and idiosyncrasies (equipment may run on different applications; personnel may need to train on specific equipment and procedures).
- Area familiarization (identify NAIs, key terrain, minefields, and boundaries; know camp locations, routes and route names, checkpoints, towns, and troubled resettlement areas).

RULES OF ENGAGEMENT

4-17. Although ROE training was presented during the plan function of the intelligence process, leaders at all levels can take the opportunity during the prepare function to ensure their subordinates completely understand the ROE. It is also during this function that commanders may need to consider exceptions to, or modifications of, the ROE to facilitate HUMINT and CI collection or to enable the placement of ISR assets.

COLLECT

4-18. Recent ISR doctrine necessitates that the entire staff, especially the G3/S3 and G2/S2, must change their reconnaissance and surveillance mind-set to conducting ISR. The staff must carefully focus ISR on the CCIR (PIR and FFIR) but also enable the quick retasking of units and assets as the situation changes. This doctrinal requirement ensures that the enemy situation not just our OPLAN "drives" ISR operations. Well-developed procedures and carefully planned flexibility to support emerging targets, changing requirements, and the need to support combat assessment is critical. The G2/S2 and G3/S3 play a critical role in this challenging task that is sometimes referred to as "fighting ISR" because it is so staff intensive during planning and execution (it is an operation within the operation). Elements of all units on the battlefield obtain information and data about enemy forces, activities, facilities, and resources as well as information concerning the environmental and geographical characteristics of a particular area.

ISR TASKS

4-19. ISR tasks are the actions of the intelligence collection effort. ISR tasks consists of three categories:

- Intelligence.
- Surveillance.
- Reconnaissance.

4-20. Intelligence tasks are included in Annex B of the OPORD under Scheme of Intelligence. They include the following:

- **Intelligence Production**. Intelligence production includes analyzing information and intelligence and presenting intelligence products, conclusions, or projections regarding the OE and enemy forces in a format that enables the commander to achieve situational understanding.
- **Request for Information**. Submitting an RFI to the next higher or lateral echelon is the normal procedure for obtaining intelligence information not available through the use of available ISR assets. Users enter RFIs into an RFI management system where every other user of that system can see it. Hence, an echelon several echelons above the actual requester becomes aware of the request and may be able to answer it. A G2/S2 who receives an RFI from a subordinate element may use intelligence reach to answer RFIs.
- **Intelligence Reach**. Intelligence reach allows the commander to access the resources of national, joint, foreign, and other military organizations and units. Requestors can acquire information through push and pull of information, databases, homepages, collaborative tools, and broadcast services. (See Chapter 2 for more information on intelligence reach.)

4-21. For information on reconnaissance or surveillance tasks, refer to FM 7-15 and FM 3-55.

SPECIAL RECONNAISSANCE

4-22. Special reconnaissance (SR) is the complementing of national and theater intelligence collection assets and systems by obtaining specific, well-defined, and time-sensitive information of strategic or operational significance. It may complement other collection methods where there are constraints of weather, terrain, hostile countermeasures, and/or other systems availability. SR places US or US-controlled personnel conducting direct observation in hostile, denied, or politically sensitive territory when authorized. SOF may conduct these missions unilaterally or in support of conventional operations. (See JP 3-05 and FM 101-5-1.)

4-23. Army Special Operations Forces (ARSOF) elements conduct SR missions to obtain information not available through other means. SR operations encompass a broad range of collection activities to include reconnaissance, surveillance, and TA. SR complements national and theater collection systems that are more vulnerable to weather, terrain masking, and hostile countermeasures. SR missions provide intelligence or information that is often not available through other means. Typical SR missions include—

- TA and surveillance of hostile C2 systems, troop concentrations, deep-strike weapons, lines of communication (LOCs), WMD systems, and other targets.
- Location and surveillance of hostage, prisoner of war, or political prisoner detention facilities.
- Post-strike reconnaissance for BDA.
- Meteorologic, geographic, or hydrographic reconnaissance to support specific air, land, or sea operations.

4-24. For more information on special reconnaissance, see FM 3-05.102.

REPORTING

4-25. The most critical information collected is worthless if not reported in a timely manner. Collectors may report information via verbal, written, graphic, or electronic means. Unit SOPs must clearly state the transmission means of different types of reports (for example, sent by voice frequency modulated [FM] radios or by automated means). In general, the transmission of reports for enemy contact and actions, CCIRs, exceptional information, and NBC reports is by voice FM, and then followed up with automated reports. Commanders and staffs must remember that timely reporting, especially of enemy activity, is critical in fast-moving operations. Collectors must report accurate information as quickly as possible. Commanders and staff must not delay reports for the sole purpose of editing and ensuring the correct format. This is particularly true for reporting information or intelligence that answers the PIR.

TIME-SENSITIVE REPORTING

4-26. Intelligence and time-sensitive combat information that affects the current operation is disseminated immediately upon recognition. Combat information is unevaluated data, gathered by or provided directly to the tactical commander which, due to its highly perishable nature or the criticality of the situation, cannot be processed into tactical intelligence in time to satisfy the user's tactical intelligence requirements. Thus combat information is provided directly to the tactical commander (see JP 1-02). The routing of combat information proceeds immediately in two directions: directly to the commander and through routine reporting channels, which include intelligence analysis and production elements.

4-27. Time-sensitive information usually includes reports concerning enemy contact and actions and CCIRs.

PROCESS

4-28. The process function converts relevant information into a form suitable for analysis, production, or immediate use by the commander. Processing also includes sorting through large amounts of collected information and intelligence (multidiscipline reports from the unit's ISR assets, lateral and higher echelon units and organizations, and non-MI elements in the battlespace). Processing identifies and exploits that information which is pertinent to the commander's intelligence requirements and facilitates situational understanding. Examples of processing include developing film, enhancing imagery, translating a document from a foreign language, converting electronic data into a standardized report that can be analyzed by a system operator, and correlating dissimilar or jumbled information by assembling like elements before the information is forwarded for analysis.

4-29. Often collection assets must collect and process their data prior to disseminating it. MI systems have their own reporting and processing systems, the details of which are in the appropriate MI system manuals and technical manuals. Some collection assets, particularly air reconnaissance and ground scouts, can report relevant information that is immediately usable by the tactical commander (for example, for targeting purposes). However, the personnel in the reporting chain still process these reports by evaluating their relevancy and accuracy. In many cases, the output of a collection asset is data, or information of limited immediate use to a commander. Also, in certain situations ROE dictate a requirement for target confirmation by other sources.

4-30. The intelligence staff processes information collected by the unit's assets as well as that received from higher echelons. The intelligence staff processes many types of information and data from intelligence reach, unmanned aerial vehicle (UAV) imagery, radar imagery, mobile target indicators (MTIs), and HUMINT and SIGINT reports.

PRODUCE

4-31. In the production step, the G2/S2 integrates evaluated, analyzed, and interpreted information from single or multiple sources and disciplines into finished intelligence products. Like collection operations, the G2/S2 must ensure the unit's information processing and intelligence production are prioritized and synchronized to support answering the CCIRs (PIRs and FFIRs).

4-32. Intelligence products must be timely, relevant, accurate, predictive, and usable. The accuracy and detail of every intelligence product has a direct effect on how well the unit plans and prepares for operations. However, the G2/S2 and unit must use intelligence (no matter what form the intelligence is in) that meets the requirements but might not be as detailed or refined as possible or in a better form. A good answer on time is better than a more refined answer that is late.

4-33. The G2/S2 produces intelligence for the commander as part of a collaborative process. The commander drives the G2/S2's intelligence production effort by establishing intelligence and IRs with clearly defined goals and criteria. Differing unit missions, environments, and situations impose numerous and varied production requirements on the G2/S2 and his staff.

4-34. The G2/S2 must employ collaborative analysis techniques and procedures that leverage intelligence production capability of higher and subordinate echelons to meet these requirements. Proficiency in these techniques and procedures enables the G2/S2 to answer the commander's and staff's requirements regardless of the mission, environment, and situation. The G2/S2 and staff intelligence products enable the commander to—

- Plan operations and employ maneuver forces effectively.
- Recognize potential COAs.
- Employ effective tactics and techniques.
- Take appropriate security measures.
- Focus ISR.

COMMON INTELLIGENCE PROCESS TASKS

ANALYZE

4-35. Analysis occurs at various stages throughout the intelligence process. Personnel conducting intelligence operations at all levels analyze intelligence, information, and problems to produce intelligence, solve problems and, most importantly, answer the PIRs. Leaders at all levels conduct analysis to assist in making many types of decisions. An example is a HUMINT collector analyzing an intelligence requirement in order to determine the best possible collection strategy to use against a specific source.

4-36. Analysis in RM is critical to ensuring the IRs receive the appropriate priority for collection. The intelligence staff analyzes each requirement to determine its feasibility, whether or not it supports the commander's intent, and to determine the best method of satisfying the IRs. The staff also analyzes collected information to determine if it satisfies requirements.

4-37. During the produce function, the intelligence staff analyzes information from multiple sources to develop all-source intelligence products. The intelligence staff analyzes information and intelligence to ensure the focus, prioritization, and synchronization of the unit's intelligence production is in accordance with the PIRs.

4-38. In situation development, the intelligence staff analyzes information to determine its significance relative to predicted ECOAs and the CCIRs (PIRs and FFIRs). Through predictive analysis, the staff attempts to identify enemy activity or trends that represent opportunities or risks to the friendly force.

They use the indicators developed for each ECOA and CCIRs (PIRs and FFIRs) during the MDMP as the basis for their analysis and conclusions.

DISSEMINATE

4-39. Successful operations at the tactical and operational levels require an increased ability to synchronize fires, have faster access to intelligence, and enhance situational understanding and effective FP. Timely and accurate dissemination of intelligence is key to the success of these and other operations. Commanders must receive combat information and intelligence products in time and in an appropriate format to support decisionmaking. Additionally, sharing the most current all-source information and intelligence at all echelons is essential for commanders to maintain situational understanding.

4-40. To achieve this, it is imperative that the commander and staff establish and support a seamless intelligence architecture—including an effective dissemination plan—across all echelons to ensure information and intelligence flow in a timely manner to all those who need them. Intelligence and communications systems continue to evolve in their sophistication, application of technology, and accessibility to the commander. Their increasing capabilities also create an unprecedented volume of information available to commanders at all echelons. Finally, the commander and staff must have a basic understanding of these systems and how they contribute to the Intelligence BOS. A dissemination plan can be a separate product, or integrated into existing products such as the ISR synchronization plan or ISM, the decision support template (DST), or decision support matrix (DSM).

Dissemination Procedures

4-41. The G2/S2 and intelligence personnel at all levels assess the dissemination of intelligence and intelligence products.

4-42. Disseminating intelligence simultaneously to multiple recipients is one of the most effective, efficient, and timely methods. This can be accomplished through various means; for example, push, broadcast. However, within the current tactical intelligence architecture, reports and other intelligence products move along specific channels. The staff helps streamline information distribution within these channels by ensuring dissemination of the right information in a timely manner to the right person or element. There are three channels through which commanders and their staffs communicate: command, staff, and technical.

4-43. **Command Channel**. The command channel is the direct chain-of-command link that commanders, or authorized staff officers, use for command-related activities. Command channels include command radio nets (CRNs), video teleconferences (VTC), and the Maneuver Control System (MCS).

4-44. **Staff Channel**. The staff channel is the staff-to-staff link within and between headquarters. The staff uses the staff channel for control-related activities. Through the staff channel, the staff coordinates and transmits intelligence, controlling instructions, planning information, provides early warning information, and other information to support C2. Examples of staff channels include the operations and intelligence radio net, telephone, the staff huddle, VTC, and the BOS-specific components of the Army Battle Command System (ABCS).

4-45. **Technical Channel**. Staffs typically use technical channels to control specific combat, CS, and CSS activities. These activities include fire direction and the technical support and sensitive compartmented information (SCI) reporting channels of intelligence and ISR operations. The SIGINT tasking and reporting radio net, intelligence broadcast communications, and the wide area networks (WANs) supporting single intelligence discipline collection, processing, and production are examples of technical channels.

Presentation Techniques and Procedures

4-46. The staff's objective in presenting information is to provide the commander with relevant information. Table 4-1 lists the three general methods that the staff uses to present information and meet its information objective. Systems within the ABCS contain standard report formats, maps, and mapping tools that assist the staff in presenting information in written, verbal, and graphic form. Audio and video systems such as large format displays and teleconferencing systems enable the staff to use a combination of the methods in multimedia presentations.

Table 4-1. Presentation Methods and Products.

Method	Products
Written Narrative	Reports, Estimates, and Studies
Verbal Narrative	Briefing (information, decision, mission, and staff)
Graphic	Charts, Overlays, and Electronic Displays

Intelligence Communications Architecture

4-47. The intelligence communications architecture transmits intelligence and information to and from various ISR elements, units, and agencies by means of automation and communication systems. With the continued development of sensors, processors, and communications systems, it is increasingly important to understand the requirements of establishing an effective communications architecture. The G2/S2 must identify the Intelligence BOS specific requirements of the unit's overall communications architecture. Refer to FM 2-33.5 for more information on intelligence communications reach. The following are some (but not all) of the questions which the staff must answer in order to establish the intelligence communications architecture:

- Where are the unit's collectors?
- What and where are the unit's processors?
- Where are the unit's intelligence production elements?
- Where are the unit's decisionmakers?
- How does the unit disseminate information from its producers to its decisionmakers and/or consumers?

- Are the systems which the unit's collectors, producers, processors, and consumers use compatible with each other? If not, what is the plan to overcome this challenge?
- How can the unit access databases and information from higher and other agencies?

ASSESS

4-48. Assessment is the continuous monitoring—throughout planning, preparation, and execution—of the current situation and progress of an operation, and the evaluation of it against criteria of success to make decisions and adjustments. Assessment plays an integral role in all aspects of the intelligence process. Assessing the situation and available information begins upon receipt of the mission and continues throughout the intelligence process. The continual assessment of intelligence operations and ISR assets, available information and intelligence, the various aspects of the battlefield environment, and the situation are critical to—

- Ensure the CCIRs (PIRs and FFIRs) are answered.
- Ensure intelligence requirements are met.
- Redirect collection assets to support changing requirements.
- Ensure operations run effectively and efficiently.
- Ensure proper use of information and intelligence.
- Identify enemy efforts at deception and denial.

4-49. During planning, the intelligence staff conducts a quick initial assessment of the unit's intelligence posture and holdings, status of intelligence estimates, and any other available intelligence products. From this assessment the commander issues his initial guidance and a WARNO.

4-50. While the majority of the unit is engaged in preparation, the ISR effort should already have begun. It is during this period when the prepare and execute activities of the operations process overlap, that the G2/S2 assesses the current situation as well as the progress of ISR operations.

4-51. During execution the intelligence staff continues assessing the effectiveness of the ISR effort while at the same time assessing the results and products derived from the collection effort. The critical aspects of assessment at this point include determining whether the PIRs have been answered, will be answered with the current ISR operations, or which ISR operations to adjust in order to answer the CCIRs (PIRs and FFIRs). This type of assessment requires sound judgment and a thorough knowledge of friendly military operations, characteristics of the AO and AOI, and the threat situation, doctrine, patterns, and projected future COAs.

PART THREE

Military Intelligence Disciplines

Part Three provides a more detailed explanation of the intelligence disciplines introduced in Part One of this manual.

Chapter 5 defines and discusses the roles and fundamentals of the all-source intelligence discipline.

Chapter 6 defines and discusses the roles, fundamentals, and generic organization of the HUMINT discipline.

Chapter 7 defines and discusses the roles, fundamentals, and generic organization of the IMINT discipline.

Chapter 8 defines and discusses the roles, fundamentals, and generic organization of the SIGINT discipline.

Chapter 9 defines and discusses the roles, fundamentals, and generic organization of the MASINT intelligence discipline.

Chapter 10 defines and discusses the roles, fundamentals, and generic organization of the TECHINT discipline.

Chapter 11 defines and discusses the roles, fundamentals, and generic organization of the CI discipline.

Chapter 5

All-Source Intelligence

DEFINITION

5-1. All-source intelligence is defined as the intelligence products, organizations, and activities that incorporate all sources of information and intelligence, including open-source information, in the production of intelligence. All-source intelligence is a separate intelligence discipline, as well as the name of the function used to produce intelligence from multiple intelligence or information sources.

ROLE

5-2. The operational environment provides an ever-growing volume of data and information available from numerous sources, from which the commander can use to achieve situational understanding. His situational understanding enables him to make decisions in order to influence the outcome of the operation; prioritize and allocate resources; assess and take risks; and understand the needs of the higher and subordinate commanders. The commander depends upon a skilled G2/S2 working within his intent to support his ISR effort and provide all-source intelligence analysis conclusions and projections of future conditions or events.

FUNDAMENTALS

5-3. All-source intelligence production satisfies intelligence requirements. It provides an overall picture of the adversary and the battlespace. It reduces the possibility of error, bias, and misinformation through the use of multiple sources of information and intelligence.

PLAN

5-4. The utilization of all-source intelligence facilitates the development of accurate and concise contingency-specific plans, orders, and intelligence products. Additionally, the G2/S2 must retrieve, update, or develop any required intelligence databases. The most important all-source products are the modified combined obstacle overlay (MCOO), event templates, ECOA sketches, and the HPT list provided during initial IPB in support of the plan function. Through all-source intelligence, the commander can make informed decisions about COAs and adversary capabilities.

5-5. The intelligence staff coordinates its efforts with other elements such as the engineer terrain team, the unit surgeon, the Air Force weather team, and other assets or elements that can support the analytical effort.

5-6. IPB plays a significant role in the planning phase of the intelligence process. It is a systematic, continuous process of analyzing the threat and environment and is designed to support staff estimates and the MDMP. The IPB is led by the G2/S2, with participation by the entire staff. IPB allows the commander and staff to make informed decisions, develop COAs, and focus ISR efforts where they are most needed. IPB continuously assists the commander and staff in focusing ISR assets on the appropriate targets at critical points in time and space.

5-7. The time available for the IPB process may not permit the luxury of conducting each step in detail. Overcoming time limitations requires the commander to identify those parts of IPB that are most important to the commander in planning and executing his mission. Applying the specific steps or degree of detail performed varies according to METT-TC. There are four steps in the IPB process: define the battlefield environment, describe the battlefield's effects, evaluate the threat, and determine ECOAs.

Define the Battlefield Environment

5-8. In this step, the G2/S2—

- Identifies characteristics of the battlefield that will influence friendly and threat operations including terrain (mobility), weather, hydrological data, infrastructure, and civilian demographics. Produces accurate, timely, and predictive IPB products that depict the aspects of the battlefield.

- Identifies the limits of the command's AO and AOI.
- Identifies gaps in current intelligence holdings, identifies IRs, and recommends CCIRs (PIRs and FFIRs).

5-9. Defining the battlefield environment includes identifying enemy forces (their location, mobility, general capabilities, and weapon ranges) and all other aspects of the environment that could have an effect on the unit's ability to accomplish the mission. Depending on the situation, these considerations may also include—

- Geography, terrain, and weather of the area.
- Information environment to include but not limited to computer and communications systems and capabilities, data acquisition systems and capabilities, media access and distribution, areas prone to electromagnetic interferences, and systems to generate electromagnetic interference.
- Population demographics (ethnic groups, religious sects, age distribution, health status, income groups).
- Medical threat to include but not limited to endemic and epidemic diseases, occupational and environmental health hazards, and poisonous and toxic plants and animals.
- Political or socio-economic factors, including factions, clans, and gangs.
- Infrastructures, such as transportation or telecommunications, and critical decisionmaking infrastructures and supporting information systems.
- ROE or legal restrictions, such as international treaties or agreements.

Describe the Battlefield's Effects

5-10. This IPB step deals with the effects of the battlefield environment on the current operations and potential enemy and friendly COAs. It begins with the assessment of existing and projected conditions of the battlefield environment, which the staff accomplishes through terrain analysis, weather analysis, and analysis of other characteristics of the battlefield. To conduct terrain analysis, the staff uses maps, reconnaissance, and other specialized terrain products (maps, overlays, databases, software). These products address such factors as wet or dry cross-country mobility, transportation systems (road and bridge information), vegetation type and distribution, surface drainage and configuration, surface materials (soils), ground water, manmade structures, and obstacles. The results of evaluating the terrain's effects should be expressed by identifying areas of the battlefield that influence each COA.

Evaluate the Threat

5-11. Evaluating the threat involves determining the threat force capabilities and the doctrinal principles and TTP that threat forces prefer to employ. The result of this evaluation produces a threat model, which portrays how threat forces normally execute operations and how they have reacted to similar situations in the past. The threat model includes an evaluation of the threat's strengths, weaknesses, and vulnerabilities, including an evaluation of typical HVTs.

Determine ECOAs

5-12. Determining ECOAs involves the identification and development of likely ECOAs that will influence accomplishing the friendly mission. Developing ECOAs is a form of predictive intelligence analysis and production. The procedures for this step include—

- Identifying the enemy's likely objectives and predicting the desired end-state.
- Identifying the full set of COAs available to the enemy.
- Predicting the enemy's most likely and most dangerous COAs.
- Evaluating and prioritizing each COA.
- Developing each COA in the amount of detail that time permits.
- Identifying initial ISR and collection requirements.

5-13. After conducting the initial IPB, the staff, primarily the G2/S2, identifies gaps in the available intelligence, develops the initial PIRs and IRs, and develops the initial input to the ISR plan based on the commander's guidance.

5-14. Indicators are the basis for situation development. Indicators are activities that will confirm or deny the event specified in an intelligence requirement. They are any positive or negative evidence of enemy activity or characteristic of the AO that points toward enemy capabilities, vulnerabilities, or intentions. Individual indicators cannot stand alone. Each indicator is integrated with other factors and indicators before patterns are detected and enemy intentions established. Indicators are developed by the analysts in the G2/S2 section. The event matrix shows the threat activities, or indicators, to look for in each NAI, and the timelines during which each NAI should be active. All indicators are developed to answer the commander's PIRs and IRs. The analyst uses indicators to correlate particular events or activities with probable ECOAs and to determine what events or activities must occur for an enemy to follow a particular COA. The ability to read indicators (including recognition of enemy deception indicators) contributes to the success of friendly operations. The analyst integrates information from all sources to confirm indicators of enemy activities. As indicators are detected and confirmed, PIRs are answered.

PREPARE

Activities

5-15. All-source activities during the prepare function include—

- Conducting rehearsals.
- Conducting communications rehearsals and verifiying communications protocols with higher, lateral, and subordinate units.
- Planning and practicing actions that support likely contingencies, branches, or sequels.
- Reviewing and updating available databases and IPB products.
- Ensuring control and coordination measures are still in effect.

- Updating the force with the most recent intelligence before mission execution.

Intelligence Estimates

5-16. The intelligence estimate is a continuous process that is the product of all actions the intelligence staff performs throughout the MDMP. The intelligence estimate provides a timely and accurate evaluation of the enemy and the AO (and often the AOIR) at a given time. It provides the background the G2/S2 uses to portray enemy actions during COA analysis.

5-17. The G2/S2 must clearly understand the weather and terrain effects and the ability to visualize the battlespace before producing the intelligence estimate. This understanding facilitates accurate assessments and projections regarding the enemy: enemy situation (including strengths and weaknesses), enemy capabilities and an analysis of those capabilities (COAs available to the enemy), and conclusions drawn from that analysis. The intelligence estimate's conclusion identifies the enemy's most likely COA and most dangerous COA. ECOAs must include a sketch of the COA with a task and purpose for the enemy's actions. Any gaps in the intelligence estimate are identified as information requirements.

5-18. The intelligence estimate may be written or oral. At the tactical level, especially during operations and exercises, intelligence estimates are usually delivered orally, supported by graphic displays and other decision support tools. During contingency planning, especially at corps level and above, estimates are usually written. During deliberate planning at joint headquarters, estimates are always written (see JP 5-00.2). However, the intelligence estimate should always be prepared as thoroughly as time and circumstances permit. A comprehensive intelligence estimate considers both the tangible (quantifiable) and the intangible aspects of the enemy's operations. It translates enemy strengths, weapon systems, training, morale, and leadership into combat capabilities and projections of future enemy actions.

5-19. Different sections of the intelligence estimate receive more emphasis during different activities within the operations process.

- During planning, the most important decision the commander makes is selecting a COA on which to base the plan. Thus, the sections of the intelligence estimate that focus on the commander's selecting the most appropriate COA are the most important at that time.
- During preparation, the intelligence estimate must focus on the commander's decisions that affect the ability of the unit to execute the upcoming operation. The intelligence estimate—and functions—that supports these decisions must focus on answering the CCIRs (PIRs and FFIRs) and guiding the ISR effort.
- During execution, the intelligence estimate focus is to support the anticipated command decisions. The most important action of the intelligence estimate—and the ISR effort— is to answer the PIR. However, it is also during the execution phase that the intelligence estimate must look ahead: anticipating branches or sequels to the

current operation, transition to other operations, changes in the ECOAs, and required adjustments to the ISR effort.

5-20. Refer to FM 2-01 for examples of intelligence estimates.

COLLECT

5-21. The collect intelligence process function within the all-source intelligence discipline is limited to gathering the necessary information, intelligence, and intelligence products required to perform analysis. Thus, all-source intelligence depends upon the other intelligence disciplines to perform collection. This is also the primary reason why OSINT is defined as a source of information and intelligence and not a separate intelligence discipline.

PROCESS

5-22. The process function converts relevant information into a form suitable for analysis, production, or immediate use by the commander. Processing also includes sorting through large amounts of collected information and intelligence to identify and exploit the information which is pertinent to the commander's intelligence requirements and facilitates situational understanding.

PRODUCE

5-23. All-source intelligence analysis and production—

- Drives collection to answer the PIR.
- Provides the enemy situation.
- Provides INTSUMs and other intelligence reports.
- Supports situational understanding.
- Provides predictive estimates of enemy actions; specifically, ECOAs.
- Provides continuously updated IPB.
- Provides all-source target packages (or folders).

5-24. As previously stated, the intelligence estimate is a continuous process. The commander and staff constantly collect, process, store, display, and disseminate information. Staff members update their estimates as they receive new information, such as—

- When they recognize new facts.
- When they replace assumptions with facts or find their assumptions invalid.
- When they receive changes to the mission or when changes are indicated.

5-25. Technological advances and NRT information allow estimates to be continuously updated; the running estimate and its intelligence component, the intelligence running estimate, exemplify this.

Running Estimate

5-26. Running estimates are continuously updated estimates based on new information as the operation proceeds. They serve as a staff technique to support the commander's visualization and decisionmaking, as well as the staff's tool for assessing during preparation and execution. In the running estimate, staff officers continuously update their conclusions and recommendations as they evaluate the impact of new facts. Current doctrine emphasizes the COP as the primary tool that provides the current situation and, when merged with the running estimates, facilitates commanders achieving situational understanding. All staff sections (BOSs) provide their respective input to the COP. The COP as used today, combined with the running estimate, is predictive and enhances our current ability to collect, process, store, display, and disseminate information.

5-27. The portion of the COP that depicts the enemy situation is currently limited to displaying the locations and dispositions of enemy and threat forces in a relatively static manner, sometimes referred to as snapshots in time. The enemy situation portion of the COP requires analysis to provide the required level of detail. The COP and running estimates sufficiently provide effective support for battle command of knowledge-based organizations.

5-28. The main differentiation between the running estimate and the old staff estimates is the emphasis on not only continuously updating the facts of the estimate but also continuously updating the conclusions and recommendations while including projections of future conditions of the entire battlespace. While the COP is primarily a display of current intelligence and information, the running estimate requires the merging of the staff's cognitive processes with automation applications. The primary focus of the staff's cognitive process is to present predictive or anticipatory intelligence in support of the commander's decisionmaking or situational understanding. See Figure 5-1.

5-29. The running estimate is a product of the entire battle staff. Just as all BOSs contribute to the COP, they will also contribute their portion of the running estimate. The running estimate integrates the running estimates from each BOS. The Intelligence BOS input to the running estimate is the intelligence running estimate.

Intelligence Running Estimate

5-30. The intelligence running estimate is a continuous flow and presentation of relevant information and predictive intelligence that, when combined with the other staff running estimates, enables the decisionmaker's visualization and situational understanding of the AOI in order to achieve information superiority. The intelligence running estimate requires constant verification to support situational understanding of the current situation as well as predictive assessments for future operations.

5-31. While it is possible to employ the concept of the intelligence running estimate today, the true seamless and continuous update feature will be achieved with technological enablers that are not yet present in the force. In the future, technology should allow the information from the running estimate to become implanted as a part of the COP. However, there will always be a place for some separate context, orally or written, separate from the COP display to add fidelity and assist in the commander's visualization and decisionmaking. Additional training must be implemented not just within MI but also throughout the force in order to institute the running estimate concept.

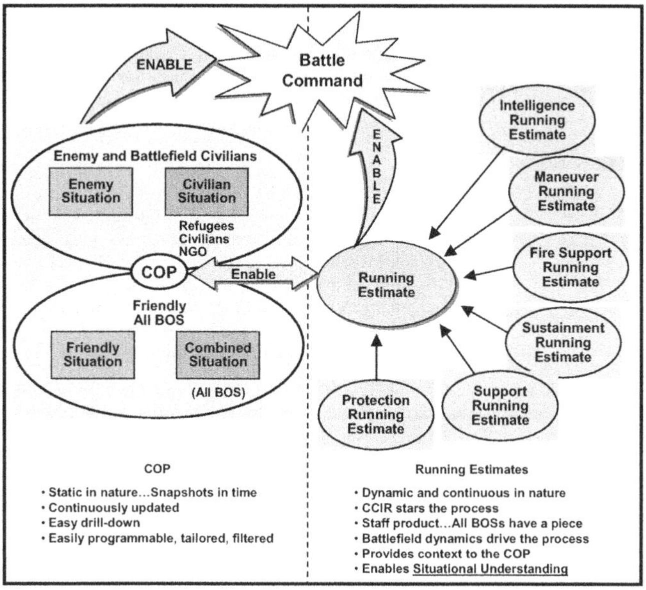

Figure 5-1. Example of the COP and a Running Estimate.

ANALYZE

5-32. Analysis is key in converting combat information and all intelligence from each discipline into all-source intelligence products and targeting information. From the multiple sources of information received, the staff analyzes and identifies critical information; determines the relevancy and accuracy of the information; and reaches conclusions about the information. These conclusions are either immediately disseminated or are used to form the basis of intelligence analysis products.

DISSEMINATE

5-33. The intelligence staff extracts and disseminates pertinent intelligence from products developed as a result of all-source analysis. The intelligence staff—

- Rapidly disseminates time-sensitive, all-source analysis intelligence and intelligence products to higher, lateral, and subordinate commands in order to keep all commanders abreast of current developments in the situation and battlespace in accordance with unit SOPs.

- Disseminates all-source analysis intelligence to other G2/S2 intelligence activities for additional processing or detailed analysis and exploitation in accordance with unit SOPs.

5-34. For intelligence reach operations all-source intelligence products are available and disseminated in a variety of forms. It is incumbent on the requestor to ensure that the all-source product can be transmitted over the available communications systems. This includes verifying the appropriate security level of the communication system.

ASSESS

5-35. The commander, the intelligence staff, and the intelligence consumers assess the analysis, production, and dissemination of all-source intelligence. The staff identifies intelligence gaps, emerging operational requirements, duplication of effort, or new targets or threats and redirects all-source activities as appropriate to meet intelligence requirements. Throughout the assessment process, the G2/S2, the intelligence staff, and intelligence users should be alert for evidence of possible enemy denial and deception efforts. The staff assesses intelligence and information for—

- Consistency with the current situation and threat trends.
- Accuracy and confirmation by other intelligence sources or disciplines.
- Source reliability and credibility.
- Pertinence to PIRs, intelligence requirements, RFIs, and other intelligence tasks or requests.

Chapter 6

Human Intelligence

DEFINITION

6-1. HUMINT is the collection by a trained HUMINT Collector of foreign information from people and multimedia to identify elements, intentions, composition, strength, dispositions, tactics, equipment, personnel, and capabilities. It uses human sources as a tool and a variety of collection methods, both passively and actively, to gather information to satisfy the commander's intelligence requirements and cross-cue other intelligence disciplines.

ROLE

6-2. The role of HUMINT Collectors is to gather foreign information from people and multiple media sources to identify adversary elements, intentions, composition, strength, dispositions, tactics, equipment, personnel, and capabilities. It uses human sources and a variety of collection methods to gather information to satisfy the commander's intelligence requirements, and cross-cue other intelligence disciplines.

HUMINT FUNCTIONS

6-3. HUMINT functions are interrelated, mutually supporting, and can be derived from one another. No single function or technical capability can provide a full understanding of our adversaries. HUMINT functions are defined below and shown in Figure 6-1.

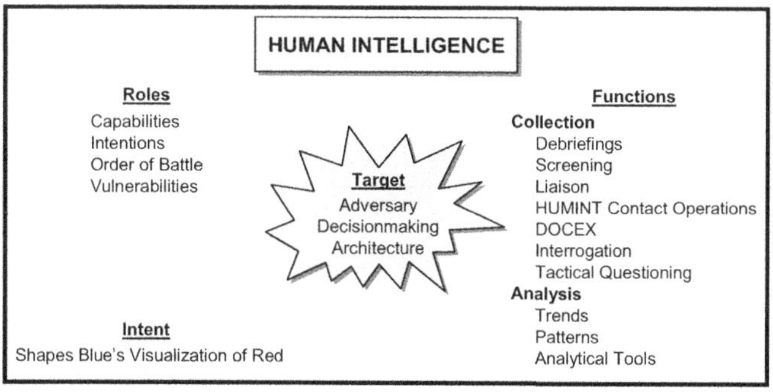

Figure 6-1. HUMINT Roles and Functions.

COLLECTION

6-4. HUMINT collection methods and operations include—

- **Debriefings.** Debriefings are the systematic questioning of individuals to procure information to answer specific collection requirements by direct and indirect questioning techniques. Sources for debriefings are categorized as friendly forces, US and non-US civilians to include refugees, displaced persons, and local inhabitants.
- **Screening.** Screening is the process of evaluating and selecting human and media sources for the prioritized collection of information in support of commander's PIRs and IRs. Screening categorizes and prioritizes sources based on the probability of a particular source having priority information and the degree of difficulty of extracting that information from the source. While screening is not in itself an information collection technique, it is vital to the rapid collection of information.
- **Liaison.** HUMINT elements conduct liaison with US, multinational, and HN military and civilian agencies, to include NGOs. Liaison is conducted to obtain information of interest and to coordinate or deconflict HUMINT activities.
- **HUMINT Contact Operations.** HUMINT contact operations are tactically oriented, overt collection activities that use human sources to identify attitude, intentions, composition, strength, dispositions, tactics, equipment, target development, personnel, and capabilities of those elements that pose a potential or actual threat to US and multinational forces. These forces provide early warning of imminent danger to US and multinational forces and contribute to the MDMP.
- **Document Exploitation (DOCEX).** DOCEX is the systematic extraction of information from all media formats in response to collection requirements.
- **Interrogation.** Interrogation is the systematic effort to procure information to answer specific collection requirements by direct and indirect questioning techniques of persons who are in the custody of the forces conducting the questioning.

UNIT SUPPORT TO HUMINT COLLECTION

6-5. Small units contribute to HUMINT collection through a number of different ways.

Tactical Questioning

6-6. Tactical questioning can provide critical information for situational understanding. Tactical questioning is the expedient initial questioning for information of immediate tactical value. Soldiers conduct tactical questioning based on the unit's SOP, ROE, and the order for that mission. Unit leaders must include specific guidance for tactical questioning in the order for appropriate missions. The unit S3 and S2 must also provide specific guidance down to the unit level to help guide tactical questioning. For a more detailed discussion of tactical questioning, see ST 2-91.6.

Combat Patrolling

6-7. Through observing and interacting with the local environment during the conduct of missions, handling EPWs/detainees, and handling captured documents, soldiers serve as the commander's "eyes and ears" whether—

- Performing traditional offensive or defensive operations.
- Performing a patrol in a stability operation.
- Manning a TCP or a roadblock in a support operation.
- Occupying an observation post.
- Passing through an area in a convoy.
- Performing any operation that involves observing and reporting on elements of the environment and activities in the AO.

ISR Operations

6-8. The information that the soldier reports as a result of completed missions up the chain of command forms a vital part of planning and operations. Careful and quick handling of EPWs/detainees and documents also helps the ISR effort. Unit headquarters must ensure that the reports they receive and forward contain all the subtle detail provided by the small unit. For tactical operations, there are four levels of reporting:

- Immediate reporting of information of critical tactical value, based either on predetermined criteria or common sense.
- Normal reporting, submitted before the unit S2 section performs the debriefing.
- Debriefing performed by the S2 section.
- Follow-up reporting, submitted after the unit S2 section performs the debriefing.

6-9. The four levels of reporting facilitate the unit S2 section capturing all the fine details of the small unit's activities for all-source analysis and future planning. The unit S2 must ensure that information of HUMINT and CI value is reported to the G2X.

ANALYSIS

6-10. IPB, all-source, and single-source analysis are used to template adversarial OB. HUMINT elements analyze operational taskings to determine the best collection methods to employ in satisfying CCIRs (PIRs and FFIRs) and IRs. As information is collected, it is analyzed for completeness and accuracy to identify significant facts for subsequent interpretation and inclusion in intelligence products, and passing to other collectors. This analysis also helps to identify collection gaps and focus or refocus collection efforts. Analysis provides the commander with situational understanding of the battlespace to execute operations. Raw information, open-source material, and finished intelligence products are analyzed in response to local and national requirements. Analysis occurs at all levels from tactical to strategic.

- At the tactical level, HUMINT teams focus their efforts on supporting mission requirements and contributing to the all-source COP.

- Operational analysis is used to assess adversary intentions, capabilities, dispositions, and their regional impacts.
- Strategic analysis supports national and Army programs through compilation of local and regional analysis into global assessments; identifies technology development trends and patterns in military activities and capabilities.

6-11. HUMINT products consist of, but are not limited to, target nomination, input to threat and vulnerability assessments, intelligence estimates, and intelligence information reports. Finalized intelligence derived from HUMINT activities is incorporated into joint and national intelligence databases, assessments, and analysis products. HUMINT products are also incorporated into the COP to support situational awareness. HUMINT production takes place at all levels.

- Operational and tactical production includes tactical alerts, spot reports, and current intelligence; input to threat and/or vulnerability assessments tailored to specific activities, units, installations, programs or geographic areas, and target studies to support contingency planning and major exercises; studies of military activities and capabilities.
- Strategic products include assessments supporting national and Army information requirements on foreign technology development; worldwide assessments of the organization, location, funding, training, operations capabilities and intentions of terrorist organizations; analyses of the capabilities of international narcotics trafficking organizations.

OPERATIONAL EMPLOYMENT

6-12. Army HUMINT must support the full spectrum of military operations. Commanders require a well-trained HUMINT force consisting of AC, RC, civilian government, and contractor personnel. The HUMINT force is focused on and dedicated to the collection of data and information relevant to the commander's PIRs and IRs. Effective employment of Army HUMINT elements in all phases of operations and at all levels from tactical to strategic will remain paramount to ensuring that commanders have the best pictures possible of their adversaries. HUMINT Collectors use the following collection methods.

LEVELS OF EMPLOYMENT

6-13. **Strategic and Departmental.** Strategic and departmental operations will be conducted by HUMINT elements supporting national, DOD, and DA required missions (for example, support to NATO and special operations and missions). The Army strategic HUMINT force resides in DHS. Trained Army HUMINT professionals must be available to augment DOD HUMINT efforts and coordinate between Army and Joint or DOD elements during crisis and war. Strategic HUMINT activities are generally carried out by departmental level HUMINT elements or Army HUMINT assets directly assigned to DOD and Joint positions. The Army overt HUMINT capability at the strategic level will be deployed and under operational control (OPCON) of the Army Service Component Commander (ASCC).

6-14. **Operational.** HUMINT operational missions must support combatant commanders, generally in geographic AORs. Operational HUMINT assets provide capabilities to support theater requirements for HUMINT collection. Operational HUMINT elements will focus on threat identification and capabilities. As with all levels of employment, HUMINT activities and functions will include DOCEX, collection, operations, analysis, and production. HUMINT elements must be capable of quickly transitioning from a peacetime mission to crisis operations to support combatant commander requirements. Operational elements may also be deployed to support or reinforce tactical forces in CONOPS.

6-15. **Tactical.** Army HUMINT play a crucial role in supporting tactical forces. HUMINT teams conduct operations throughout the battlespace during CONOPS. CONOPS support activities include debriefings, screenings, contact operations, and interrogations. HUMINT activities in CONOPS focus on the threat and assisting the senior intelligence officer (SIO) and commander in understanding the threat's decisionmaking process. During peacetime, organic tactical HUMINT teams conduct activities pursuant with approved regulations and command guidance.

SUPPORT TO CONTINGENCY OPERATIONS

6-16. **Initial Phase.** The initial phase of operations from peacetime military engagement (PME) to major theater war (MTW) lays the foundation of future team operations. In general, the priority of effort focuses inward on security of operating bases, areas of troop concentration, and C2 nodes to identify the collection threat to US forces that could be used by adversarial elements to plan hostile acts against US activities and locations.

6-17. **Continuation Phase.** Once security of the operating bases has been established, the operational focus of HUMINT teams shifts outside the operating base to continue to detect, identify, and neutralize the collection threat to US forces as well as to provide I&W of hostile acts targeting US activities. The HUMINT team uses several collection methods, to include HUMINT contact operations, elicitation, and liaison, to answer the supported commander's requirements.

6-18. **HUMINT Team.** A key element to the HUMINT team's success is the opportunity to spot, assess, and develop relationships with potential sources of information. Operating as independent teams, without being tied to combat assets, enables the HUMINT team's maximum interaction with the local population, thereby maximizing the pool of potential sources of information. A second key element of a HUMINT team's success is its approachability to the local population. A soft posture enables a HUMINT team to appear as non-threatening as possible. Experience has shown that the local population in general is apprehensive of fully and openly armed patrols and soldiers moving around population centers. During some operations civilian attire or non-tactical vehicles may be used to lower the

HUMINT team profile. NOTE: In some special situations, these measures are taken to make the operation less visible to the casual observer. Also, in some cultures, sharing food and beverages among friends is expected; exceptions to restrictions or general orders should be considered to facilitate successful HUMINT team operations, many of which are geared towards developing relationships with potential sources of information.

SUPPORT TO INSTALLATIONS AND OPERATING BASES

6-19. Commanders may restrict personnel to base camps and installations during the initial stages of operations (PME to MTW), when the operational environment is being assessed, or as a temporary expedient when the threat level exceeds the ability to provide reasonable FP. Operational restrictions minimize the risk to the HUMINT team, but minimizing its collection potential may increase the risk to the force as a whole. While confined to an installation or a base camp, the HUMINT team can maintain a limited level of information collection by—

- Screening locally employed personnel.
- Debriefing combat and ISR patrols.
- Debriefing friendly force personnel who are in contact with the local population.
- Conducting limited local open-source information collection.
- Contributing to the threat and vulnerability assessments of the base camp.

TACTICS, TECHNIQUES, AND PROCEDURES

6-20. At the HUMINT team level, team members conduct mission analysis and planning specific to their AO. Backwards planning and source profiling are used extensively to choose HUMINT targets. To verify adequate area coverage, the HUMINT team may periodically develop and use HUMINT target overlays and other HUMINT analytical tools that illustrate the HUMINT situation, identify gaps, and help refocus the collection effort.

6-21. The HUMINT team is also in constant contact with the supported S2 and the other ISR assets (Scouts, PSYOP, CA, and MPs) in order to coordinate and deconflict operations and to cross-check collected information. The supported unit S2, with the help of the HUMINT team, regularly and systematically debriefs all ISR assets.

6-22. The HUMINT team must be integrated into the supported unit's ISR plan. The HUMINT operational management team (OMT) chief will advise the supported unit on the specific capabilities and requirements of the team to maximize mission success.

OPERATIONAL RISK MITIGATION

6-23. The employment of HUMINT teams includes varying degrees of contact with the local population. As the degree of contact with the population increases, both the quantity and quality of collection increases. In many instances, however, there is a risk to the team inherent with increased exposure to the local population. The decision at what level to employ a team is METT-TC dependent. The risk to HUMINT assets must be balanced with the need to collect priority information and to protect the force as a whole. ROE, SOFA, direction from higher headquarters, and the overall threat level may also restrict the deployment and use of HUMINT teams. The commander should consider exceptions to the ROE to facilitate HUMINT collection.

6-24. Risks are minimized through the situational awareness of HUMINT team members. They plan and rehearse to readily react to any situation and carry the necessary firepower to disengage from difficult situations. If it becomes necessary to call for assistance, adequate and redundant communications equipment is critical. These scenarios and actions should be trained prior to deployment into a contingency area and rehearsed continuously throughout the deployment.

6-25. A supported unit commander is often tempted to keep the HUMINT team "inside the wire" when the threat condition (THREATCON) level increases. The supported commander must weigh the risk versus potential information gain when establishing operational parameters of supporting HUMINT teams. This is necessary especially during high THREATCON levels when the supported unit commander needs as complete a picture as possible of the threat arrayed against US and multinational forces.

6-26. When it is not expedient to deploy the HUMINT team independently due to threat levels or other restrictions, the team can be integrated into other ongoing operations. The HUMINT team may be employed as part of a combat, ISR, or MP patrol or used to support CA, PSYOP, engineer, or other operations. This method reduces the risk to the team while allowing a limited ability to collect information. It has the advantage of placing the team in contact with the local population and allowing it to spot, assess, and interact with potential sources of information. However, this deployment method restricts collection by subordinating the team's efforts to the requirements, locations, and timetables of the unit or operation into which it is integrated and does not allow for the conduct of sensitive source operations. **This method of employment should be considered a last resort.**

HUMINT EQUIPMENT

6-27. Basic C2, transportation, and weapons requirements do not differ significantly from most soldier requirements and are available as unit issue items. However, HUMINT teams have unique communications, collection, processing, and mission-specific requirements.

COMMUNICATIONS

6-28. Dedicated and secure long-range communications are key to the success of the HUMINT team mission. HUMINT team operations require a secure, three-tiered communications architecture consisting of inter/intra-team radios, vehicle-based communications, and a CI and HUMINT base station.

6-29. The HUMINT team must have access to existing communications networks such as the tactical local area network (LAN). The HUMINT team must also be equipped with its own COMSEC devices. It is imperative that the HUMINT team acquire access to the public communication system of the HN. This can be in the form of either landlines or cellular telephones. Such access enables the HUMINT team to develop leads which can provide early indicators to US forces.

Interoperability

6-30. Communications systems must be equipped with an open-ended architecture to allow for expansion and compatibility with other service elements, government organizations, NGOs, and multinational elements to effectively communicate during CONOPS. All ISR systems must be vertically and horizontally integrated to be compatible across all BOSs and with Legacy and Interim Force elements.

Satellite Communications On-The-Move (SOTM)

6-31. To provide real time or NRT information reporting, HUMINT elements must have the capability to transmit voice, data, imagery, and video while on the move. HUMINT teams must be able to transmit while geographically separated from their parent unit while operating remotely. This broadband requirement can only be achieved through a SATCOM capability and must be achievable while mobile.

HUMINT COLLECTION AND PROCESSING SYSTEMS

6-32. The HUMINT team must rely on automation to achieve and maintain information dominance in a given operation. With time, effective collection planning and management at all echelons, the HUMINT team can collect a wealth of information. The sorting and analysis of this information in a timely and efficient manner is crucial to operations. Automation helps the HUMINT team to report, database, analyze, and evaluate the collected information quickly and to provide the supported unit with accurate data in the form of timely, relevant, accurate, and predictive intelligence.

6-33. Automation hardware and software must be user friendly as well as interoperable among different echelons and services. They must interface with the communications equipment of the HUMINT team as well as facilitate the interface of audiovisual devices. Technical support for hardware and software must be available and responsive.

6-34. The demand for accurate and timely HUMINT reporting, DOCEX, and open-source information has grown tremendously. Biometric (physiological, neurological, thermal analysis, facial and fingerprint recognition) technologies will allow rapid identification; coding and tracking of adversaries and human sources; and cataloging of information concerning enemy prisoners of war (EPWs), detainees, and civilians of HUMINT interest on the battlefield. Biometrics will also provide secure authentication of individuals seeking network or facility access.

6-35. HUMINT teams work with multinational forces, and other foreign nationals, and require the ability to communicate in their respective languages. Often HUMINT personnel have little or no training in the target language, and lack of skilled interpreters can hinder HUMINT activities. HUMINT teams require textual and voice translation devices, source verification, and deception detection machines (biometrics) to improve collection capability and accuracy.

6-36. HUMINT teams require dynamic machine language translation (MLT) tools that provide both non-linguists and those with limited linguist skills a comprehensive, accurate means to conduct initial screenings and basic interviews in a variety of situations. HUMINT elements will focus on in-depth interviews and communications with persons of higher priority. MLT tools will minimize reliance on contract linguists and will allow soldiers to concentrate on mission accomplishment.

MISSION SPECIFIC

6-37. The HUMINT team may conduct night operations and must be equipped with NVDs for its members and photographic and weapons systems. The HUMINT team also may operate in urban and rural areas, where the threat level can vary from semi-hostile to hostile. The safety of the HUMINT team can be enhanced with equipment that can detect, locate, suppress, illuminate, and designate hostile optical and E-O devices. In addition, high power, gyro-stabilized binoculars, which can be used from a moving vehicle, also increase the survivability of the HUMINT team. It also gives the team another surveillance and collection device.

6-38. Some of the HUMINT team missions may require the documentation of incidents. The teams can use the following equipment in their open-source collection efforts.

- Small, rugged, battery-operated digital camcorders and cameras which are able to interface with the collection and processing systems as well as communication devices.
- GPSs that can be mounted and dismounted to move in the AO efficiently.
- Short-range multichannel RF scanning devices that can also identify frequencies which enhance their security.

6-39. In some cases HUMINT teams require a stand-off, high resolution optical surveillance and recording capability that can provide target identification at extended ranges to protect the intelligence collector while avoiding detection by the adversary target. An advanced optical capability provides intelligence collectors the ability to locate and track adversarial targets (passive and hostile) for identification, collection, and target exploitations.

INTEGRATION OF LINGUISTS

6-40. Integrating linguists into the HUMINT team should take place as soon as possible. Security clearances and contractual agreements will help the team determine the level of integration.

6-41. Along with the basic briefing of what is expected of the civilian linguists as interpreters, HUMINT teams should be informed about the civilians' chain of command and the scope of their duties beyond interpreting. The HUMINT team leader must ensure that linguists are trained and capable of completing all tasks expected of them.

BATTLE HAND-OFF

6-42. HUMINT teams are always engaged. A good battle hand-off is critical to smooth transition and mission success. The battle hand-off can directly contribute to mission success or failure of the outgoing team, but especially of the incoming team. The battle hand-off begins the first day the HUMINT team begins to operate in an AO. Regardless of how long the team believes it will operate within the AO, it must ensure there is a seamless transition to an incoming team, other US unit, or agency. The HUMINT team accomplishes this transition by establishing procedures for source administration, database maintenance, and report files.

6-43. Teams must plan and implement a logical and systematic sequence of tasks to enable an incoming team to assume the operations in the AO. Adequate time must be allotted for an effective battle hand-off. In some environments, a few weeks may be necessary to accomplish an effective battle hand-off. Introductions to sources of information, especially HUMINT contact operations sources, are critical and teams must prioritize their time. During this time the outgoing HUMINT team must familiarize the new HUMINT team with all aspects of the operation, which includes past, present, and planned activities within the AO. Area orientation is critical. These include major routes, population centers, potential hot spots, and other points of interest (such as police stations, political centers, and social centers).

ORGANIZATION

6-44. HUMINT activities require a complex C2 relationship to ensure that the requirements of the supported commander are fulfilled while balancing the need for strict integrity and legality of HUMINT operations. This complex relationship balances the role of the SIO as the requirements manager and the *2X as the mission manager with the MI commander as the asset manager. ("*2X" indicates 2X functions at all levels.)

COMMAND VERSUS CONTROL

6-45. ARFOR will normally deploy as part of a joint and/or multinational operation. In all cases, commanders at each echelon will exercise command over the forces assigned to their organization. Command includes the authority and responsibility for effectively using resources, planning for and employment of forces, and ensuring that forces accomplish assigned missions. Leaders and staffs exercise control to facilitate mission accomplishment.

6-46. While the MI commander supervises subordinates and produces reports, the *2X synchronizes activities between intelligence units and provides single-source processing and limited analysis. While the MI commander takes care of the operators executing missions, the *2X obtains the data and reports from higher echelons required to execute the missions.

STAFF RESPONSIBILITIES AND FUNCTIONS

*2X Staff

6-47. The *2X staff is responsible for the integration, correlation, and fusion of all Human Sensor information into the Intelligence BOS within the *2X

AOIR. The *2X is also responsible for analyzing adversary collection, terrorist and sabotage activities, developing countermeasures to defeat threat collection activities, identifying and submitting collection requirements to fill collection gaps, and providing input to the all-source picture regarding adversary intelligence activities.

*2X Staff Officer

6-48. The *2X Staff Officer provides CI and HUMINT collection expertise. The *2X—

- Is the single focal point for all matters associated with CI and HUMINT in the AOIR.
- Is the CI and HUMINT advisor to the G2 and commander.
- Is an extension of the collection manager and ensures that the best asset or combinations of assets are used to satisfy information requirements.
- Along with his subordinate elements (CICA, HOC, OSC, CI Analysis Cell [CIAC], and HAC), exercises technical control over his assigned Army CI and HUMINT elements in the designated AOIR.
- Is the principal representative of the G2 and the commander when coordinating and deconflicting CI and HUMINT activities with national or theater agencies operating in the AOIR.
- Supports specific RM efforts in conjunction with the requirements manager through—
  - Planning and coordinating CI and HUMINT operations.
  - Reviewing and validating HUMINT requirements.
  - Recommending assignment of tasks to specific collectors.
  - Conducting liaison with non-organic HUMINT collection. This liaison includes national level and multinational force assets for source deconfliction and special activities outside the *2X AOIR.
- Will provide OMTs with capability to reach back to current database information, technical information and guidance, and source deconfliction necessary to monitor the collection activities of the HUMINT teams.

The HUMINT Analysis Cell (HAC)

6-49. The HAC is the single-source fusion point for all HUMINT reporting and operational analysis. It determines gaps in reporting and coordinates with other analysis teams and technical controllers to cross-cue other collection sensor systems. The HAC—

- Uses analytical tools to develop long-term collection plans and provides reporting feedback that supports all HUMINT and CI entities in the supported command's AOIR.
- Produces and disseminates HUMINT products and provides input to intelligence summaries.
- Uses analytical tools found at the ACE or JISE to develop long-term analyses and provides reporting feedback that supports the J/G/S2X, HUMINT operations section, OMTs, and HUMINT teams.
- Produces country and regional studies tailored to HUMINT collection.

- Compiles target folders to assist J/G/S2X assets in focusing collection efforts.
- Analyzes and reports on trends and patterns found in HUMINT reporting.
- Analyzes source reliability and credibility as reflected in reporting and communicating that analysis to the collector.
- Develops and maintains databases specific to HUMINT collection activities that directly support the collection efforts of HUMINT teams and are directly accessible by HUMINT teams.
- Provides collection requirements input to the HOC.
- Supports RM through the development of HUMINT SIRs based on command PIRs.
- Answers HUMINT-related RFIs.

6-50. For intelligence reach operations, HUMINT products are available and disseminated in a variety of forms. It is incumbent on the requestor to ensure that the HUMINT product can be transmitted over the available communications systems. This includes verifying the appropriate security level of the communications systems.

HUMINT Operations Cell (HOC)

6-51. The HOC in the *2X coordinates and synchronizes all HUMINT activities in the AOIR. The HOC exercises technical control over all HUMINT entities in the designated AOIR and deconflicts HUMINT activities with higher, lower, and adjacent HUMINT elements. The HOC accomplishes all responsibilities through coordination with the operational units and the CICA and operations support cell (OSC). The HOC tracks all HUMINT activities in the AOR. The J/G2X uses this information to advise the SIO on all HUMINT activities conducted within the AO. The HOC—

- Exercises technical control of all HUMINT assets and coordinates and deconflicts HUMINT activities in the deployed AO.
- Establishes and maintains a HUMINT source database.
- Coordinates and supervises HUMINT FP source operations conducted by all services and components in the AO.
- Develops and manages collection requirements for HUMINT in coordination with the requirements manager.
- Develops and provides the HUMINT portion of the intelligence synchronization plan to the J/G/S2X and requirements manager for inclusion in the intelligence synchronization plan.
- Coordinates the activities of HUMINT collectors assigned or attached to interrogations and debriefing facilities.
- Expedites preparation of intelligence information reports and their distribution to consumers at all levels.
- Performs liaison with HN and US national HUMINT organizations.

Operations Support Cell

6-52. The OSC in the *2X staff maintains the source registry for all HUMINT activities in the designated AOIR. The OSC provides management of intelligence property book operations, source incentive programs, and ICFs.

6-53. Integrating linguists and DOD emergency-essential civilians, such as technical support contractors, into the HUMINT team should take place as soon as possible. Security clearances and contractual agreements will help the team determine the level of integration.

6-54. Along with the basic briefing of what is expected of the civilian linguists as interpreters and the emergency-essential civilians as support personnel, HUMINT teams should be informed about the civilians' chain of command and the scope of their duties beyond interpreting and technical support. The HUMINT team leader must ensure that linguists and emergency-essential civilians are trained and capable of completing all tasks expected of them.

Counterintelligence Coordinating Authority

6-55. The CICA coordinates all CI activities for a deployed force. There can be only one CICA in a theater of operations. When multiple echelons exist, the highest echelon has the CICA and subordinate G2X offices have a CICA. Depending on the size and scope of the operation, the CICA could be the unified command's CI staff officer; the CI or HUMINT staff officer from corps or division; or a senior warrant officer or branch CI officer designated by the unified command CI staff officer or task force commander. For more information about the CICA functions, see Chapter 11.

HUMINT TEAM STRUCTURE

Operational Management Team

6-56. The OMT is a four-person team consisting of a warrant officer (WO), two noncommissioned officers (NCOs), and a junior enlisted soldier. Civilians may be inserted into this structure when appropriate. Rank structure and standards of grade for OMTs vary depending upon the skill sets required and mission focus. HUMINT OMTs provide operational guidance for two to four HUMINT teams, depending on mission focus and operational tempo. When two or more HUMINT teams are deployed in DS of a maneuver element, an OMT also deploys to provide technical control. The OMT works closely with the supported S2 and analysis and control team (ACT) to furnish current threat information and answer the supported commander's PIRs and IRs. OMTs coordinate with the supported 2X and manage subordinate HUMINT team echelons to—

- Provide guidance and technical control of operational activity.
- Provide the collection and operational focus for HUMINT teams.
- Provide quality control and dissemination of reports for subordinate HUMINT teams.
- Conduct single-discipline HUMINT analysis, and assist in mission analysis for the supported commander.

- Act as a conduit between subordinate HUMINT teams, the HOC, and supported unit headquarters.
- Provide administrative support for subordinate HUMINT teams, to include reporting mission and equipment status to the HOC and the supported unit headquarters.
- Educate the supported commander on the capabilities of the HUMINT teams.
- Integrate the HUMINT teams directly into the maneuver commander's ISR planning.

HUMINT Team

6-57. The HUMINT team is a four-person team consisting of two NCOs and two junior enlisted personnel. Civilians may be inserted into this structure when appropriate. Rank structure and standards of grade for HUMINT teams vary depending upon the skill sets required and the mission focus. HUMINT teams are trained to execute the full range of HUMINT functions as defined in **HUMINT FUNCTIONS** section. However, they may be assigned to mission-focused elements such as DOCEX, interrogation, debriefing, or contact operations.

Chapter 7

Imagery Intelligence

DEFINITION

7-1. IMINT is intelligence derived from the exploitation of imagery collected by visual photography, infrared, lasers, multi-spectral sensors, and radar. These sensors produce images of objects optically, electronically, or digitally on film, electronic display devices, or other media.

ROLE

7-2. The role of imagery is to assist the commander in focusing and protecting his combat power. Imagery often enhances the commander's situational understanding of the battlespace. Other than direct human observation, imagery is the only intelligence discipline that allows the commander to see the battlefield in real time as the operation progresses. In those cases where maps are not available, digital imagery in hardcopy or softcopy can be used as a substitute. Imagery can also be used to update maps or produce grid-referenced graphics. Detailed mission planning often requires imagery, to include three-dimensional stereo images, in order to provide the degree of resolution necessary to support such specialized planning.

FUNDAMENTALS

7-3. Some imagery assets are very responsive to the individual commander's intelligence requirements. Some imagery systems can directly transmit imagery into the tactical operations center (TOC); examples include imagery from UAVs and the Joint Surveillance Target Attack Radar System (JSTARS). This direct downlink enables the G2/S2 to use the imagery as soon as possible instead of having to wait for finished imagery products. A note of caution is required, however, if not a trained imagery analyst because the G2/S2 could incorrectly interpret the imagery.

7-4. Imagery-related equipment has undergone a reduction in size as well as a reduction in the time it takes to provide products, particularly softcopy imagery. The modularity and size reduction of imagery analysis, processing, and display systems make transport easier; they also allow the commander to bring lesser amounts than in the past, while still retaining those systems (or subsystems) required to complete the mission. Additionally, data compression allows faster transmission of imagery products directly to the warfighter.

SOURCES OF IMAGERY

7-5. There are three general sources of imagery: national, civil, and commercial. National imagery traditionally refers to imagery collected by DOD imagery systems. However, there are other sources of imagery provided by non-national sources such as handheld photography (film, digital, and video), UAV imagery, and gun-camera images.

National

7-6. National systems are developed specifically for supporting the President of the United States, the Secretary of Defense, other national agencies, and US military forces. These systems respond to the needs of the nation and those of the combatant commands.

Civil

7-7. Civil imagery systems are usually government funded in terms of building, launching, and operating the system. In many, but not all, cases the agencies operating these civil imagery systems also process, distribute, and archive the imagery data or images.

Commercial

7-8. Commercial companies build, launch, and operate imagery systems for profit. In times of crises, license agreements with the US Government obligate US commercial satellite imaging systems to provide data only to the US Government at the market value. This protects information concerning US operations from adversarial exploitation from commercial systems such as the ACE Imagery Company. However, foreign commercial imagery systems are not bound to this arrangement, and thus may be used by our nation's adversaries. Commercial imagery has become increasingly valuable for many reasons:

- Due to its unclassified nature, civil and commercial imagery is useful in an open environment, may be released to multinational partners, and can be made available to the press. They are especially useful for geospatial products. Their use allows national systems more time to focus on other intelligence functions.
- Civil and commercial imagery offer radar and multi-spectral. Some offer large area collection useful for broad area coverage purposes.
- Commercial satellite imagery resolution varies from less than one meter to several kilometers.

7-9. The Central Imagery Tasking Office (CITO) is responsible for ordering commercial imagery. The Commercial Satellite Imagery Library is available to research DOD purchased commercial imagery. The G2/S2 should consult the CITO when forming commercial imagery requests. NGA will deliver the imagery primarily on CD-ROM media via courier or mail service. Limited digital or electronic delivery is available as well.

TYPES OF IMAGERY SENSORS

7-10. There are four types of imagery sensors: visible (optical), infrared, radar, and multi-spectral. Each sensor has a unique capability, with distinct advantages and disadvantages. The G2/S2 must understand each sensor's capability in order to select the best sensor for the mission and thus enable the user to better understand the intelligence received. Certain sensors are better suited for military operations than others. (See Table 7-1.)

Table 7-1. Sensor Characteristics Matrix.

SENSORS	ADVANTAGES	DISADVANTAGES
Visible (Optical) Best tool for daytime, clear weather, detailed analysis. Includes video and electro-optical.	• Affords a familiar view of a scene. • Offers system resolution that cannot be achieved in other optical systems or in thermal images and radars. • Preferred for detailed analysis and mensuration. • Offers stereoscopic viewing.	• Restricted by terrain and vegetation. • Limited to daytime use only. • Reduced picture size.
Infrared Best tool for nighttime, clear weather, detailed analysis. Includes Overhead Non-Imaging Infrared (ONIR).	• A passive sensor and is impossible to jam. • Offers camouflage penetration. • Provides good resolution. • Nighttime imaging capability.	• Not effective during thermal crossover periods. • Product not easily interpretable. • Requires skilled analysis. • Cannot penetrate clouds.
Radar Useful for detecting presence of objects at night and in bad weather. Includes synthetic aperture radar (SAR), coherent change detection (CCD), and MTI.	• All weather; can penetrate fog, haze, clouds, smoke. • Day or night use. • Does not rely on visible light nor thermal radiation. • Good standoff capability. • Large area coverage. • Allows moving target detection. • Foliage and ground penetration.	• Product not easily interpretable. • Requires skilled analysis. • Terrain masking inhibits use.
Multi-Spectral Imagery (MSI) Best tool for mapping purposes and terrain analysis.	• Large database available. • Band combinations can be manipulated to display desired requirements. • Images can be merged with other digital data to provide higher resolution.	• Product not easily interpretable. • Requires skilled analysis. • Computer manipulation requires large amounts of memory and storage; requires large processing capabilities.

PLAN

7-11. The first step in planning for IMINT is determining the need for IMINT products based on the PIRs and the initial IPB. The G2/S2 should research targets using online imagery databases early and request those imagery products that are not perishable for contingency planning. National and COCOM imagery databases may hold recently imaged targets that could meet the commander's immediate needs instead of requesting new imagery. The staff must clearly articulate their intelligence requirements to include communicating what the mission is and how the requested product will aid in mission accomplishment. The G2/S2 should submit the imagery requirement using established procedures such as those in the unit's SOP or as established by the COCOM.

7-12. The G2/S2 must also determine the specific imagery requirements so as not to burden the system with unnecessary requests. The desire for imagery products often exceeds the capabilities of the IMINT system. Therefore, it is imperative that the G2/S2 consider what type of analysis they require, and request only that which they require. The specifications of the request for IMINT products often affect the timeliness of the response. For example, determining if vehicles are tanks takes less time and requires less resolution than determining if a tank is a T-64 or a T-72.

7-13. Here are some of the roles IMINT can perform that the G2/S2 may consider when determining IMINT requirements:

- Imagery can detect and/or identify and locate specific unit types, equipment, obstacles, potential field fortifications, etc., from which intelligence analysts are able to analyze enemy capabilities and develop possible COAs.
- Imagery can also update maps and enhance the interpretation of information from maps. Detailed mission planning uses imagery to include stereo images for three-dimensional viewing of the terrain and many other geospatial uses.
- Imagery can be used as a substitute for maps when maps are not available. The most common application of this technique is by constructing imagery mosaics: a combination of two or more overlapping photographic prints that form a single picture.
- MTI displays or products can provide an NRT picture of an entity's movement by indicating its speed, location, and direction of travel. MTI systems do not differentiate friendly from enemy. Imagery assets, particularly MTI systems, are useful in cueing other ISR systems.
- Imagery can be used to support protection of the force by helping the commander visualize how his forces look—including their disposition, composition, and vulnerabilities—as exploited by enemy IMINT systems.
- Imagery analysts use combat assessment imagery to confirm destruction, determine the percentage of destruction, or whether the target was unaffected.

PREPARE

7-14. The G2/S2 IMINT-related actions during the prepare function of the intelligence process include establishing or verifying the portion of the intelligence communications architecture that supports IMINT display and analysis functions properly. Additionally, the G2/S2 must ensure that required IMINT analytical assets and resources are prepared to provide support or are available through intelligence reach. Lastly, the G2/S2 must also ensure IMINT reporting and dissemination channels and procedures are in place and rehearsals are conducted with all pertinent IMINT elements to ensure interoperability.

COLLECT

7-15. As previously mentioned, there are four types of imagery sensors. Depending on the type of sensor, it can record hardcopy or softcopy single frame or continuous (video). A given imagery target will not necessarily receive continuous coverage due to the possible conflict between the number and priority of targets and the number and availability of imagery assets. However, a commander may decide to have continuous surveillance of certain targets, for specified periods of time, usually using his own imagery assets (for example, UAV) even though this detracts from the commander's ability to use these assets for other imagery targets within his AOI.

PROCESS

7-16. The process function regarding IMINT involves converting imagery data into a form that is suitable for performing analysis and producing intelligence. Examples of IMINT processing include developing film, enhancing imagery, converting electronic data into visual displays or graphics, and constructing electronic images from IMINT data.

PRODUCE

7-17. The IMINT producer must ensure the IMINT product satisfies the associated intelligence requirements and that the product is in the required format. The quality and resolution of the product is highly dependent upon the type of sensor, the time of day, and the weather conditions, as well as the imagery analyst's ability to identify items, vehicles, equipment, and personnel within the images. Specific IMINT products are discussed in the **ANALYZE** section below.

ANALYZE

7-18. Timeliness is critical not only to IMINT collection but also to IMINT analysis and reporting. It is difficult to separate IMINT reporting from IMINT analysis in this discussion. This is demonstrated by the three phases of IMINT reporting presented below; all are dependent upon the timeliness requirements. Each phase represents a different degree of analysis and period of time available to accomplish the exploitation of the imagery.

- First Phase imagery analysis is the rapid exploitation of newly acquired imagery and reporting of imagery-derived information within a specified time from receipt of imagery. This phase satisfies priority requirements of immediate need and/or identifies changes or activity of immediate significance. First Phase imagery analysis results in an Initial Phase Imagery Report (IPIR).
- Second Phase imagery analysis is the detailed exploitation of newly acquired imagery and the reporting of imagery-derived intelligence and information while meeting the production and timeliness requirements. Other intelligence discipline source material may support phase two imagery as appropriate. Second phase imagery analysis results in a Supplemental Imagery Report (SUPR).
- Third Phase imagery analysis is the detailed analysis of all available imagery pertinent to an SIR, and the subsequent production and reporting resulting from this analysis within a specified time. This

phase provides an organized detailed analysis of an imagery target or topic, using imagery as the primary data source but incorporating data from other sources as appropriate.

7-19. The two types of imagery exploitation are national and direct support (DS).

- National exploitation is imagery exploitation that supports presidential requirements, National Security Council (NSC) requirements, congressional requirements, or requirements of a common concern to the intelligence community.
- DS exploitation is imagery exploitation that supports assigned missions of a single agency, department, or command (the warfighter).

7-20. IMINT assets will complete DS exploitation in order to satisfy (First Phase) requirements and report the results as soon as possible, but not later than 24 hours after receipt of the imagery. Collectors will complete national exploitation in order to satisfy (Second and Third Phases) requirements and report the results within the time specifications of each individual requirement.

DISSEMINATE

7-21. IMINT products are distributed or disseminated in hardcopy, softcopy, or direct viewing such as through remote terminals. The distribution of hardcopy products will be via couriers or other type of mail system. The dissemination of softcopy products will be either as hardcopy products (for example, CD-ROM and 3.5-inch disks) or electronically. The requestor must ensure that the requested product is transmittable over the available communications systems.

ASSESS

7-22. The requestor should immediately assess the imagery product upon receipt for accuracy and relevance to the original request. The requestor must then notify the producer and inform him of the extent to which the product answered the PIR. Providing feedback to the producer regarding the product helps ensure the producer will provide the required information in the correct format. The following are some of the questions which the requestor should consider when providing feedback to the producer.

- Is the format of the product acceptable?
- Is additional information needed on the product or future products?
- Is excess information included on the product?

Chapter 8

Signals Intelligence

DEFINITION

8-1. SIGINT is a category of intelligence comprising either individually or in combination all COMINT, ELINT, and FISINT, however transmitted; intelligence is derived from communications, electronics, and foreign instrumentation signals. SIGINT has three subcategories:

- COMINT – The intelligence derived from foreign communications by other than the intended recipients.
- ELINT – The technical and geo-location intelligence derived from foreign non-communications electromagnetic radiations emanating from other than nuclear detonations or radioactive sources.
- FISINT – Technical information and intelligence derived from the intercept of foreign electromagnetic emissions associated with the testing and operational deployment of non-US aerospace, surface, and subsurface systems. Foreign instrumentation signals include but are not limited to telemetry, beaconry, electronic interrogators, and video data links. (See JP 1-02)

8-2. SIGINT provides intelligence to the commander based upon intercepted communications and provides transmitter location data.

ROLE

8-3. SIGINT provides intelligence on threat capabilities, disposition, composition, and intentions. In addition, SIGINT provides targeting information for the delivery of lethal and non-lethal effects.

FUNDAMENTALS

8-4. It is important that the G2/S2 understand how SIGINT assets are organized not only within the Army but also throughout the DOD. The majority of SIGINT assets are located at EAC. SIGINT assets from all the armed services, combined with national SIGINT assets, work together to support commanders from the tactical to the strategic level. Only by understanding the SIGINT structure that transcends traditional service component boundaries can the G2/S2 understand how to use SIGINT effectively.

TECHNICAL CONTROL AND ANALYSIS ELEMENT (TCAE) ARCHITECTURE

8-5. The TCAE architecture supports Army SIGINT collectors and analysts by providing SIGINT-specific intelligence and guidance to SIGINT personnel. The SIGINT technical architecture complements existing C2 relationships; it does not replace the commander's authority or chain of command.

8-6. There is a TCAE at each Army operational echelon. The TCAE at each echelon provides a single POC and resource dedicated to supporting the commander at that level. TCAEs or SIGINT elements at varying echelons draw information from each other in order to provide a more complete and detailed intelligence picture for their respective commanders.

ARMY TCAE (ATCAE)

8-7. The ATCAE, established at the national level, plays a significant role in TCAE operations by providing technical support oversight and providing access to national databases—in coordination with the NSA and other intelligence organizations. Additionally, the ATCAE works closely with NSA's IO element assisting in operations, monitoring technical capabilities, and providing liaison with Army's 1st Information Operations Command (Land). (This unit was previously called Land Information Warfare Activity.)

8-8. The ATCAE supports the Army's Quick Reaction Capability (QRC) SIGINT and signals research and target development (SRTD) operations. The QRC systems are responsive to the ground force commander's requirements and allow the ability to conduct SIGINT operations against modern communications systems.

REGIONAL TCAE (RTCAE)

8-9. The Army established RTCAEs to allow supported units at all echelons access to SIGINT regarding their respective AOI. The Army placed an RTCAE within each Regional Security Operations Center (RSOC). An RSOC is a joint service facility established by the Director, NSA, to conduct continuous security operations on selected targets in support of national and warfighter intelligence requirements using remoting technologies. The RSOC embeds a focus on the tactical commander's SIGINT requirements at a national level. It also creates a framework for enhanced interoperability among service SIGINT activities, especially within the context of a deployed JTF.

8-10. In addition to each RSOC, the RTCAE works closely with Theater TCAEs as well as Corps and Division SIGINT elements with the same regional focus. It provides analytic support and tailored products and answers RFIs. During peacetime and real-world operations RTCAEs facilitate intelligence reach support. They provide tailored technical support packages, support for surge and survey operations, analytic expertise, and signal data access and digital file transfers of live collection for Army tactical units.

THEATER TCAE

8-11. The Theater TCAE performs SIGINT and EW technical control and analysis and management. It provides SIGINT technical support for assigned, OPCON, and lower echelon SIGINT resources deployed in the theater. This includes mission tasking, processing, analyzing, and reporting of SIGINT data, information, and intelligence. The TCAE provides direction for the Theater Collection and Exploitation Battalion's SIGINT mission and for other theater tactical SIGINT assets.

SIGINT THEATER INTELLIGENCE BRIGADE/GROUP (TIB/TIG)

8-12. The TIB/TIG conducts operational level multidiscipline SIGINT operations. It provides timely intelligence to the commander throughout the full spectrum of operations. The TIB/TIG is structured to provide ground and aerial SIGINT support. It executes the full range of SIGINT missions or tasks.

SIGINT AT ECHELONS CORPS AND BELOW

8-13. There is a varying mixture of SIGINT assets within each corps. A reference listing the types of SIGINT assets found at echelons corps and below is listed in ST 2-50.

PLAN

8-14. An important SIGINT planning consideration is that if at all possible, SIGINT collection should be employed in conjunction with another intelligence discipline collection system. SIGINT is often used to cue, and be cued by, other ISR assets.

8-15. During planning, the G2/S2 should retrieve, update, or develop any required SIGINT databases. This includes effecting coordination with other SIGINT assets or elements that can support the operation.

8-16. SIGINT ground collection assets are usually placed in proximity to enemy signal sources due to the limited height of collector antennas, the low power output of threat or enemy emitters, and the line-of-sight (LOS) constraints imposed by terrain. Ground-based SIGINT teams are most effective when positioned to—

- Maximize threat emitter interception. This allows teams to overcome the constraints of threat emitter characteristics usually allowed by proximity to the threat.
- Minimize system receiver interference. This increases the potential capability of the team to acquire threat emitters in a timely manner.
- Optimize overlapping areas of intercept coverage. This ensures coverage of the AOIR and allows targets to be handed off from team to team if necessary.

8-17. Aerial SIGINT assets have additional planning requirements. Coordination must be effected with Army Airspace Command and Control (A2C2) for ingress and egress routes, restricted operating zones (ROZs) where the aircraft can travel on an intercept track, as well as with the supporting air operations center to determine the availability of asset types and times.

PREPARE

8-18. The G2/S2 ensures the SIGINT unit and asset leaders have effected all necessary coordination and conducted rehearsals. This includes establishing or verifying the operation of the SIGINT technical architecture. The G2/S2 also ensures all required SIGINT assets and resources are available, SIGINT reporting and dissemination channels and procedures are in place, and connectivity and interoperability exist with all pertinent SIGINT elements.

8-19. SIGINT OPCON is the authoritative direction of SIGINT activities, including tasking and allocation of effort, and the authoritative prescription of those uniform techniques and standards by which SIGINT information is collected, processed, and reported.

8-20. SIGINT operational tasking is the authoritative operational direction of and direct levying of SIGINT information needs by a military commander on designated SIGINT resources. These requirements are directive, regardless of other priorities, and are conditioned only by the capability of those resources to produce such information. Operational tasking includes authority to deploy all or part of the SIGINT resources for which SIGINT operational tasking authority (SOTA) has been delegated.

8-21. SOTA is the military commander's authority to operationally direct and levy SIGINT requirements on designated SIGINT resources; it includes authority to deploy and redeploy all or part of the SIGINT resources for which SOTA has been delegated.

COLLECT

8-22. SIGINT performs two major collection activities: signals intercept and direction finding (DF).

Signals Intercept

8-23. Signals intercept are those SIGINT actions used to search for, intercept, and identify threat electromagnetic signals for the purpose of immediate threat recognition. Signals intercept provides information required to answer PIRs, and other intelligence requirements in support of the ISR effort.

Direction Finding

8-24. Even when threat radio operators use COMSEC procedures, SIGINT teams can often intercept and approximate the location of the threat's signals. Specifically, SIGINT teams can use DF to determine—

- Movement of threat personnel or equipment.
- Locations of emitters associated with weapon systems and units.
- New emitter locations and confirm known emitter locations.
- Possible friendly targets the enemy intends to attack (lethal and non-lethal).

8-25. In addition to finding threat forces, DF operations can assist the (radio-equipped) friendly force by—

- Locating and vectoring assets or units during limited visibility.
- Locating downed aircraft and personnel radio beacons.
- Conducting signal security assessments.
- Locating sources of communication interference and jamming.

PROCESS

8-26. SIGINT processing involves converting intercepts of SIGINT into written and verbal reports, automated message, graphic displays, recordings, and other forms suitable for analysis and intelligence production. Since US forces routinely conduct operations against adversaries who speak languages other than English, SIGINT processing often also includes translation of these intercepts.

8-27. Due to the complexity of many SIGINT systems, automated processing may occur several times before SIGINT data or information receives any human interaction.

PRODUCE

8-28. The SIGINT producer must ensure the SIGINT product satisfies the associated intelligence requirements and that the product is in the required format. The quality, fidelity, and timeliness of SIGINT products are highly dependent upon the type of intercept, the collection system, the system's position in relation to the threat emitter, the weather (including space-based weather), as well as the SIGINT operator's ability to identify the appropriate threat signal activity.

8-29. SIGINT production results in some of the reports and formats mentioned in the process section above; however, the objective for SIGINT is to be used in an all-source analytical approach.

ANALYZE

8-30. The intelligence staff analyzes intelligence and information about the enemy's communications capabilities to determine appropriate SIGINT collection strategies. Conversely, a corresponding analysis of the friendly forces' SIGINT capabilities must be conducted to ensure the continued effectiveness of, or to improve upon, SIGINT collection.

8-31. SIGINT analysts also sort through large amounts of SIGINT and information and intelligence to identify and use only that which is pertinent to the CCIRs (PIRs and FFIRs).

DISSEMINATE

8-32. SIGINT of critical importance to the force, including answers to the CCIRs (PIRs and FFIRs), is disseminated via the most expeditious means possible. Due to the usually highly perishable nature of SIGINT, the most expeditious reporting means is often immediately augmented with a follow-up report or augmented by a report transmitted through additional means, enhancing the probability of receipt. Sometimes the most expeditious means of reporting critical SIGINT information to the commander is face to face.

8-33. For intelligence reach operations, SIGINT products are available and disseminated in a variety of forms: hardcopy, softcopy, direct viewing, or listening (television or radio). It is incumbent on the requestor to ensure that the SIGINT product can be transmitted over the available communications systems. This includes verifying the appropriate security level of the communications system.

ASSESS

8-34. The primary goal of the assess function when applied to SIGINT is to determine whether the results of SIGINT collection meet the requirements of the unit's ISR effort. SIGINT producers must assess all facets of SIGINT operations, from receipt of the ISR task to the dissemination of SIGINT, in an effort to determine their effectiveness. This assessment is not only directed at each SIGINT asset individually but also throughout the supporting SIGINT architecture and the unit's entire ISR effort.

8-35. The G2/S2 immediately assesses SIGINT products upon receipt for accuracy and relevance. He must inform the SIGINT producer of the extent to which the product answered the PIR or intelligence requirement. Providing feedback to the SIGINT producer—and collector—helps improve the effectiveness and efficiency of SIGINT.

Chapter 9

Measurement and Signatures Intelligence

DEFINITION

9-1. MASINT is technically derived intelligence that detects, locates, tracks, identifies, and/or describes the specific characteristics of fixed and dynamic target objects and sources. It also includes the additional advanced processing and exploitation of data derived from IMINT and SIGINT collection.

9-2. MASINT collection systems include but are not limited to radar, spectroradiometric, E-O, acoustic, RF, nuclear detection, and seismic sensors, as well as techniques for gathering NBC and other material samples.

9-3. It requires the translation of technical data into recognizable and useful target features and performance characteristics. Computer, communication, data, and display processing technologies now provide MASINT in support of commanders throughout the full spectrum of operations.

9-4. The subdisciplines within MASINT include, but are not limited to, the following:

- **Radar Intelligence (RADINT).** The active or passive collection of energy reflected from a target or object by LOS, bistatic, or over-the-horizon radar systems. RADINT collection provides information on radar cross-sections, tracking, precise spatial measurements of components, motion and radar reflectance, and absorption characteristics for dynamic targets and objectives. A SAR system, coupled with advanced MASINT processing techniques—
  - Provides a high resolution, day and night collection capability.
  - Can produce a variety of intelligence products that identify or provide change detection, terrain mapping, underwater obstacles, dynamic sensing of targets in clutter, and radar cross-section signature measurements.
- **Frequency Intelligence.** The collection, processing, and exploitation of electromagnetic emissions from a radio frequency weapon (RFW), an RFW precursor, or an RFW simulator; collateral signals from other weapons, weapon precursors, or weapon simulators (for example, electromagnetic pulse signals associated with nuclear bursts); and spurious or unintentional signals.
  - *Electromagnetic Pulses.* Measurable bursts of energy that result from a rapid change in a material or medium, resulting in an explosive force, produces RF emissions. The RF pulse emissions associated with nuclear testing, advanced technology devices, power and propulsion systems, or other impulsive events can be used to detect, locate, identify, characterize, and target threats.

  - *Unintentional Radiation Intelligence (RINT).* The integration and specialized application of MASINT techniques against unintentional radiation sources that are incidental to the RF propagation and operating characteristics of military and civil engines, power sources, weapons systems, electronic systems, machinery, equipment, or instruments. These techniques may be valuable in detecting, tracking, and monitoring a variety of activities of interest.
- **E-O Intelligence.** The collection, processing, exploitation, and analysis of emitted or reflected energy across the optical portion (ultraviolet, visible, and infrared) of the EMS. MASINT E-O provides detailed information on the radiant intensities, dynamic motion, spectral and spatial characteristics, and the materials composition of a target. E-O data collection has broad application to a variety of military, civil, economic, and environmental targets. E-O sensor devices include radiometers, spectrometers, non-literal imaging systems, lasers, or laser radar (LIDAR).
  - *Infrared Intelligence (IRINT).* A subcategory of E-O that includes data collection across the infrared portion of the EMS where spectral and thermal properties are measured.
  - *LASER Intelligence (LASINT).* Integration and specialized application of MASINT E-O and other collection to gather data on laser systems. The focus of the collection is on laser detection, laser threat warning, and precise measurement of the frequencies, power levels, wave propagation, determination of power source, and other technical and operating characteristics associated with laser systems—strategic and tactical weapons, range finders, and illuminators.
  - *Hyperspectral Imagery (HSI).* A subcategory of E-O intelligence produced from reflected or emitted energy in the visible and near infrared spectrum used to improve target detection, discrimination, and recognition. HSI can detect specific types of foliage—supporting drug-crop identification; disturbed soil—supporting the identification of mass graves, minefields, caches, underground facilities or cut foliage; and variances in soil, foliage, and hydrologic features—often supporting NBC contaminant detection.
- **Spectroradiometric Products.** Include E-O spectral (frequency) and radiometric (energy) measurements. A spectral plot represents radiant intensity versus wavelength at an instant in time. The number of spectral bands in a sensor system determines the amount of detail that can be obtained about the source of the object being viewed. Sensor systems range from multispectral (2 to 100 bands) to hyperspectral (100 to 1,000 bands) to ultraspectral (1,000+ bands). More bands provide more discrete information, or greater resolution. The characteristic emission and absorption spectra serve to fingerprint or define the makeup of the feature that was observed. A radiometric plot represents the radiant intensity versus time. An example is the radiant intensity plot of a missile exhaust plume as the missile is in flight. The intensity or brightness of the object is a function of several conditions including its temperature, surface properties or material, and how fast it is moving. For each point along a time-intensity

radiometric plot, a spectral plot can be generated based on the number of spectral bands in the collector.

- **Geophysical Intelligence.** Geophysical MASINT involves phenomena transmitted through the earth (ground, water, atmosphere) and manmade structures including emitted or reflected sounds, pressure waves, vibrations, and magnetic field or ionosphere disturbances.
    - *Seismic Intelligence.* The passive collection and measurement of seismic waves or vibrations in the earth surface.
    - *Acoustic Intelligence.* The collection of passive or active emitted or reflected sounds, pressure waves or vibrations in the atmosphere (ACOUSTINT) or in the water (ACINT). ACINT systems detect, identify, and track ships and submarines operating in the ocean.
    - *Magnetic Intelligence.* The collection of detectable magnetic field anomalies in the earth's magnetic field (land and sea). An example is a Remotely Emplaced Battlefield Surveillance System (REMBASS) sensor detection indicating the presence and direction of travel of a ferrous object.
- **Nuclear Intelligence (NUCINT).** The information derived from nuclear radiation and other physical phenomena associated with nuclear weapons, reactors, processes, materials, devices, and facilities. Nuclear monitoring can be done remotely or during onsite inspections of nuclear facilities. Data exploitation results in characterization of nuclear weapons, reactors, and materials. A number of systems detect and monitor the world for nuclear explosions, as well as nuclear materials production.
- **Materials Intelligence.** The collection, processing, and analysis of gas, liquid, or solid samples. Materials intelligence is critical to collection against NBC warfare threats. It is also important to analyzing military and civil manufacturing activities, public health concerns, and environmental problems. Samples are both collected by automatic equipment, such as air samplers, and directly by humans. Samples, once collected, may be rapidly characterized or undergo extensive forensic laboratory analysis to determine the identity and characteristics of the sources of the samples.

ROLE

9-5. MASINT provides intelligence to the commander throughout the full spectrum of operations to facilitate situational understanding. MASINT can thwart many of the camouflage, concealment, and deception techniques currently used to deceive ISR systems.

9-6. MASINT is perceived as a "strategic" discipline with limited "tactical" support capabilities. But, by application of real-time analysis and dissemination, MASINT has a potential ability to provide real-time situation awareness and targeting not necessarily available to the classic disciplines. Specifically, MASINT "sensors" have unique capabilities to detect missile launch, detect and track aircraft, ships, and vehicles; do non-cooperative target identification (NCTI), combat assessment, and BDA; and detect and track fallout from nuclear detonations. Often, these contributions are the first indicators of hostile activities. For example, two EXOCET-equipped

Mirage F-1s were shot down during the Operation DESERT STORM due to MASINT collection and analysis. As evidenced by Operation IRAQI FREEDOM (OIF), MASINT will play a decisive role in the targeting of smart munitions with the signatures (fingerprint) of the targets they are seeking (for example, infrared signatures).

9-7. The MASINT systems most familiar on today's battlefield are employed by ground surveillance and NBC reconnaissance elements.

9-8. MASINT spans the entire EMS and its capabilities complement, rather than compete with, the other intelligence disciplines. MASINT provides, to varying degrees, the capability to—

- Use automatic target recognition (ATR) and aided target recognition (AiTR).
- Penetrate manmade and/or natural camouflage.
- Penetrate manmade and/or natural cover, including the ability to detect subterranean anomalies or targets.
- Counter stealth technology.
- Detect recently placed mines.
- Detect natural or manmade environmental disturbances in the earth's surface not discernible through other intelligence means.
- Provide signatures (target identification) to munitions and sensors.
- Enhance passive identification of friend or foe.
- Detect the presence of NBC agents to include prior to, during, or after employment.
- Detect signature anomalies that may affect target-sensing systems.

FUNDAMENTALS

9-9. Before discussing the functions of the intelligence process within a MASINT, the following paragraph provides an overview of organizational structure of MASINT.

9-10. Within DOD, the DIA provides central coordination for MASINT collection efforts through the Central MASINT Office. Each service, in turn, has a primary command or staff activity to develop requirements and coordinate MASINT effort. Army responsibility currently resides with INSCOM. Army weapons systems programs that require MASINT information to support system design or operations submit requests through the Army Reprogramming Analysis Team (ARAT) or INSCOM channels for data collection and processing. The S&TI community also performs MASINT collection and processing primarily to support R&D programs and signature development. Every S&TI center has some involvement in MASINT collection or production that reflects that center's overall mission (for example, NGIC does work on armored vehicles and artillery). Service R&D centers such as the Communications-Electronics Command (CECOM) Research, Development, and Engineering Center (RDEC), and Night Vision and Electronic Systems Laboratory are also involved in developing sensor systems for collecting and processing MASINT.

9-11. In addition to supporting the S&TI mission, INSCOM units also execute limited ground-based operational collection to support Theater and ASCC PIRs. This capability will expand upon the standup of INSCOM TIBs and TIGs.

PLAN

9-12. Some MASINT sensors can provide extremely specific information about detected targets, whereas other sensors may only be capable of providing an indication that an entity was detected. Additionally, there are varying capabilities of detection, identification, and classification among MASINT sensors. It is these varying capabilities that require synchronizing the employment of MASINT sensors both within the MASINT discipline and within the ISR effort as a whole. See FM 2-01 for more specific information on ISR synchronization.

9-13. As previously mentioned, there are many types of MASINT sensors. Depending on the type of sensor employed, a given MASINT collection target or NAI may not necessarily receive continuous coverage due to the possible conflict between the number and priority of targets and the number and availability of MASINT assets. However, a commander may decide to have continuous surveillance of certain targets by using his own MASINT assets (for example, REMBASS or Improved-REMBASS).

9-14. Another consideration when planning MASINT missions is whether to use active, passive, or a combination of both when planning MASINT coverage.

PREPARE

9-15. The G2/S2 MASINT related actions during the prepare function of the intelligence process include properly establishing or verifying the MASINT portion of the intelligence communications architecture functions. Additionally, the G2/S2 must ensure that required MASINT analytical assets and resources are prepared to provide support or are available through intelligence reach. Since the products of MASINT are not as well known as products from other intelligence disciplines, the G2/S2 must be aware of the types of MASINT products available to support the operation, and then educate the rest of his unit's staff on the use of these MASINT products. Lastly, the G2/S2 must also ensure MASINT reporting and dissemination channels and procedures are in place and rehearsals are conducted with all pertinent MASINT elements to ensure interoperability.

COLLECT

9-16. MASINT provides information required to answer PIRs and other intelligence requirements in support of the ISR effort. As stated earlier in this chapter, MASINT collection must not only be synchronized within its own discipline but also be synchronized and integrated into the unit's overall ISR effort in order to be effective.

9-17. MASINT sensors are employed throughout the full spectrum of operations from a variety of platforms—sub-surface, ground, marine, and aerospace.

PROCESS

9-18. Just as in the other intelligence disciplines, MASINT involves dealing with huge volumes of data that have to be processed before beginning analysis and production. The process function regarding MASINT involves converting esoteric data into a form that is suitable for performing analysis and producing intelligence. MASINT processing can include relatively simple actions such as converting a REMBASS sensor activation into a report, to a complex task such as processing HSI into a report identifying the composition and concentrations of carcinogenic chemicals present in the emissions from a factory upwind from a US forces encampment.

PRODUCE

9-19. Effective and timely MASINT requires personnel with diverse skill sets. The MASINT producer must ensure the MASINT product satisfies the associated intelligence requirements and that the product is in the required format. The quality, fidelity, and timeliness of MASINT products are highly dependent upon the type of target, the collection system, the system's position in relation to the target or NAI, and the weather, as well as the MASINT system operator's ability to identify the appropriate threat activity.

9-20. The objective of MASINT production is to be used in an all-source analytical approach.

ANALYZE

9-21. The intelligence staff analyzes intelligence and information about the enemy's equipment, doctrine, and TTP to determine appropriate MASINT collection strategies. Conversely, a corresponding analysis of the friendly force's MASINT capabilities must be conducted to ensure the continued effectiveness of, or to improve upon, MASINT collection.

DISSEMINATE

9-22. MASINT of critical importance to the force, including answers to the PIRs, is disseminated via the most expeditious means possible.

9-23. For intelligence reach operations, MASINT products are available and disseminated in a variety of forms. The requestor must ensure that the MASINT product can be transmitted over the available communications systems. This includes verifying the appropriate security level of the communications system.

ASSESS

9-24. The primary goal of the MASINT assess function is to determine whether the results of MASINT collection and production meet the requirements of the unit's ISR effort. MASINT producers must assess all facets of MASINT operations, from receipt of the ISR task to the dissemination of MASINT, in an effort to determine the effectiveness of MASINT. This assessment is not only directed at each MASINT asset individually but also throughout the supporting intelligence communications architecture, to include intelligence reach and the unit's entire ISR effort.

9-25. Also, the G2/S2 immediately assesses MASINT products upon receipt for accuracy and relevance. He must inform the MASINT producer of the extent to which the product answered the PIR or intelligence requirement. Providing feedback to the MASINT producer—and collector—helps improve the effectiveness and efficiency of MASINT.

Chapter 10

Technical Intelligence

DEFINITION

10-1. TECHINT is intelligence derived from the collection and analysis of threat and foreign military equipment and associated materiel.

ROLE

10-2. The strength of the US military lies, in part, to the diversity and extent of its technology base. While the US aspires to be the leader in integrating technology, the threat can achieve temporary technological advantage in certain areas by acquiring modern systems or capabilities. The world arms market is willing to provide these advanced systems to countries or individuals with the resources to pay for them. A concerted TECHINT program is vital to providing precise direction and purpose within the US R&D process to ensure quick and efficient neutralization of this advantage.

10-3. The role of TECHINT is to ensure that the warfighter understands the full technological capabilities of the threat. With this understanding, the warfighter can adopt appropriate countermeasures, operations, and tactics.

10-4. TECHINT has two goals within its role:

- To ensure the US armed forces maintain technological advantage against any adversary.
- To provide tailored, timely, and accurate TECHINT support to the warfighter throughout the entire range of military operations. This includes providing US forces intelligence, information, and training on foreign weapons systems to an extent that allows their use of CEE.

FUNDAMENTALS

10-5. The G2/S2 must understand how TECHINT assets are organized in order to properly apply the intelligence process, as the majority of TECHINT assets are located at EAC.

DEFENSE INTELLIGENCE AGENCY

10-6. DIA manages and reviews overall TECHINT activities. The S&TI Directorate within DIA is the action element for TECHINT. This directorate coordinates with external TECHINT agencies on non-policy matters concerning the production of S&TI. The following organizations provide TECHINT support under the control of DIA:

- **Armed Forces Medical Intelligence Center (AFMIC).** AFMIC, based at Fort Detrick, MD, is a DOD intelligence production center under DIA control. AFMIC is responsible for exploiting foreign medical materiel. The director supports the Army Foreign Materiel Exploi-tation

Program (FMEP) and Army medical R&D requirements. The director coordinates planning, programming, and budgeting with the Army Deputy Chief of Staff, Intelligence (DCS, G2).

- **Missile and Space Intelligence Center (MSIC).** MSIC, based at Redstone Arsenal, AL, is a DOD intelligence production center under DIA control and supports the FMEP. The MSIC acquires, produces, maintains, and disseminates S&TI pertaining to missile and space weapons systems, subsystems, components, and activities. The S&TI produced at MSIC also covers foreign state-of-the-art technology and research applicable to missiles.
- **Defense HUMINT Service.** DHS conducts worldwide HUMINT operations in support of foreign materiel acquisition (FMA) and foreign materiel exploitation (FME).

10-7. The organizations and agencies discussed below constitute the Army TECHINT structure.

ARMY DCS, G2

10-8. The Army DCS, G2 exercises general staff responsibility for all Army TECHINT activities. The Army DCS, G2 forms policies and procedures for S&TI activities, supervises and carries out the Army S&TI program, coordinates DA staff and MSC requirements for TECHINT, and is responsible for the Army Foreign Materiel Program (FMP).

INSCOM

10-9. Under the direction of Headquarters, Department of the Army (HQDA), INSCOM is responsible for peacetime TECHINT operations. HQ, INSCOM, fulfills its responsibilities through its TECHINT oversight function and manages the Army's Foreign Materiel for Training (FMT) Program and FMEP. It provides the interface with strategic S&TI agencies in support of FME and organizes, trains, and equips EAC TECHINT organizations during peacetime. TECHINT exploitation within INSCOM is performed by the following elements:

- **National Ground Intelligence Center.** HQ, INSCOM, exercises direct OPCON over the NGIC. NGIC produces and maintains intelligence on foreign scientific developments, ground force weapons systems, and associated technologies. NGIC analysis includes but is not limited to military communications electronics systems, types of aircraft used by foreign ground forces, NBC systems, and basic research in civilian technologies with possible military applications.
- **203d Military Intelligence Battalion.** The 203d MI Battalion is a multi-component unit headquartered at Aberdeen Proving Ground, MD, and is the Army's sole TECHINT battalion. It performs the following functions:
  - Conducts TECHINT collection and reporting in support of validated S&TI objectives.
  - Acts as the HQDA executive agent for foreign materiel used for training purposes.

- Conducts TECHINT training for DOD analysts and RC TECHINT personnel.
- Supports INSCOM's FMA and FME operations as directed.
- Analyzes and exploits foreign captured enemy documents (CEDs), equipment, weapon systems, and other war materiel.
- Reports on the capabilities and limitations of enemy combat materiel.
- Provides reports alerting the command to the tactical threat posed by technical advances in new or recently discovered foreign or enemy materiel.
- Provides countermeasures to any enemy technical advantage.
- Provides foreign or enemy equipment for troop familiarization and training.
- Provides recommendations on the reuse of CEM.
- Supervises evacuating items of TECHINT interest.
- Provides task-organized battlefield TECHINT teams to support a subordinate command's TECHINT effort.

US ARMY MATERIEL COMMAND (AMC)

10-10. AMC plays a significant support role in TECHINT. Among AMC elements are a series of RDECs, the Army Research Laboratory System, and the Test and Evaluation Command (TECOM). Each conducts highly technical evaluations of foreign equipment. In peacetime, the AMC conducts FME on equipment purchased by each laboratory and RDEC for the intelligence community and for DOD as part of the International Materiel Evaluation Program (IMEP). AMC elements include—

- The Foreign Ordnance Exploitation Team, which is located at the Fire Support Armaments Center (FSAC) in the Picatinny Arsenal. This team exploits foreign ground ordnance and develops render safe procedures (RSPs) for foreign ordnance. It also prepares detailed intelligence reports to support explosive ordnance disposal (EOD), MI, and US munitions developers.
- The Science and Technology Center Europe and the Science and Technology Center Far East, which have the responsibility of collecting information on foreign technical developments by attending arms shows and technology exhibitions.
- The Soldier Biological Chemical Defense Command, which is the headquarters for the US Army Technical Escort Unit (TEU). The mission of the TEU is to collect and escort (transport) chemical, biological, radiological, and nuclear (CBRN) samples for testing and evaluation.

10-11. There are many other agencies with TECHINT responsibilities within the DOD. Refer to FM 34-54 for more information on TECHINT.

PLAN

10-12. TECHINT collection usually begins when an organization or individual reports the recovery or acquisition of unusual, new, or newly employed threat materiel. However, there are often indications that the threat may be using materiel not yet associated with the threat among the myriad intelligence products available. Conversely, it may be known that the threat is using a particular item, the capabilities of which are unknown to US forces. It is in these cases that a unit may receive a TECHINT related ISR task. It is conceivable that such a task may be reflected in the PIR. An example of such a task is the identification or verification of suspected external modifications on a particular model of an enemy's main battle tank; the results of which are linked to a commander's decision.

10-13. TECHINT related ISR tasks should include a mission or target folder. At a minimum, this folder should include a description of the item, with its associated major combat systems, as well as handling instructions, reporting instructions, and a photograph or sketch of the item if available.

10-14. Additionally, the G2/S2 is responsible for ensuring the staff coordinates and establishes a plan for evacuating the desired materiel.

PREPARE

10-15. The G2/S2 must ensure that required TECHINT analytical assets, resources, and evacuation means are prepared to provide support. This includes verifying coordination effected with the task-organized battlefield TECHINT teams from the 203d MI Battalion. The G2/S2 must also ensure the means to report and disseminate TECHINT results to the unit and its soldiers are in place so that they can immediately adopt appropriate countermeasures, operations, or tactics in order to enhance their survival and mission accomplishment.

COLLECT

10-16. TECHINT collection includes capturing, reporting, and evacuating CEM. TECHINT collection begins when an organization or individual reports the recovery or acquisition of threat materiel. An item of materiel is exploited at each level, and continues on through succeeding higher levels until an appropriate countermeasure to neutralize the item's capabilities is identified or developed.

10-17. Army personnel (soldiers and civilian) and units will normally safeguard CEM and report it through intelligence channels to the first TECHINT element in the reporting chain. The location of this TECHINT element will be in accordance with the METT-TC factors; however, there will usually be TECHINT representation at the Corps G2 or the COCOM J2. The TECHINT representative or element will verify if the type of materiel is of intelligence value and determine its further disposition in conjunction with the unit's staff.

PROCESS

10-18. TECHINT processing starts (simultaneously with collection) with the capture of a piece of equipment of TECHINT value. This confirms that the enemy is indeed employing this materiel. In accordance with METT-TC factors, a TECHINT team may move to the location of the item at the capture site or wait until the item is evacuated before conducting a hasty exploitation. After hasty exploitation, the team decides if further processing is required. If it is, the items are sent to the first (or nearest) Captured Materiel Exploitation Center (CMEC). If the item is deemed to yield no immediate tactical intelligence value, it may still be evacuated to the S&TI centers in CONUS for further analysis if the systems represent a change in the technological posture of an enemy.

PRODUCE

10-19. Battlefield TECHINT teams normally report initial and secondary examinations of CEM using either a preliminary technical report or a complementary technical report.

- A preliminary technical report—
  - Includes a general description of the item reported and recommended RSP.
  - Alerts others to information that can be used immediately by tactical units.
- A complementary technical report is more in-depth and—
  - Follows a secondary or an in-depth initial examination.
  - Allows the CMEC to compare new information with intelligence holdings.

10-20. At each successive echelon of exploitation, TECHINT analysts add to the overall body of information on an item by either adding to previous reports or by preparing new reports. The CMEC or other national level S&TI activities prepare more advanced technical reports and analyses. These reports include—

- Detailed technical reports.
- Translation reports.
- Special technical reports.

10-21. Other TECHINT products include—

- CMEC publications such as operator manuals, maintenance manuals, TECHINT bulletins, and tactical user bulletins.
- S&TI analysis bulletins.
- Foreign materiel exploitation reports.

ANALYZE

10-22. TECHINT analysts use checklists established by S&TI agencies and the CMECs to analyze each type of the adversary's equipment for which requirements exist. Analysis always begins with what is, and what is not, known about the piece of equipment. TECHINT units maintain procedures and plans for sampling, analyzing, and handling materiel.

DISSEMINATE

10-23. TECHINT of critical importance to the force, including answers to the PIR, is disseminated via the most expeditious means possible.

10-24. Routine TECHINT reports and products are usually transmitted through the unit's existing intelligence communications architecture in the format of an intelligence information report (IIR) format. For intelligence reach operations, TECHINT products are available and disseminated in a variety of forms. The requestor must ensure that the TECHINT product can be transmitted over the available communications systems. This includes verifying the appropriate security level of the communications systems.

ASSESS

10-25. The primary goal of the TECHINT assess function is to determine whether the results of TECHINT production meet the unit's PIR or intelligence requirements. The G2/S2 immediately assesses TECHINT products upon receipt for accuracy and relevance. He must inform the TECHINT producer of the extent to which the product answered the PIR or intelligence requirement. Providing feedback to TECHINT analysts helps improve the effectiveness and efficiency of TECHINT.

10-26. The G2/S2 also assesses the success of the unit's ISR effort in accomplishing any TECHINT associated ISR task and shares his assessment with the staff and the pertinent units or personnel.

Chapter 11

Counterintelligence

DEFINITION

11-1. CI counters or neutralizes intelligence collection efforts through collection, CI investigations, operations, analysis and production, and functional and technical services. CI includes all actions taken to detect, identify, exploit, and neutralize the multidiscipline intelligence activities of friends, competitors, opponents, adversaries, and enemies. It is the key intelligence community contributor to protect US interests and equities. Figure 11-1 shows the CI overview.

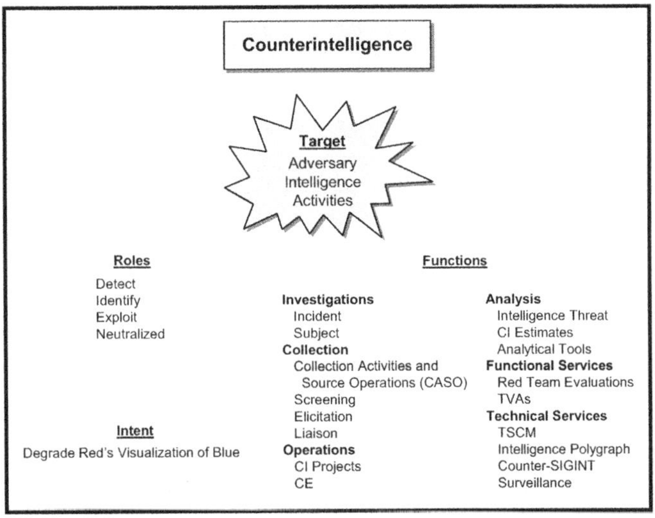

Figure 11-1. Counterintelligence Overview.

ROLE

11-2. The role of CI is to detect, identify, exploit, and neutralize all adversary intelligence entities targeting US and multinational interests. CI will focus on countering adversary intelligence collection activities targeting information or material concerning US personnel, activities, operations, plans, equipment, facilities, publications, technology, or documents—either classified or unclassified—without official consent of designated US release authorities, for any purpose that could cause damage or otherwise adversely impact the interests of national security of the US ability to fulfill national policy and objectives. Adversary intelligence threats include but are not limited to any (US, multinational, friendly, competitor, opponent, adversary, or recognized enemy) government or NGOs, companies, businesses, corporations, consortiums, groups, agencies, cells or persons, terrorists, insurgents, guerrilla entities, and persons whose demonstrated actions, views, or opinions are inimical to US interests.

11-3. CI elements are instrumental in contributing to situational awareness in the AOI. CI elements may corroborate other intelligence discipline information as well as cue other intelligence assets through the CI core competencies and through CI technical services. The CI core competencies are collection, investigations of national security crimes within the purview of CI, operations, and analysis and production. CI technical services include computer network operations, technical surveillance countermeasures (TSCM), and polygraph. CI focuses on combating adversary intelligence activities targeting Army personnel, plans, operations, activities, technologies, and other critical information and infrastructure.

COUNTERINTELLIGENCE FUNCTIONS

11-4. CI functions are interrelated, mutually supporting, and can be derived from one another. No single function or technical capability can defeat adversary intelligence efforts to target US interests. CI functions are discussed below.

CI INVESTIGATIONS

11-5. Investigative activity is essential to countering the adversary intelligence threat to Army interests. CI places emphasis on investigative activity to support force, infrastructure and technology protection, homeland defense, information assurance, and security programs. CI investigations focus on resolving allegations of known or suspected acts that may constitute National Security Crimes under US law. The primary objective in any CI investigation is the detection, identification, exploitation, and/or neutralization of adversary intelligence threats directed against the US Army. CI investigations are also conducted to identify systemic security problems that may have damaging repercussions to Army operations and national security interests. All CI investigations are conducted within guidelines established in AR 381-10, AR 381-12, AR 381-20, applicable DOD policy and directives, and US laws.

CI OPERATIONS

11-6. CI operations are characterized as those activities that are not solely associated with investigative, collection, analysis, or production functions. CI operations can be either offensive or defensive in nature; they are derived from or transition to a collection or investigative activity depending on the scope, objective, or continued possibility for operational exploitation. CI operations fall into two categories: CI support operations and CI sensitive operations.

- **CI Support Operations**. These are defensive operations used to support ARFOR and technology protection, security projects, and programs. They include technical services support, support to acquisition, FP, special access, international security, foreign visitor or contact, treaty verification, information assurance, homeland defense, and other approved projects and programs.
- **CI Sensitive Operations**. These operations are generally offensive in nature and involve direct or indirect operations against a known adversary intelligence threat. These operations include counter-espionage (CE) and CI projects and are conducted by designated units.

CI COLLECTION

11-7. **Collection Activities**. CI elements conduct collection activities focused on adversary intelligence threats that target US and multinational interests. CI collection is conducted through the use of sources, assets, official contacts, and other human or multimedia sources to obtain information that impacts the supported unit. CI will not be used as a substitute for Army HUMINT collection. These activities are designed to collect specific information or develop leads concerning adversary intelligence collection requirements, capabilities, efforts, operations, structure, personalities, and methods of operation targeting US and multinational interests. CI collection can result from ongoing CI investigations or operations or serve to initiate CI investigations and/or operations.

11-8. **Liaison**. CI elements conduct liaison with US, multinational, and HN military and civilian agencies, to include NGO, for the purpose of obtaining information of CI interest and coordinating or deconflicting CI activities. Liaison activities are designed to ensure a cooperative operating environment for CI elements and to develop CI leads for further exploitation.

11-9. **CI Collection Activities and Source Operations (CASO)**. CASO is used to collect information on direct threats to US Army installations, organizations, activities, and personnel. The CASO program is not intended to be used as a substitute for tactical HUMINT contact operations. CASO can be used to initiate CI investigations, identify potential leads for offensive operations, or develop additional CASO leads. **Only designated units, in accordance with applicable policy, will be authorized to pursue investigative and offensive operational leads from ongoing CASO operations.**

11-10. **Screening**. CI Special Agents work jointly with HUMINT Collectors during screening operations to identify civilians on the battlefield, EPWs, detainees, and other noncombatants who may have information of CI interest or to develop CI leads. CI screening is also conducted during the process of hiring HN citizens for Army and DOD employment. Information obtained during screening operations may be used to initiate CI investigations and operations or to cue other intelligence collection disciplines such as HUMINT, IMINT, SIGINT, and MASINT.

11-11. **Debriefing**. CI Special Agents conduct debriefings of friendly force, HN, or the local population who may have information of CI interest regarding adversary intelligence collection or targeting efforts focused on US and multinational interests.

11-12. **Functional Services**. CI elements conduct functional services to assist traditional CI activities of investigations, collection, operations, analysis, and production; they also provide tailored support to US, DOD, and Army protection and security programs for commanders at all echelons. CI elements may use one or several of the functional services simultaneously to provide tailored support to a particular CI mission or supported program. CI functional services consist of—

- CI threat vulnerability assessments (TVAs).
- Adversary intelligence simulation (Red Team Evaluation).
- Covering agent support.

11-13. **CI Technical Services**. CI technical services are used to assist the CI functions of investigations, collections, and operations or to support functional services conducted by CI elements. For additional information on CI technical services, refer to AR 381-20 (SECRET/NOFORN) and FM 34-5. CI technical services consist of—

- Surveillance.
- Intelligence polygraphs.
- TSCM.
- Computer Network Operations (CNO).
- IO.
- Counter-Signals Intelligence (C-SIGINT).

ANALYSIS

11-14. IPB, all-source, and single-source analysis are used to template adversary intelligence activities. Analysis is also used to make recommendations to the commander on how to counter adversary intelligence efforts and to refine CI activities to potentially neutralize and/or exploit those efforts and to continually focus the efforts of CI teams. CI analysis focuses on the multidiscipline adversary intelligence collection threat targeting information on US multinational personnel, operations, activities, technology, and intentions. Raw information, open-source material, and finished intelligence products are analyzed in response to local and national requirements. Analysis occurs at all levels from tactical to strategic.

- At the tactical level, CI teams focus their efforts on supporting mission requirements and contributing to the all-source COP.
- Operational analysis is used to assess how adversary intelligence views and targets US interests; identify US vulnerabilities that could be exploited or targeted; and to determine adversary intelligence targeting methods of operation (MOs).
- CI collection priorities; assessing adversary intelligence technical options for countering US weapons and intelligence systems; and

assessing the impact of technology transfer activities on the US military's technological overmatch.

PRODUCTION

11-15. CI products consist of, but are not limited to, target nomination, CI input to TVAs, CI estimates, and investigative and intelligence information reports. Finalized intelligence derived from CI activities are incorporated into joint and national intelligence databases, assessments, and analysis products. CI products are also incorporated into the COP to support battlefield situational awareness. CI production takes place at all levels.

- Operational and tactical production includes tactical alerts, spot reports, and current intelligence; CI threat and/or vulnerability assessments tailored to specific activities, units, installations, programs, or geographic areas; CI studies to support contingency planning and major exercises; studies of adversary intelligence organization, MO, personnel, activities, and intentions that pose a current or potential threat to the supported command.
- Strategic products include assessments supporting national and Army programs including SAPs and acquisition programs; worldwide assessments of the organization, location, funding, training, operations capabilities, and intentions of terrorist organizations; global trends in adversary intelligence MO; after-action studies of individual espionage cases; analyses of the intelligence collection capabilities of international narcotics trafficking organizations; and multimedia threat products to support Army CI awareness programs.

TECHNICAL SERVICES

11-16. CI organizations with technically trained CI Special Agents are chartered with providing unique technical capabilities to augment CI investigations, collection, and operations. These technical capabilities are not used as substitutes for CI activities, but support traditional CI techniques employed to counter and neutralize adversary intelligence activities targeting US interests.

11-17. Selected CI Special Agents are trained to identify human deception indicators during the course of investigations, operations, and collection missions. Required training and skills include the use of polygraph and emerging biometric technologies that can recognize indications of deception by human sources, contacts, and subjects of investigations, as well as analysis and reporting of results.

11-18. TSCM CI Special Agents are trained in sophisticated electronic and sensing equipment to identify technical collection activities carried out by adversary intelligence entities. The use of TSCM is critical to ensuring that sensitive and/or restricted areas are clear of any adversary intelligence-placed active or passive electronic sensing, or eavesdropping or collection devices. These areas include SCIF, secure working and planning areas, and C2 facilities.

11-19. CNO consist of computer network attack (CNA), computer network defense (CND), and computer network exploitation (CNE). CI Special Agents are specially trained in the areas of computer operation, network theory and administration, and forensics, along with IO in order to ensure US information dominance. The reliance on networked systems will result in greater emphasis being placed on information assurance. Specially selected CI agents will be trained in CNO in order to assist in protecting US information and information systems while exploiting and/or attacking adversary information and information systems.

11-20. For more information on technical services, refer AR 381-20 and FM 34-5.

OPERATIONAL EMPLOYMENT

11-21. Army CI supports the full spectrum of military operations. The Army requires a well-trained CI force consisting of AC, RC, civilian government employee, and contractor personnel. CI elements are focused on and dedicated to detecting, identifying, neutralizing, and/or exploiting adversary intelligence elements attempting to collect information on US forces. Effective employment of Army CI elements in all phases of operations and at all levels from tactical to strategic is paramount to countering any adversary intelligence threat to US interests and resources. Moreover, Army CI protects special programs that provide R&D, acquisition, and integration of technologies leading to future technological overmatch. CI elements have the capability and authority to conduct complex, nontraditional operations in this Information Age to CE, to both protect critical technologies and satisfy Army, DOD, and national level CI objectives.

LEVELS OF EMPLOYMENT

Strategic and Departmental

11-22. Strategic and departmental operations will be conducted by CI elements supporting national, DOD, and DA required missions (for example, support to NATO and special operations and missions). Strategic and departmental CI will conduct compartmented investigations and operations to affect the knowledge of adversary intelligence regarding CONOPS and defense information. Army CI will execute the full range of CI functions and missions at the strategic and departmental level including CI investigations and operations, CE, technology protection, SAP support, treaty verification, and technical CI services (polygraph, TSCM, and computer forensics). Strategic and departmental CI will also support SOF and special mission units (SMUs) within the scope of applicable national, DOD, and DA security policies and regulations.

Operational

11-23. Operational missions of CI elements will support combatant commanders, generally in geographic AOIR. Operational CI elements will focus on threat identification and countering regional adversary intelligence threats. Operational level CI activities and functions include CI investigations and operations, CE, technology protection, SAP support, treaty verification, and technical CI services (polygraph, TSCM, and computer forensics). CI elements must be capable of quickly transitioning from a peacetime mission to crisis operations to support combatant commander requirements. Theater CI assets will conduct unilateral to multinational operations in designated theaters. Operational elements may also be deployed to support or reinforce tactical forces in CONOPS.

Tactical

11-24. CI teams will conduct operations throughout the battlespace during CONOPS. CONOPS support activities include conduct of CASO, limited CI investigative capability, personnel security investigations, screenings, and debriefings. CI activities in CONOPS focus on countering the adversary intelligence threat and assisting in conducting TVAs. TVAs will be conducted in conjunction with MPs, Engineers, and Medical Service personnel to provide the commander with a comprehensive FP assessment. During peacetime, organic tactical CI teams conduct activities in accordance with approved regulations and command guidance.

SUPPORT TO CONTINGENCY OPERATIONS

11-25. The initial phase of operations from PME to MTW lays the foundation of future team operations. In general, the priority of effort focuses inward on security of operating bases, areas of troop concentration, and C2 nodes to identify the collection threat to US forces that could be used by adversary elements to plan hostile acts against US activities and locations.

11-26. Once security of the operating bases has been established, the operational focus of CI teams shifts outside the operating base to continue to detect, identify, and neutralize the collection threat to US forces as well as to provide I&W of hostile acts targeting US activities. The CI team uses several collection methods, to include CASO, elicitation, and liaison, to answer the supported commander's requirements. This is referred to as the continuation phase. The CI team conducts CI investigations to identify, neutralize, and exploit reported threat intelligence collection efforts.

11-27. A key element to the CI team's success is the opportunity to spot, assess, and develop relationships with potential sources of information. Operating as independent teams, without being tied to ISR or combat assets, enables the CI team's maximum interaction with the local population, thereby maximizing the pool of potential sources of information. Along with the opportunity to spot, assess, and interact with potential sources of information, a second key element of a CI team's success is its approachability to the local population. A soft posture enables a CI team to appear as non-threatening as possible. Experience has shown that the local population in general is apprehensive of fully and openly armed patrols and soldiers moving around population centers.

11-28. During some operations, civilian attire or nontactical vehicles may be used to lower the CI team profile. In some special situations, these measures are taken to make the operation less visible to the casual observer. Also, in some cultures, sharing food and beverages among friends is expected; exceptions to restrictions or general orders should be considered to facilitate successful CI team operations, many of which are geared towards developing relationships with potential sources of information.

SUPPORT TO INSTALLATIONS AND OPERATING BASES

11-29. CI teams, as part of a multi-agency team consisting of MPs, CA, medical, and EOD, support the conduct of TVAs of installations and operating bases to identify the intelligence threat to the operating locations. Detailed TVAs identify weaknesses in operational and physical security procedures and recommend countermeasures to mitigate intelligence collection on friendly forces limiting the ability to plan hostile acts on US activities and locations. CI activities supporting installations and operating bases include—

- Interviewing walk-in sources and locally employed personnel.
- Screening local national (LN) hires. Commanders, staff planners, and SIOs should always provide input to personnel assigned to establish and negotiate contracts using local national (LN) hires. This requirement ensures that LN hires can be screened, interviewed, and in some instances used as CI sources or assets in order to provide intelligence information that impacts the security of the base camp.
- Debriefing friendly force personnel who are in contact with the local population, such as—
  - ISR patrols.
  - MP patrols.
  - Combat patrols.
  - Liaison personnel.
  - CA and PSYOP teams.
- Conducting limited local open-source information collection.
- Providing support to TVAs of the base camp.

TACTICS, TECHNIQUES, AND PROCEDURES

11-30. At the CI team level, team members conduct mission analysis and planning specific to their AO. Backwards planning and source profiling are used extensively to choose CI targets. To verify adequate area coverage, the CI team may periodically develop and use CI target overlays and other CI analytical tools that illustrate the CI situation, identify CI gaps, and help refocus the collection effort.

11-31. The CI team is also in constant contact with the supported S2 and the other ISR assets (Scouts, PSYOP, CA, and MP) in order to coordinate and deconflict operations and to cross-check collected information. The supported unit S2, with the help of the CI team, regularly and systematically debriefs all ISR assets.

11-32. The CI team must be integrated into the supported unit's ISR plan. The CI OMT chief will advise the supported unit on the specific capabilities and requirements of the team to maximize mission success.

OPERATIONAL RISK MITIGATION

11-33. The employment of CI teams includes varying degrees of contact with the local population. As the degree of contact with the population increases, both the quantity and quality of CI collection increases. In many instances, however, there is a risk to the CI team inherent with increased exposure to the local population. The decision at what level to employ a CI team is METT-TC dependent. The risk to the CI assets must be balanced with the need to collect priority information and to protect the force as a whole. ROE, SOFA, direction from higher headquarters, and the overall threat level may also restrict the deployment and use of CI teams. The commander should consider exceptions to the ROE to facilitate CI collection.

11-34. Risks are minimized through the situational awareness of CI team members. They plan and rehearse to readily react to any situation and carry the necessary firepower to disengage from difficult situations. If it becomes necessary to call for assistance, adequate and redundant communications equipment is critical. These scenarios and actions should be trained prior to deployment into a contingency area and rehearsed continuously throughout the deployment.

11-35. A supported unit commander is often tempted to keep the CI team "inside the wire" when the THREATCON level increases. The supported commander must weigh the risk versus potential information gain when establishing operational parameters of supporting CI teams. This is necessary especially during high THREATCON levels when the supported unit commander needs as complete a picture as possible of the threat arrayed against US and multinational forces.

11-36. When it is not expedient to deploy the CI team independently due to threat levels or other restrictions, the team can be integrated into other ongoing operations. The CI team may be employed as part of a combat, ISR, or MP patrol or used to support CA, PSYOP, engineer, or other operations. This method reduces the risk to the team while allowing a limited ability to collect information. It has the advantage of placing the team in contact with the local population and allowing it to spot, assess, and interact with potential sources of information. However, this deployment method restricts collection by subordinating the team's efforts to the requirements, locations, and timetables of the unit or operation into which it is integrated and does not allow for the conduct of sensitive source operations. **This method of employment should be considered a last resort.**

COUNTERINTELLIGENCE EQUIPMENT

11-37. Basic C2, transportation, and weapons requirements do not differ significantly from most soldier requirements and are available as unit issue items. However, CI teams have unique communications, collection, processing, and mission-specific requirements.

COMMUNICATIONS

11-38. **Dedicated and Secure Long-Range Communications.** These are keys to the success of the CI team mission. CI team operations require a secure, three-tiered communications architecture consisting of inter/intra-team radios, vehicle-based communications, and a CI and HUMINT base station.

11-39. **Communications Network.** The CI team must have access to existing communications networks such as the tactical LAN. The CI team must also be equipped with its own COMSEC devices. It is imperative that the CI team acquire access to the public communication system of the HN. This can be in the form of either landlines or cellular telephones. Such access enables the CI team to develop leads which can provide early indicators to US forces.

11-40. **Interoperability.** Communications systems must be equipped with an open-ended architecture to allow for expansion and compatibility with other service elements, government organizations, NGOs, and multinational elements to effectively communicate during CONOPS. All ISR systems must be vertically and horizontally integrated to be compatible across all BOSs and with Legacy and Interim Force elements.

11-41. **SOTM.** To provide real-time and NRT information reporting, CI elements must have the capability to transmit voice, data, imagery, and video while on the move. CI teams must be able to transmit while geographically separated from their parent unit while operating remotely. This broadband requirement can only be achieved through a SATCOM capability and must be achievable while mobile.

CI COLLECTION AND PROCESSING SYSTEMS

11-42. The CI team must rely on automation to achieve and maintain information dominance in a given operation. With time, effective collection planning and management at all echelons, the CI team can collect a wealth of information. The sorting and analysis of this information in a timely and efficient manner is crucial to operations. Automation helps the CI team to report, database, analyze, and evaluate the collected information quickly and to provide the supported unit with accurate data in the form of timely, relevant, accurate, and predictive intelligence.

11-43. Automation hardware and software must be user friendly as well as interoperable among different echelons and services. They must interface with the communications equipment of the CI team as well as facilitate the interface of audiovisual devices. Technical support for hardware and software must be available and responsive.

11-44. The demand for accurate and timely CI reporting, DOCEX, and open-source information has grown tremendously. Biometric (physiological, neurological, thermal analysis, facial and fingerprint recognition) technologies will allow rapid identification, coding, and tracking of adversaries and human sources; as well as cataloging of information concerning EPWs, detainees, and civilians of CI interest on the battlefield. Biometrics will also provide secure authentication of individuals seeking network or facility access.

11-45. CI teams work with multinational forces and other foreign nationals and require the ability to communicate in their respective languages. Often CI personnel have little or no training in the target language, and lack of skilled interpreters can hinder CI activities. CI teams require textual and voice translation devices, source verification, and deception detection machines (biometrics) to improve collection capability and accuracy.

11-46. CI teams require dynamic MLT tools that provide both non-linguists and those with limited linguist skills a comprehensive, accurate means to conduct initial CI screenings and basic interviews in a variety of situations. CI elements will focus on in-depth interviews and communications with persons of higher priority. MLT tools minimize reliance on contract linguists and allow soldiers to concentrate on mission accomplishment.

MISSION SPECIFIC

11-47. The CI team may conduct night operations and must be equipped with NVDs for its members, and photographic and weapons systems. The CI team also may operate in urban and rural areas, where the threat level can vary from semi-hostile to hostile. The safety of the CI team can be enhanced with equipment that can detect, locate, suppress, illuminate, and designate hostile optical and E-O devices. In addition, high power, gyro-stabilized binoculars, which can be used from a moving vehicle, increases the survivability of the CI team and also gives the team another surveillance and collection device.

11-48. Some of the CI team missions may require the documentation of incidents. The CI teams can use the following equipment in their open-source collection efforts.

- Small, rugged, battery-operated digital camcorders and cameras which are able to interface with the collection and processing systems as well as communication devices.
- GPSs that can be mounted and dismounted to move in the AO efficiently.
- Short-range multichannel RF scanning devices that can also identify frequencies which enhance their security.
- In some cases CI teams require a stand-off, high resolution optical surveillance and recording capability that can provide target identification at extended ranges to protect the intelligence collector while avoiding detection by the adversary target. An advanced optical capability provides intelligence collectors the ability to locate and track adversary targets (passive and hostile) for identification, collection, and target exploitations.

INTEGRATION OF LINGUISTS

11-49. Integrating linguists into the CI team should take place as soon as possible. Security clearances and contractual agreements will help the team determine the level of integration.

11-50. Along with the basic briefing of what is expected of the civilian linguists as interpreters, CI teams should be informed about the civilians' chain of command and the scope of their duties beyond interpreting. The CI team leader must ensure that linguists are trained and capable of completing all tasks expected of them.

BATTLE HAND-OFF

11-51. CI teams are always engaged. A good battle hand-off is critical to smooth transition and mission success. The battle hand-off can directly contribute to mission success or failure of the outgoing team, but especially of the incoming team. The battle hand-off begins the first day the CI team begins to operate in an AO. Regardless of how long the team believes it will operate within the AO, it must ensure there is a seamless transition to an incoming team, other US unit, or agency. The CI team accomplishes this transition by establishing procedures for source administration, database maintenance, and report files.

11-52. Teams must plan and implement a logical and systematic sequence of tasks that enables an incoming team to assume the operations in the AO. Adequate time must be allotted for an effective battle hand-off. In some environments, a few weeks may be necessary to accomplish an effective battle hand-off. Introductions to sources of information, especially CASO sources, are critical, and teams must prioritize their time. During this time the outgoing CI team must familiarize the new CI team with all aspects of the operation, which include past, present, and planned activities within the AO. Area orientation is critical. These include major routes, population centers, potential hot spots, and other points of interest (such as police stations, political centers, and social centers).

ORGANIZATION

11-53. CI activities require a complex C2 relationship to ensure that the requirements of the supported commander are fulfilled while balancing the need for strict integrity and legality of CI operations. This complex relationship balances the role of the SIO as the requirements manager and the 2X as the mission manager with the MI commander as the asset manager.

COMMAND VERSUS CONTROL

11-54. ARFOR will normally deploy as part of a joint, multinational, and/or combined operation. In all cases, commanders at each echelon will exercise command over the forces assigned to their organization. Command includes the authority and responsibility for effectively using resources, planning for and employment of forces, and ensuring that forces accomplish assigned missions. Leaders and staffs exercise control to facilitate mission accomplishment.

11-55. While the MI commander supervises subordinates and produces reports, the *2X synchronizes activities between intelligence units and provides single-source processing and limited analysis. **(*2X " indicates 2X functions at all levels.)** While the MI commander takes care of the operators executing missions, the *2X obtains the data and reports from higher echelons required to execute the missions.

STAFF RESPONSIBILITIES AND FUNCTIONS

11-56. The *2X staff is responsible for the integration, correlation, and fusion of all Human Sensor information into the Intelligence BOS within the *2X AOIR. The *2X is also responsible for analyzing adversary intelligence collection, terrorist and sabotage activities, developing countermeasures to defeat threat collection activities, identifying and submitting collection requirements to fill CI collection gaps, and providing input to the all-source picture regarding adversary intelligence activities.

11-57. The *2X Staff Officer provides CI and HUMINT collection expertise. The *2X—

- Is the single focal point for all matters associated with CI and HUMINT in the AOIR.
- Is the CI and HUMINT advisor to the G2 and commander.
- Is an extension of the collection manager and ensures that the best asset or combinations of assets are used to satisfy information requirements.
- Along with his subordinate elements—CICA, HOC, OSC, CIAC, and HAC—exercises technical control over his assigned Army CI and HUMINT elements in the designated AOIR.
- Is the principal representative of the G2 and the commander when coordinating and deconflicting CI and HUMINT activities with national or theater agencies operating in the AOIR.
- Supports specific RM efforts in conjunction with the requirements manager through the planning and coordination of CI and HUMINT operations; the review and validation of CI requirements; the recommendation for assignment of tasks to specific collectors; and the conduct of liaison with non-organic HUMINT collection. This liaison includes national level and multinational force assets for source deconfliction and special activities outside the *2X AOIR.
- Will provide OMTs with capability to reach back to current database information, technical information and guidance, and source deconfliction necessary to monitor the collection activities of the CI teams.
  - *CICA.* The CICA is responsible for coordinating and synchronizing all CI activities in the designated AOIR. The CICA exercises technical control over all CI entities in the designated AOIR and deconflicts CI activities with higher, lower, and adjacent CI elements. The CICA accomplishes all responsibilities through coordination with the operational units and other *2X staff elements.
  - *OSC.* The OSC in the *2X staff maintains the source registry for all CI activities in the designated AOIR. The OSC provides management of intelligence property book operations, source incentive programs, and ICFs.
  - *CIAC.* The CIAC analyzes adversary intelligence collection capabilities. The CIAC leverages all intelligence discipline reporting and analysis to counter threat collection capabilities against the deployed force. CIAC analysis provides information and analysis to the COP.

11-58. The ACE and JISE CI analysis team analyzes threat intelligence collection and the intelligence collection efforts of foreign organizations involved in terrorism and sabotage in order to develop countermeasures against them. CI analysis cross-cues IMINT, SIGINT, MASINT, and TECHINT resources in addition to CI-related HUMINT reporting and analysis to counter threat collection capabilities against the deployed force. While the HAC supports the positive collection efforts of the force, the CI analysis team supports the "defend" aspects of the commander's FP program.

11-59. CI analysis is the analysis of the adversary's HUMINT, IMINT, SIGINT, and MASINT capabilities in support of intelligence collection, terrorism, and sabotage in order to develop countermeasures against them. It involves a reverse IPB process in which the analyst looks at US forces and operations from the threat's perspective. CI analytical products are an important tool in the COA development in the MDMP. This analytical tool supports the commander's FP program and facilitates the nomination of CI targets for neutralization or exploitation. (See FM 2-01.2 (FM 34-60) for more information on CI analysis.) Specifically, CI analysis—

- Produces and disseminates CI products and provides input to INTSUMs.
- Provides collection requirements input to the CICA.
- Analyzes source reliability and credibility as reflected in reporting and communicating that analysis to the collector.
- Nominates CI targets for neutralization or exploitation.
- Identifies and submits CI-related requirements to fill collection gaps.
- Assists HAC personnel in focusing the CI aspects of the HUMINT collection program.
- Presents CI analysis products such as CI estimates, target lists, reports, and graphics that support the commander.

11-60. For intelligence reach operations, CI products are available and disseminated in a variety of forms. It is incumbent on the requestor to ensure the CI product can be transmitted over the available communications systems. This includes verifying the appropriate security level of the communications systems.

CI TEAM STRUCTURE

11-61. **OMT.** The OMT is a four-person team consisting of a WO, two NCOs, and a junior enlisted soldier. (Civilians may be inserted into this structure as appropriate.) Rank structure and standards of grade for OMTs will vary depending upon the skill sets required and mission focus. CI OMTs will provide operational guidance for 1 to 4 CI teams, depending on mission focus and operational tempo. When two or more CI teams are deployed in a DS role, an OMT is also deployed to provide technical control. The OMT works closely with the supported S2 and ACT to furnish current threat information and to answer the supported commander's PIRs and IRs. OMTs coordinate with the supported 2X and manage subordinate CI teams to—

- Provide guidance and technical control of operational activity.
- Provide the collection and operational focus for CI teams.

- Provide quality control and dissemination of reports for subordinate CI teams.
- Conduct single-discipline CI analysis and assist in mission analysis for the supported commander.
- Act as a conduit between subordinate CI teams, the CICA, and supported unit headquarters.
- Provide administrative support for subordinate CI teams to include reporting mission and equipment status to the CICA and the supported unit headquarters.
- Educate the supported commander on the capabilities of the CI teams.
- Integrate the CI teams directly into the maneuver commander's ISR planning.

11-62. **CI Team.** The CI team is a four-person team consisting of two NCOs and two junior enlisted personnel. Rank structure and standards of grade for CI teams will vary depending upon the skill sets required and mission focus. CI teams are trained to execute the full range of CI functions; however, they may be assigned to mission-focused elements (for example, CE, CI projects). Assignment to a TSCM, polygraph, or information warfare team requires additional, specialized technical training.

Appendix A

Intelligence and Information Operations

THE INFORMATION ENVIRONMENT

A-1. The information environment is the aggregate of individuals, organizations, and systems that collect, process, store, display, and disseminate information; also included is the information itself. (FM 3-0) The information environment, it should be noted, is not an exclusively military one; in fact, the military applications of information are almost obscured in today's universal usage of the information spectrum by national, international, and non-state players.

THE COMMANDER AND INFORMATION

A-2. Information is facts, data, or instructions in any medium or form; the meaning that a human assigns to data by means of the known conventions used in their representation. (JP 1-02) Information provides the key to battlefield success in the 21st century. Commanders must have detailed information to command. Information is the medium that allows the commander's decisionmaking and execution cycle to function. Information gives direction to actions by the force, identifies the enemy's centers of gravity, provides COAs for force activity, and enables the force to accomplish its operational mission.

INFORMATION SUPERIORITY

A-3. Information superiority is the operational advantage derived from the ability to collect, process and disseminate an uninterrupted flow of information while exploiting or denying an adversary's ability to do the same. (FM 3-0) Relevant information drawn from intelligence supports the creation and development of situational understanding that contributes directly to information superiority during decisive operations. The requirement for information superiority is not new; what is new is that today's information technologies are creating a base of knowledge for military planning and execution that is unprecedented in scope, volume, accuracy, and timeliness. This means commanders receive accurate, timely information that enables them to make better decisions and act faster than their adversaries.

A-4. Information superiority, however, is neither a staple in today's battlespace nor is it necessary in military operations as a constant condition; rather, at specific times during operations, information superiority becomes a key enabler for assuring military success. At the operational level of predominant interest to the land component commander (LCC), information superiority is realized through the integration of operational level, interdependent ISR, information management, and IO to gain and maintain operational initiative and to achieve an operational advantage.

INFORMATION OPERATIONS

A-5. IO are the employment of core capabilities of electronic warfare, computer network operations, PSYOP, military deception, and operations security, in concert with specified supporting and related capabilities to affect or defend information and information systems, and to influence decisionmaking.... Information systems are...the equipment and facilities that collect, process, store, display, disseminate information. These include computers, hardware, software, and communications, as well as policies and procedures for their use. (FM 3-0). Offensive IO are capable of degrading an adversary's will to resist and ability to fight. Defensive IO measures passively and actively protect friendly information and C2 systems and limit their vulnerability.

A-6. Offensive and defensive IO are conducted as a fully coordinated effort to ensure the complementary, asymmetric, and reinforcing effects to attack enemy forces, influence others, and protect friendly forces. Relevant information assures that the right person has the right information at the right time for decisionmaking and execution.

THE ELEMENTS OF IO

A-7. Full-spectrum IO incorporates, integrates, and synchronizes traditionally independent capabilities and activities in support of the commander's mission. (See FM 3-0 and FM 3-13.) The Army's doctrinal view of full-spectrum IO as evolving core capabilities are—

- PSYOP.
- OPSEC.
- EW.
- Military deception.
- CNO.
  - CNA.
  - CND.
  - CNE.

A-8. Supporting IO capabilities are—

- Physical destruction.
- Physical security.
- Information assurance.
- Counterpropaganda.
- Counterdeception.
- Counterintelligence.

IO-RELATED ELEMENTS

A-9. The IO-related activities of PA and CMO are closely associated and integrated with the elements of IO as key contributors to information superiority.

INTELLIGENCE SUPPORT TO IO

A-10. All-source intelligence support, encompassing all of its forms, is the principal enabler for successful IO. The commander's capability to synchronize military operations can be heavily influenced by the ability to identify the threat and understand the adversary's capabilities and intentions in the information environment.

A-11. It is essential that the elements of the command's IO capabilities and vulnerabilities are integrated into the command intelligence plan. This leads to COAs that synchronize the elements of IO and IO-related activities into the commander's warfighting plans. Integration of the full-spectrum aspects of IO ensures that the relative importance of information is recognized in the development of a synchronized OPLAN. Further, integration of IO into the planning process provides a methodology for analyzing the threat from a knowledge base that enhances protection of friendly systems and assets while exposing windows of opportunity for attack or exploitation. IPB is the best process we have for understanding the battlefield and the options it presents to friendly and threat forces.

A-12. Provide Intelligence support to IO includes, but is not limited to, the following:

- **Provide Intelligence Support to Offensive IO**. The Intelligence BOS supports offensive IO by providing information to identify critical enemy C2 nodes. Intelligence also helps to identify enemy systems and procedures that may be vulnerable for offensive IO. Additionally, intelligence plays a key role in evaluating and assessing the effectiveness of offensive IO.
  - *Provide Intelligence Support to PSYOP.* This task identifies the cultural, social, economic, and political environment of the AO. It identifies target groups and subgroups and their location, conditions, vulnerabilities, susceptibilities, cultures, attitudes, and behaviors. PSYOP influence foreign target audiences in the AO to support achieving the commander's goals in the AO.
    - Identify the cultural, social, economic, and political environment of the AOI; for example, adversary mechanisms for political control, adversary communication and broadcast systems used to elicit support from the populace, and current and past adversary propaganda activities and their effectiveness.
    - Identify target groups and subgroups and their location, conditions, vulnerabilities, susceptibilities, cultures, attitudes, and behaviors. This includes determining the audience demographics of popular radio and television programs and periodicals; groups influenced by media personalities and political cartoons.
    - Identify impact of planned PSYOP on individuals outside the targeted group (for example, multinational partners, neighboring populations).
  - *Provide Intelligence Support to Military Deception.* This task identifies the capabilities and limitations of the adversary's

intelligence-gathering systems and identifies adversary biases and perceptions.

  - Profiles of key adversary leaders.
  - Cultural, religious, social, and political characteristics of the country and region.
  - Sources of military, economic, or political support.
  - Adversary decisionmaking processes, patterns, and biases.
  - Adversary perceptions of the military situation in the AO.
  - Capabilities and limitations of adversary CI and security services.
- *Provide Intelligence Support to Electronic Attack.* This task supports electronic attack employing jamming, electromagnetic energy, or directed energy against personnel, facilities, or equipment. It identifies critical adversary information systems and C2 nodes. It includes determining and presenting the adversary's electronic OB and their information system infrastructure. The enemy's C2 system vulnerabilities and the means they use to protect their C2 systems is part of the electronic OB.

- **Provide Intelligence Support to Defensive IO**. The Intelligence BOS supports defensive IO by providing information to identify threat IO capabilities and tactics. Intelligence provides information relating to CND, physical security, OPSEC, counterdeception, and counterpropaganda. The Intelligence BOS supports defensive IO by providing information to identify threat IO capabilities and tactics. Intelligence provides information relating to CND. Provide Intelligence Support to OPSEC—identify capabilities and limitations of the adversary's intelligence system to include adversary intelligence objectives and the means, methods, and facilities used by the threat to collect, process, and analyze information—supports the identification of indicators that could be interpreted or pieced together to penetrate EEFI in time to be useful to adversaries.
- **Provide Intelligence Support to Activities Related to IO**. The Intelligence BOS when operating outside US territories supports activities related to IO under some circumstances.
  - *Provide Intelligence Support to CMO.* This task allows military intelligence organizations to collect and provide information and intelligence products concerning foreign cultural, social, economic, and political elements within an IO in support of CMO. Identify cultural, social, economic, and political environment of the AOI, including—
    - Population demographics.
    - Civilian populace attitudes, alliances, and behavior.
    - Availability of basic necessities (food, clothing, water, shelter, medical care) and the ability of the populace to care for itself.
    - Access to medical care.
    - Locations and potential routes, destinations, and assembly areas or sites of displaced persons.
    - Local government type, status, organization, and capabilities.

- Availability of local material and personnel to support military operations.
- NGOs in the AOI, their agenda, resources, and capabilities.

- *Provide Intelligence Support to Public Affairs.* This task identifies the multinational and foreign public physical and social environment, as well as world, HN national, and HN local public opinion, in addition to the propaganda and misinformation capabilities, activities, targets, themes, and dissemination means of the adversary. Identify world, national, and local public opinion (location, biases or predispositions, and agenda of national and international media representatives in the AOI, and trends reflected by the national and international media).

Appendix B
Linguist Support

ROLE OF LINGUISTS

B-1. Military operations are highly dependent on foreign language support. The requirement to communicate with and serve on multinational staffs, communicate with local populations, and exploit enemy forces necessitates the use of linguists. The growing focus on multinational operations increases the competition for limited linguist resources that are vital for mission success. This appendix establishes the framework and process to access, prioritize, and employ the Army's limited organic linguist resources.

LINGUISTIC SUPPORT CATEGORIES

B-2. Foreign language support requirements of US Armed Forces typically fall into one of four broad categories:

- Intelligence and Information Gathering. This category includes the traditional SIGINT and HUMINT disciplines, as well as foreign language support to FP and exploitation of open-source information.
- CMO. This category encompasses all functions relating to military interaction with the civilian population. Foreign language support is critical to CMO in areas such as government liaison, legal agreements, medical support and operations, law enforcement, engineering projects, public safety, security and population control, CA, and PSYOP.
- Logistics. This category consists of foreign language support to sustainment or transportation functions. These include logistical contracting, port, railhead, airhead, or transshipment operations and convoy operations.
- Multinational Operations and Liaison. This category includes the coordination of military operations and liaison with multinational partners, previously unaffiliated nations, and at times adversary or former adversary nations. Multinational operations are becoming more common and increasingly important.

DETERMINING LINGUIST REQUIREMENTS

B-3. To identify linguist requirements, the staff conducts mission analysis and identifies specified or implied tasks requiring foreign language support. Other critical factors are the organization or echelon of command and the location of the mission. The staff uses these criteria to determine the allocation of linguists, such as one linguist team per echelon of command, one linguist per piece of equipment, or one linguist team per location where the function is to be performed. The staff then applies task organization and scheme of maneuver to determine the number of linguists needed for an operation.

B-4. The staff must analyze each linguist assignment to determine the minimum level of foreign language proficiency needed. While interpretation for a peace negotiation requires not only outstanding linguistic capability but also cultural acumen, the translation of routine documents (with the aid of a dictionary) requires a much different skill set. Poor identification of linguist proficiency requirements can tie up the best linguists in less effective roles, creating linguist shortfalls in other areas.

B-5. The relative importance of each of the four linguist support categories is mission dependent. For example, during a NEO civil and military coordination would probably not be as critical as intelligence and information gathering. However, the situation is reversed for a humanitarian assistance mission in which CMOs have a significant impact on mission success. Identifying these "dynamics" helps the commander and staff prioritize linguist requirements.

B-6. Determining linguist requirements for any operation can be difficult because each operation is unique. However, commanders and staffs with a basic knowledge of organic Army linguistic assets, foreign language resource alternatives, and MI skills can successfully assess, prioritize, and employ linguists in support of their military operations.

PLANNING AND MANAGING LINGUIST SUPPORT

B-7. Commanders must consider the linguist requirements as part of their MDMP for every CONPLAN and OPLAN assigned to their commands. Prior staff planning and identification of linguist requirements should prompt commanders to initiate linguist support requests and identify command relationships prior to actual operations. If the mission analysis reveals requirements for linguistic support, the commander must identify what foreign languages are needed, the foreign language proficiency levels needed for each assignment, and the best source of linguists. In addition, if the mission includes intelligence and information collection, the commander must identify MI collection skills required. During mission analysis, the commander should consider linguist requirements for every CONPLAN and OPLAN assigned to his command.

LINGUIST CATEGORIES

B-8. The commander and staff must identify linguist requirements by category:

- Category I – Have native proficiency in the target language (level 4-5) and an advanced working proficiency (Interagency Language Round Table [ILRT] level 2+) in English. They may be locally hired or from a region outside the AO. They do not require a security clearance. They must be screened by the Army CI support team.
- Category II – Are US citizens screened by Army CI personnel and are granted access to SECRET by the designated US government personnel security authority. Have native proficiency in the target language (level 4-5) and an advanced working proficiency (ILRT 2+) in English.
- Category III – Are US citizens screened by Army CI personnel and are granted either TS/SCI clearance or an interim TS/SCI clearance by the

designated US government personnel security authority. Meet a minimum requirement of ILRT level 3. They are capable of understanding the essentials of all speech in a standard dialect. They must be able to follow accurately the essentials of conversation, make and answer phone calls, understand radio broadcasts and news stories, and oral reports (both of a technical and non-technical nature).

PRIMARY STAFF RESPONSIBILITIES

B-9. Primary staff at each echelon has responsibilities for evaluating requirements and managing linguist support. The responsibilities include but are not limited to those discussed below. In addition, each staff section is responsible for determining its linguist support required to meet its operational missions.

Assistant Chief of Staff, G1 (S1):

- Identify linguist requirements needed to support G1/S1 functions in all contingency areas. G1/S1 requirements for linguist support include but are not limited to the following:
  - Coordinate with local authorities on matters of civilian hire, finance, and recordkeeping.
  - Contract for local hire personnel.
  - Coordinate for local morale support and community activities.
  - Coordinate with local authorities for postal operations.
  - Support for administration, counseling, personal affairs, and leave for LN and third-country national (TCN) personnel.
  - Coordinate for local medical support.
  - Liaison with multinational counterparts.
- Linguist staffing and linguist replacement management.
- Identify foreign language skill identifiers for all assigned, attached, or OPCON Army linguists.
- Identify all Army foreign language skilled soldiers not identified on electronic Military Personnel Office System (eMILPO) and Defense Integrated Management Human Resource System (DIMHRS). The Standard Installation Division Personnel System (SIDPERS) was replaced by eMILPO.
- Deploy and provide administrative support of DA and DOD civilian linguists.
- Hire, contract for, and provide administrative support of LN linguists.
- Procure Army foreign language support personnel for screening local labor resources.

Assistant Chief of Staff, G2 (S2):

- Identify linguist requirements needed to support G2/S2 functions in all contingency areas. G2/S2 requirements for linguist support include but are not limited to—
  - Evaluate and/or use local maps and terrain products in operations.
  - Process for MI purposes material taken from EPWs or civilian internees.

  - At lower echelons, conduct tactical questioning of refugees, detainees, and EPWs.
  - Assess local open-source information for intelligence value.
  - Coordinate intelligence and liaison with multinational and HN counterpart.
- Determine, during the initial IPB, all foreign languages (spoken and written) and dialects needed for mission accomplishment.
- Collect, process, produce, and disseminate information derived from linguist sources.
- Provide intelligence training for MI linguists employed in AOs.
- Coordinate for security investigations, as necessary, for local hire linguists.
- Provide support to CI screening of contracted linguists and LN labor force.

Assistant Chief of Staff, G3 (S3):

- Identify linguist requirements needed to support G3/S3 functions in all contingency areas. G3/S3 requirements for linguist support include but are not limited to—
  - Operational coordination and liaison with multinational and HN counterparts.
  - Translate OPORDs and OPLANs for use by multinational counterparts.
- Consolidate unit linguistic requirements and establish priorities.
- Develop linguist deployment and employment plans.
- Develop plans to train linguists and to use linguists for training the force in AO's foreign language survival skills. In addition to global language skills, linguists must have training in specific vocabulary used in the AO; for example, terms used for military, paramilitary, civilian or terrorist organizations, and ethnic groups within the area, nomenclatures of equipment used, and other military or technical vocabulary. Training in the specific dialect used in the AO would also be beneficial.
- Assign, attach, and detach linguists and linguist teams.
- Integrate additional or replacement linguists through operational channels.
- Recommend modernization and development of linguist systems and methods.
- Coordinate mobilization and demobilization of RC linguist support.
- Plan linguist usage for deception operations.
- Plan linguist support for movement of EPWs, detainees, and refugees.
- Coordinate evaluation of linguist support by all staff elements.

Assistant Chief of Staff, G4 (S4):

- Identify linguist requirements needed to support G4/S4 functions in all contingency areas. G4/S4 linguist requirements for linguist support include but are not limited to—

- Procure local supply, maintenance, transportation, and services.
- Coordinate logistics at air and seaports of debarkation.
- Contract with local governments, agencies, and individuals for sites and storage.

- Provide logistical, supply, maintenance, and transportation support to attached linguists.

Assistant Chief of Staff, G5 (S5):

- Identify linguist requirements needed to support G5/S5 functions in all contingency areas. G5/S5 linguist requirements for linguist support include but are not limited to—
  - Determine civilian impact on military operations.
  - Minimize civilian interference with combat operations.
  - Inform civilians of curfews, movement restrictions, and relocations.
  - Provide assistance to liaison with HN and multinational agencies, dignitaries, and authorities.
  - Promote positive community programs to win over support.
  - Determine if multinational operations PSYOP efforts are mutually planned and synchronized.
  - Interpret support to assist resolution of civilian claims against the US Government.
  - Solicit linguistic and cultural knowledge support to protect culturally significant sites.
  - Use linguistic and cultural support to identify cultural and religious customs.
- Assist the G1 in the contracting of local hire linguists.
- Identify foreign language requirements for CMOs.

Assistant Chief of Staff, G6 (S6):

- Identify linguist requirements needed to support G6/S6 functions in all contingency areas. G6/S6 linguist requirements for linguist support include but are not limited to—
  - Coordinate suitable commercial information systems and services.
  - Coordinate with multinational forces on command frequency lists.
  - Coordinate signal support interfaces with HN and multinational forces.
- Manage RF assignments for supporting SIGINT linguist elements.
- Support linguist operations with internal document reproduction, distribution, and message services.
- Integrate automation management systems of linguist units.

SPECIAL STAFF OFFICER RESPONSIBILITIES

B-10. If no special staff officer is assigned the duties below, the corresponding coordinating staff officer should assume those responsibilities. Linguist requirements for special staff officers include but are not limited to the following staff officers.

Liaison Officer:

- Should speak the required foreign language. If not, he requires a translator or interpreter for all aspects of his duties.
- Request interpreters to assist when representing the multinational operations.
- Translate orders, maps, traces, overlays, and documents into multinational foreign languages.

Civilian Personnel Officer:

- Recruit, interview for suitability, and hire civilian local labor force if required.
- Negotiate host country on labor agreements.

Dental Surgeon:

- Administer dental care to support humanitarian mission requirements.
- Rehabilitate, construct, and gain usage of existing dental facilities as required.

Finance Officer:

- Support the procurement process of local goods and services not readily available through normal logistical channels.
- Ensure limited non-US and US pay functions to foreign national, HN, civilian internees, and EPWs are provided.
- Ensure all necessary banking functions are performed in theater.

Surgeon:

- Support medical humanitarian assistance and disaster relief operations.
- Provide medical care of EPWs and civilians within the command's AO.
- Coordinate medical laboratory access in AO.
- Determine the nature of local health threats to the force through populace interviews.
- Determine the identity of local or captured medical supplies.

Veterinary Officer:

- Determine source and suitability of local foods.
- Assist the local population with veterinary service needs.

Chemical Officer:

- Identify enemy force chemical weapons and equipment.
- Communicate NBC risks to supported populations.

Engineer Coordinator:

- Procure proper local materials to support engineering missions.
- Communicate engineering project requirements to contracted local work force.

- Communicate engineering project impact on local landowners and other affected parties.
- Determine, in coordination with G2/S2, suitability of local topographic maps and terrain products.
- Assess environmental concerns of HN and local populations in combined operations.

Provost Marshal:

- Support dislocated and civilian straggler control activities.
- Support internment and resettlement operations, to include displaced civilians.
- Support weapons buy-back programs, as required, and work closely with civil-military liaisons for payments to local officials.
- Support counter-drug and customs activities.
- When authorized, help foreign civil authorities maintain control.
- Conduct liaison with local LEAs.

PSYOP Officer:

- Produce approved PSYOP propaganda and counter-propaganda media.
- Evaluate PSYOP impact on target audience.

Air Defense Coordinator:

- Identify enemy air defense artillery (ADA) weapons and radars.
- Communicate air defense warnings to supported populations.
- Communicate air defense project requirements to contracted local work force.

Safety Officer:

- Provide safety training to local labor force.
- Communicate warnings of dangerous military operations and other hazards to local populace.

Transportation Officer:

- Coordinate commercial and local transportation needs.
- Coordinate movement scheduling and routes with multinational forces and/or HN.

PERSONAL STAFF OFFICER RESPONSIBILITIES

B-11. Personal staff officers are under immediate control of the commander and have direct access to the commander. Most personal staff officers also perform special staff officer duties, working with a coordinating staff officer. These assignments are on a case-by-case basis, depending on the commander's guidance and the nature of the mission; they are very common in stability operations and support operations. Linguist requirements for special staff officers include but are not limited to the following staff officers.

Chaplain:

- Coordinate religious support with multinational partners.
- Determine the impact of local population religious group faiths and practices on military operations.
- Provide religious support to the community to include hospital patients, EPWs, refugees, and civilian detainees.
- Conduct liaison with local population religious leaders in close coordination with the G5.

Public Affairs Officer:

- Act as the commander's spokesman for all communication with external media.
- Assess the accuracy of foreign media interpretation of Public Affairs Office (PAO) releases.
- Assess and recommend news, entertainment, and other information (assisting G5) for contracted services foreign nationals.

Staff Judge Advocate:

- Translate and interpret foreign legal codes, SOFAs, and international laws.
- Determine local environmental laws and treaties through translation services.
- Assess the treatment of EPWs and civilian internees.
- Translate documents to support G4 in local contracts.

SOURCES OF LINGUISTS

B-12. There are various sources that a commander can use to obtain the linguists necessary to support operations. It is vital to know the advantages and disadvantages of each type of linguist and to carefully match the available linguists to the various aspects of the operation.

ARMY LANGUAGE-QUALIFIED MOSs

B-13. The AC MI language-dependent military occupational specialities (MOSs) are 98G with a skill qualification identifier (SQI) of L/352G (Cryptologic Communications Interceptor/ Locator), and their related WO fields. Some soldiers in MOS 96B (All-Source Intelligence Analyst), MOS 97B (CI Agent), MOS 97E/351E (HUMINT Collector) and MOS 98C (SIGINT Analyst), and their related WO fields are trained in foreign languages. Using soldiers in the MOSs mentioned above has many advantages. They are already trained in the military system, are not subject to deployment restrictions (a limiting factor with civilian linguists), have a security clearance and, as US personnel, support the command's interests. The major disadvantage to utilizing these individuals for general foreign language support is that in doing so, they are removed from their primary MI functions. They should be used only in linguistic duties that include intelligence potential. For example, a HUMINT collector (97E) provides linguist support to a medical assistance team as a method to provide access to the local population to determine their attitudes toward US Forces.

B-14. Non-MI Army language qualified MOSs include some enlisted and WOs in career management fields 18 (Special Forces), 37 (PSYOP), 180A (Special Forces); and commissioned officers with a branch code 18 (Special Forces); and functional areas 39 (PSYOP and CA) and 48 (Foreign Area Officer). Particular attention must be paid to the recorded language proficiency and test date of these individuals since the standards vary by field. The same advantages and disadvantages apply as with the AC MI linguists.

B-15. RC language-dependent MOSs include those listed above in the AC. RC linguists have the same set of advantages and disadvantages as listed above for AC language-dependent MOSs. The RC also includes linguists in MOS 97L (translator/interpreter). The 97Ls are specifically trained to be a translator and interpreter. They have the same advantages as the AC linguists. An added advantage is that since their sole job is translation and interpretation, they do not have to be removed from another job in order to be used as a linguist. Their major disadvantage is that they have no additional skill that gives them dual functionality.

Army Linguists Not DOD Trained

B-16. The Army also includes numerous soldiers of all grades who are proficient in a foreign language and are receiving Foreign Language Proficiency Pay (FLPP) but whose primary duties do not require foreign language proficiency. They may have attended a civilian school to learn a foreign language, or they may have acquired proficiency through their heritage. They have the advantage of being trained soldiers and are therefore readily deployable to all areas of the battlefield.

B-17. These soldiers may have the specific vocabulary and military skill knowledge for certain linguist support missions. For example, a supply sergeant who speaks the local language would be an invaluable asset to the G4. There are disadvantages in that they already have another job and units are reluctant to give up personnel especially if they are in key positions. Their capabilities are difficult to assess. Since they are not required to take the Defense Language Proficiency Test (DLPT) if they are not receiving FLPP, it is often difficult for the G1/S1 to identify them as a linguist or for a non-linguist to judge the level of their foreign language capability.

Other Service Linguists

B-18. Other service linguists have the advantage of deployability, loyalty, and clearance, but must often learn the Army system and specific Army vocabulary. They are also difficult to obtain since their parent service probably also lacks a sufficient number of trained linguists. Other service linguists, however, will be valuable in joint operation centers and joint activities. When serving the JTF headquarters, Army commanders and staffs must be aware of the linguists in the other services in order to plan for the participation and optimize their employment.

US Contract Linguists

B-19. US civilians can be contracted to provide linguist support. They have an advantage over LN hires in that their loyalty to the US is more readily evaluated, and it is easier for them to be granted the necessary security clearance. However, there are usually severe limitations on the deployment and use of civilians. A careful assessment of their language ability is important because, in many cases, they use "old fashioned" terms, or interject US idioms. If the linguists are recent émigrés, the use of the language in their country of origin could be dangerous to them, or their loyalty may reside with their own country when at odds with US interests.

Multinational Linguists

B-20. Multinational linguists have their own set of advantages and disadvantages. These linguists may be unfamiliar with the US military system unless they have previously participated in a multinational operation with US forces. They may have a security clearance, but clearances are not necessarily equal or reciprocal, automatically guaranteeing access to classified or sensitive information between nations. They support the command's interest but may have differing priorities or responsibilities within their assigned AOs. These linguists also are already fulfilling specific duties for their own nation, which may also have a shortage of linguists. The major disadvantage to acquiring and maintaining multinational linguist support is that they are outside the C2 (via military authority or military contract) of the US forces. These linguists will be valuable in multinational operations centers and activities.

Local National Contract Linguists

B-21. LN hires will provide the bulk of your linguist support. They are usually less expensive to hire than US civilians and will know the local dialect, idioms, and culture. The expertise of these linguists in particular areas or subject matters can be an asset. However, there are several potential problems with using LN hires, to include limited English skills and loyalty considerations. Therefore, a screening interview or test is necessary to determine their proficiency in English. These individuals must also be carefully selected and screened by CI personnel (with US linguist support) initially and periodically throughout their employment. Their loyalty is always questionable. Local prejudices may influence them, and they may place their own interests above those of the US.

EVALUATING LINGUIST PROFICIENCY

B-22. Commanders and staffs must understand the Army linguist proficiency evaluation system in order to effectively plan for and employ linguists. Evaluation and reevaluation of linguist proficiency is covered in detail in AR 611-6, Section III. Language testing is required for all Army personnel in a language-dependent MOS, who have received foreign language training at government expense, who are receiving FLPP, or who are in a language-required position regardless of MOS. Other Army personnel who have knowledge of a foreign language are encouraged to take the proficiency test and may work as linguists.

B-23. The Army uses the DLPT to determine foreign language proficiency levels. DLPTs are listed by foreign language in DA Pam 611-16. In foreign languages where no printed or recorded test exists, oral interview tests are arranged. The DLPT is an indication of foreign language capability, but it is not the definitive evaluation of an individual's ability to perform linguist support.

B-24. AR 611-6, Appendix D, Sections 1 through 4, describes the proficiency levels for the skills of speaking, listening, reading, and writing a foreign language based on the interagency roundtable descriptions. The plus-level designators, shown as a "+" symbol, are used to designate when a linguist is above a base level, but not yet to the capability of the next level. For example, 2+ would indicate a better than limited working proficiency in the foreign language. The six "base levels" of proficiency, as established by DLPT and/or oral exam, are—

- Level 0 (No proficiency). The soldier has no functional foreign language ability. Level 0+. The minimum standard for Special Forces personnel indicates a memorized proficiency only.
- Level 1 (Elementary proficiency). The soldier has limited control of the foreign language skill area to meet limited practical needs and elementary foreign language requirements.
- Level 2 (Limited working proficiency). The linguist is sufficiently skilled to be able to satisfy routine foreign language demands and limited work requirements.
- Level 3 (General professional proficiency). The linguist is capable of performing most general, technical, formal, and informal foreign language tasks on a practical, social, and professional level.
- Level 4 (Advanced professional proficiency). The linguist is capable of performing advanced professional foreign language tasks fluently and accurately on all levels.
- Level 5 (Functionally native proficiency). The linguist is functionally equivalent to an articulate and well-educated native in all foreign language skills; and reflects the cultural standards of the country where the foreign language is natively spoken.

B-25. The above proficiency base levels designate proficiency in any of the four language skills: listening, reading, speaking, and writing. The most evaluated skills on the DLPT are reading and listening. These tests are not for use in evaluating linguists above the 3 proficiency level. Most Army linguist DLPT scores show only two skill levels: listening and reading (for example, 2+/3, or 3/1+).

SUSTAINING MILITARY LINGUIST PROFICIENCY

B-26. Language proficiency diminishes with lack of use and absence of exposure to the foreign language. To ensure combat readiness, commanders should require all military linguists receive periodic language training. In-country language immersion training, in-garrison contracted language instructors, on-line foreign newspapers, and foreign radio broadcasts are all examples of language training resources. Funding for language training is available through MACOM language training program funds.

Glossary

The glossary lists acronyms and terms with Army or joint definitions, and other selected terms. Where Army and joint definitions are different, (Army) follows the term. Terms for which FM 2-0 is the proponent FM (authority) are marked with an asterisk (*) and followed by the number of the paragraph (¶) where they are discussed. For other terms, the number of the proponent FM follows the definition. JP 1-02 and FM 1-02 are posted in the Joint Electronic Library, which is available online and on CD-ROM.

- Use this URL to access JP 1-02 online: http://www.dtic.mil/doctrine/jel/doddict/.
- Use this URL to access FM 1-02 online: http://www.dtic.mil/doctrine/jel/service_pubs/101_5_1.pdf.
- Follow this path to access JP 1-02 on the Joint Electronic Library CD-ROM: Mainmenu>Joint Electronic Library>DOD Dictionary.
- Follow this path to access FM 1-02 on the Joint Electronic Library CD-ROM: Mainmenu>Joint Electronic Library>Service Publications>Multiservice Pubs> FM 101-5-1.

A2C2	Army Airspace Command and Control
AA	avenue of approach
AAMDC	Area Air and Missile Defense Command
ABCS	Army Battle Command System
AC	Active Component
ACE	analysis and control element
ACINT	acoustic intelligence (water)
ACOUSTINT	acoustic intelligence (atmosphere)
ACR	armored cavalry regiment
ACT	analysis and control team
ADA	air defense artillery
ADCON	administrative control
ADDO	assistant deputy director for operations
ADO	air defense officer
ADP	automated data processing
AFACSI	Air Force Assistant Chief of Staff for Intelligence
AFIWC	Air Force Information Warfare Center
AFMIC	Armed Forces Medical Intelligence Center

AFOSI Air Force Office of Special Investigation

AGM attack guidance matrix

AIA Air Force Intelligence Agency

AITR aided target recognition

ALOC administrative and logistics operations center

AMC US Army Materiel Command

AMD Air Missile Defense

AMHS Automated Message Handling System

AMIB Allied Military Intelligence Battalion

***analysis** Determination of the significance of the information, relative to information and intelligence already known, and drawing deductions about the probable meaning of the evaluated information. (¶1-58)

AO area of operations

AOC area of concentration

AOI area of interest

AOIR area of intelligence responsibility

AOR area of responsibility

AR Army Regulation

ARAT Army Reprogramming Analysis Team

ARCENT US Army Central Command

ARFOR Army forces

ARISC Army Reserve Intelligence Support Center

ARL airborne reconnaissance low

ARNG US Army National Guard

ARSOF Army Special Operations Forces

ASCC Army Service Component Commander

ASPO Army Space Program Office

AT antiterrorism

ATCAE Army Technical Control Analysis Element

ATCCS Army Tactical Command and Control System

ATR automatic target recognition

ART Army tactical task

BDA battle damage assessment

BDU battle-dress uniform

BFACS	Battlefield Functional Area Control System
C/J2	combined J2
C2	command and control
CA	civil affairs
CASO	Collection Activities and Source Operations
CAT	crisis action team
CBRN	chemical, biological, radiological, and nuclear
CBRNE	chemical, biological, radiological, nuclear, and high-yield explosive
CCD	coherent change detection
CCIR	commander's critical information requirement
CD-ROM	compact disc-read only memory
CE	counterespionage
CECOM	Communications-Electronics Command
CED	captured enemy document
CEE	captured enemy equipment
CEM	captured enemy materiel
CFSO	counterintelligence force protection source operations
CGS	common ground station
CI	counterintelligence
CIA	Central Intelligence Agency
CIAC	counterintelligence analysis cell
CICA	Counterintelligence Coordinating Authority
CITO	Central Imagery Tasking Office
CJCS	Chairman, Joint Chiefs of Staff
CMEC	Captured Materiel Exploitation Center
CMF	career management field
CMISE	Corps Military Intelligence Support Element
CMO	civil-military operations
CNA	computer network attack
CND	computer network defense
CNE	computer network exploitation
CNN	Cable News Network
CNO	computer network operations

COA course of action

COCOM combatant command (command authority)

COG center of gravity

COMINT communications intelligence

COMSEC communications security

CONOPS contingency operations

CONPLAN contingency plan

CONUS continental United States

COP common operational picture

COSCOM Corps Support Command

***counterintelligence** **1.** Information gathered and activities conducted to protect against espionage, other intelligence activities, sabotage, or assassinations conducted by or on behalf of foreign governments or elements thereof, foreign organizations, or foreign persons, or international terrorist activities. (JP 1-02) **2.** (Army) Counterintelligence counters or neutralizes intelligence collection efforts through collection, CI investigations, operations, analysis and production, and functional and technical services. CI includes all actions taken to detect, identify, exploit, and neutralize the multidiscipline intelligence activities of friends, competitors, opponents, adversaries, and enemies; and is the key intelligence community contributor to protect US interests and equities. (FM 2-0) (¶1-111)

CP command post

CRN command radio net

CS combat support

CSG cryptologic support group

CSS combat service support

CSSCS Combat Service Support Control System

CT Counterterrorism

***cueing** The use of one or more sensor systems to provide data that directs collection by other systems. (¶2-53)

CWT combat weather team

DA Department of the Army

DAC Department of the Army Civilian

DAPDM Department of the Army Production and Dissemination Management

DCI Director of Central Intelligence

DCS, G2 Deputy Chief of Staff, Intelligence

***debriefing** The systematic questioning of individuals to procure information to answer specific collection requirements by direct and indirect questioning techniques. (¶6-4)

DF direction finding

DHS Defense HUMINT Service

DIA Defense Intelligence Agency

DIAC Defense Intelligence Analysis Center

DIMHRS Defense Integrated Management Human Resource System

DIO Defensive Information Operations

DISCOM Division Support Command

DISUM daily intelligence summary

DLPT Defense Language Proficiency Test

DMS Director of Military Support

DOCEX document exploitation

***document exploitation** The systematic extraction of information from all media formats in response to collection requirements. (¶6-4)

DOD Department of Defense

DOJ Department of Justice

DOT Department of Transportation

DP decision point

DPM dissemination program manager

DS direct support

DSM decision support matrix

DST decision support template

EA electronic attack

EAC echelons above corps

ECOA enemy course of action

EEFI essential elements of friendly information

ELINT electronic intelligence

EM electromagnetic

eMILPO electronic Military Personnel Office System

EMS electromagnetic spectrum

E-O electro-optical

EOB electronic order of battle

EOD explosive ordinance disposal

EP	electronic protection
EPW	enemy prisoner of war
ES	electronic warfare support (Army)
***evaluating**	**1.** Comparing relevant information (RI) on the situation or operation against criteria to determine success or progress. (FM 6-0) **2.** In intelligence usage, appraisal of an item of information in terms of credibility, reliability, pertinence, and accuracy. (FM 2-0) (¶1-26 and 1-98)
EW	electronic warfare
EWS	electronic warfare support (joint)
FBI	Federal Bureau of Investigation
FDA	functional damage assessment
FDU	Force Design Update
FEMA	Federal Emergency Management Agency
FFIR	friendly force information requirement
FISINT	foreign instrumentation signals intelligence
FISS	Foreign Intelligence Security Service
FLPP	Foreign Language Proficiency Pay
FM	Field Manual
FMA	foreign materiel acquisition
FME	foreign materiel exploitation
FMEP	Foreign Materiel Exploitation Program
FMP	Foreign Materiel Program
FMT	Foreign Materiel for Training
FP	force protection
FRAGO	fragmentary order
FS	fire support
FSAC	Fire Support Armaments Center
FSE	fire support element
FSO	fire support officer
G1	Assistant Chief of Staff, Personnel
G2	Assistant Chief of Staff, Intelligence
G3	Assistant Chief of Staff, Operations
G4	Assistant Chief of Staff, Logistics
G5	Assistant Chief of Staff, Civil/Military Affairs

G6 Assistant Chief of Staff, Command, Control, Communications, and Computer Operations

G7 Assistant Chief of Staff, Information Operations

GEOINT geospatial intelligence

GMI general military intelligence

GPS Global Positioning System

GS general support

GS-R general support-reinforcing

HN host nation

HOC HUMINT operations cell

HPT high-payoff target

HQ headquarters

HQDA Headquarters, Department of the Army

HSI hyperspectral imaging

HUMINT human intelligence

***human intelligence** Collection by a trained HUMINT collector of foreign information from people and multimedia to identify elements, intentions, composition, strength, dispositions, tactics, equipment, personnel, and capabilities. It uses human sources and a variety of collection methods, both passively and actively, to gather information to satisfy the commander's intelligence requirements and cross-cue other intelligence disciplines. (¶1-106 and ¶6-1)

HVT high-value target

I/R internment and resettlement (operations)

I&W indications and warnings

ICC Intelligence Coordination Center (US Coast Guard)

ICF intelligence contingency fund

ICL intelligence coordination line

IDC Information Dominance Center

IIR intelligence information report

IM information management

IMA individual mobilization augmentee

IMEP International Materiel Evaluation Program

IMETS Integrated Meteorological System

IMINT imagery intelligence

***indications and warning** (joint) Those intelligence activities intended to detect and report time-sensitive intelligence information on foreign developments that could involve a threat to the United States or allied and/or coalition military, political, or economic interests or to US citizens abroad. It includes forewarning of enemy actions or intentions; the imminence of hostilities; insurgency; nuclear or non-nuclear attack on the United States, its overseas forces, or allied and/or coalition nations; hostile reactions to US reconnaissance activities; terrorists attacks; and other similar events. (¶2-19)

***indicator** (Army) Positive or negative evidence of threat activity or any characteristic of the AO which points toward threat vulnerabilities or the adoption or rejection by the threat of a particular capability, or which may influence the commander's selection of a COA. Indicators may result from previous actions or from threat failure to take action. (FM 2-0) (¶5-14)

INFOSEC information security

***infrastructure** **1.** (joint) All building and permanent installations necessary for support, redeployment, and military forces operations (e.g., barracks, headquarters, airfields, communications facilities, stores, port installations, and maintenance stations). (JP 4-01.8) **2.** (Army) In intelligence usage, the basic underlying framework or feature of a thing; in economics, basic resources, communications, industries, and so forth, upon which others depend; in insurgency, the organization (usually hidden) of insurgent leadership. (FM 2-0) (¶5-8)

INR Bureau of Intelligence and Research, Department of State

INS Immigration and Naturalization Service

INSCOM US Army Intelligence and Security Command

intelligence **1.** (joint) The product resulting from the collection, processing, integration, analysis, evaluation, and interpretation of available information concerning foreign countries or areas. **2.** Information and knowledge about an adversary obtained through observation, investigation, analysis, or understanding. (¶1-3)

***intelligence coordination line** A line that designates the boundary between AOIRs. The G2/S2 establishes ICLs to facilitate coordination between higher, lateral, and subordinate units; coordinates with the G3/S3 to direct subordinates to track enemy units and HPTs in their areas; and hands over intelligence responsibility for areas of the battlefield. The establishment of ICLs ensures that there are no gaps in the collection effort; that all echelons are aware of the location, mission, and capabilities of other assets; facilitates asset cueing, and provides timely exchange of information between assets. The G2/S2 keeps abreast of collection activities in progress (all echelons) and battlefield developments through the ICLs. (¶3-13)

intelligence discipline — (joint) A well-defined area of intelligence collection, processing, exploitation, and reporting using a specific category of technical or human resources. There are seven major disciplines: human intelligence, imagery intelligence, measurement and signature intelligence, signals intelligence (communications intelligence, electronic intelligence, and foreign instrumentation signals intelligence), all-source intelligence, technical intelligence, and counterintelligence. (¶1-104)

***intelligence process** — A theoretical model used to describe intelligence operations. The intelligence process is not a framework for actual operations because the functions and common tasks occur in parallel as opposed to sequentially. (¶1-91)

***intelligence products** — Intelligence products are generally placed in one of six categories: I&W, current, general military, target, S&T, and CI. The categories are distinguished from each other primarily by the purpose for which the intelligence was produced. The categories often overlap, and the same intelligence data can be used in each of the categories. (¶1-5)

INTREP — intelligence report

INTSUM — intelligence summary

***intelligence requirements** — (Army) Those requirements generated from the staff's IRs regarding the enemy and environment that are not a part of the CCIR (PIR and FFIR). Intelligence requirements require collection and can pro-vide answers in order to identify indicators of enemy actions or intent, which reduce the uncertainties associated with an operation. Significant changes (i.e., branches and sequels) with an operation usually lead to changes in intelligence requirements. (¶2-7)

***intelligence running estimate** — The intelligence running estimate is a continuous flow and presentation of relevant information and predictive intelligence that, when combined with the other staff running estimates, enable the decisionmaker's visualization and situational understanding of the AOI in order to achieve information superiority. The intelligence running estimate requires constant verification to support situational understanding of the current situation as well as predictive assessments for future operations. (¶5-30)

***intelligence synchronization** — The task that ensures ISR operations are linked to the commander's requirements and respond in time to influence decisions and operations. The intelligence officer, with staff participation, synchronizes the entire collection effort to include all assets the commander controls, assets of lateral units and higher echelons units and organizations, and intelligence reach to answer the commander's CCIR (PIR and FFIR). (¶1-24)

***intelligence synchronization plan**	The plan the intelligence officer uses, with staff input, to synchronize the entire collection effort to include all assets the commander controls, assets of lateral units, and higher echelon units and organizations, and intelligence reach to answer the commander's CCIR (PIR and FFIR). (¶1-38)
intelligence warfighting function	The related tasks and systems that facilitate understanding of the operational environment
IO	information operations
IPB	intelligence preparation of the battlefield
IPIR	Initial Phase Imagery Report
IPTF	international police task force
IR	information requirement
IRINT	infrared intelligence
IRR	Individual Ready Reserve
ISM	intelligence synchronization matrix
ISR	intelligence, surveillance, and reconnaissance
J2	Joint Staff Directorate, Intelligence
J2A	Joint Staff Directorate, Intelligence (Administrative)
J2J	Joint Staff Directorate, Intelligence (Joint)
J2M	Joint Staff Directorate, Intelligence (Management)
J2O	Joint Staff Directorate, Intelligence (Operations)
J2P	Joint Staff Directorate, Intelligence (Assessments, Doctrine)
J2T	Joint Staff Directorate, Intelligence (Targeting)
J2X	Joint Staff Directorate, Intelligence (CI & HUMINT)
J3	Joint Staff Directorate, Operations
J5	Joint Staff Directorate, Civil Affairs
JAC	Joint Analysis Center
JCMEC	Joint Captured Materiel Exploitation Center
JCO	Joint Commission Observers
JCS	Joint Chiefs of Staff
JDEC	Joint Document Exploitation Center
JDISS	Joint Deployable Intelligence Support System
JFC	joint force commander
JIAWG	joint interagency working group
JIC	Joint Intelligence Center

JIF	Joint Interrogation Facility
JISE	joint intelligence support element
JOA	joint operations area
JP	Joint Publication
JSTARS	Joint Surveillance Target Attack Radar System
JTF	joint task force
JTMD	joint table of mobilization and distribution
JWICS	Joint Worldwide Intelligence Communications System
LAN	local area network
LASINT	laser intelligence
LCC	land component commander
LEA	law enforcement agency
LIDAR	laser radar
LN	local national
LNO	liaison officer
LOC	line of communication
LOS	line of sight
LPT	logistics preparation of the theater
LTIOV	latest time information is of value
MACOM	major Army command
MASINT	measurement and signature intelligence
***materiel**	(Army) In intelligence usage, the all-encompassing term for the weapons systems, equipment, apparatus, documents, and supplies of a foreign military force or nonmilitary organization. (¶2-24)
MCIA	Marine Corps Intelligence Agency
MCOO	modified combined obstacle overlay
MCS	Maneuver Control System
MDMP	military decisionmaking process
MEA	munitions effects assessment
METL	mission-essential task list
METT-TC	mission, enemy, terrain and weather, troops and support available, time available, civil considerations
MI	military intelligence
MLT	machine language translation

MNFC	multinational force commander
MO	method of operation
MOPP	mission-oriented protective posture
MOS	military occupational specialty
MP	military police
MSC	major subordinate command
MSE	mobile subscriber equipment
MSI	multi-spectral imagery
MSIC	Missile and Space Intelligence Center
MTI	moving target indicator
MTOE	modified table of organization and equipment
MTW	major theater war
NAI	named area of interest
NAIC	National Air Intelligence Center
NATO	North Atlantic Treaty Organization
NBC	nuclear, biological, and chemical
NCA	National Command Authorities
NCO	noncommissioned officer
NCR	national cryptologic representative
NCTI	non-cooperative target identification
NEO	noncombatant evacuation operation
NFH	National Defense Headquarters, Canada
NGA	National Geospatial-Intelligence Agency
NGO	non-governmental organization
NIC	National Intelligence Center
NIIRS	National Imagery Interpretability Rating Scale
NIMA	National Imagery and Mapping Agency
NIST	National Intelligence Support Team
NLT	not later than
NMCC	National Military Command
NMIC	National Maritime Intelligence Center
NMJIC	National Military Joint Intelligence Center
NRO	National Reconnaissance Office
NRT	near-real time

NSA	National Security Agency
NSC	National Security Council
NUCINT	Nuclear Intelligence
NVD	night vision device
OB	order of battle
OE	operational environment
OMA	Office of Military Affairs
OMB	Office of Management and Budget
OMT	operational management team
ONDCP	Office of National Drug Control Policy
ONI	Office of Naval Intelligence
ONIR	overhead non-imaging infrared
OPCON	operational control
OPLAN	operation plan
OPORD	operation order
OPSEC	operations security
OSC	operations support cell
OSCE	Organization for Security and Cooperation in Europe
OSD	Office of the Secretary of Defense
OSINT	open-source intelligence
PAO	Public Affairs Office
PDA	physical damage assessment
PIAP	Police Information Assessment Program
PIO	police intelligence operations
PIR	priority intelligence requirement
PME	peacetime military engagement
POC	point of contact
POD	port of debarkation
POE	port of embarkation
***predictive intelligence**	Intelligence analysis conclusions, assessments or products that attempt to define, describe, present, or portray the future situation or conditions of the enemy, +terrain, weather and civil considerations. (¶1-3)
***priority intelligence requirements**	(joint) Those intelligence requirements for which a commander has an anticipated and stated priority in his task of planning and decision-making. (FM 2-0) (¶1-32)

***process/ processing** A function of the intelligence process that involves converting collected data, which is not already in a comprehensible form when it is reported, into a form that is understandable and suitable for analysis and production of intelligence. Examples of processing include developing film, enhancing imagery, translating a document from a foreign language, converting electronic data into a standardized report that can be analyzed by a system operator, and correlating dissimilar or jumbled information by assembling like elements before the information is forwarded for analysis. (FM 2-0) (¶1-96)

***production** In intelligence usage, conversion of information into intelligence through the integration, analysis, evaluation, and interpretation of all source data and the preparation of intelligence products in support of known or anticipated user requirements. (¶4-20)

PSYOP psychological operations

QRC quick reaction capability

QRF quick reaction force

R&D research and development

RADINT radar intelligence

RC reserve component

RDEC Research, Development, and Engineering Center

***reconnaissance, surveillance, and target acquisition/intelligence surveillance and reconnaisssance** A full spectrum combined arms mission that integrates ground and air capabilities to provide effective, dynamic, timely, accurate and assured combat information and multi-discipline actionable intelligence for lethal and non-lethal effects/decisions in direct support of the ground tactical commander.

REMBASS Remotely Monitored Battlefield Sensor System

***requirements management** The intelligence task that develops a prioritized list of what information needs to be collected and produced into intelligence, dynamically updates and adjusts those requirements in response to mission adjustments/changes, and places a latest time intelligence is of value to ensure intelligence and information is reported to meet operational requirements. (¶2-83)

RF radio frequency

RFI request for information

RFW radio frequency weapon

RI relevant information

RM requirements management

ROE rules of engagement

ROK Republic of Korea

ROZ restricted operating zone

RSO&I	reception, staging, onward movement, and integration
RSOC	Regional Security Operations Center
RSTA	reconnaissance, surveillance, and target acquisition
RSP	render safe procedure
RTCAE	Regional Technical Control and Analysis Element
running estimate	A staff estimate, continuously updated, based on new information as the operation proceeds. (FM 6-0) (¶5-28)
S&T	scientific and technical
S&TI	scientific and technical intelligence
S1	Staff Officer, Personnel
S2	Staff Officer, Intelligence
S2X	Staff Officer, Intelligence (CI & HUMINT)
S3	Staff Officer, Operations
S4	Staff Officer, Logistics
S5	Staff Officer, Civil Affairs
S6	Staff Officer, Command, Control, Communications, and Computer Operations
SAP	special access program
SAR	synthetic aperture radar
SATCOM	satellite communications
SBCCOM	Chemical and Biological Defense Command
SCI	sensitive compartmented information
SCIF	sensitive compartmented information facility
SIDPERS	Standard Installation Division Personnel System
SIGINT	signals intelligence
SIGO	signal officer
SII	statement of intelligence interest
SIO	senior intelligence officer
SIPRNET	Secret Internet Protocol Router Network
SIR	specific information requirement
SITMAP	situation map

***situation template**	(Army) A depiction of a potential adversary course of action as part of a particular adversary operation. Situation templates are developed on the adversary's current situation (for example, training and experience levels, logistic status, losses, and disposition), the environment, and adversary doctrine or patterns of operations. The commander dictates the level to depict the adversary based on the factors of METT-TC (at minimum two levels of command below the friendly force) as a part of his guidance for mission analysis. (FM 2-0) (¶1-58)
SJA	staff judge advocate
SMU	special mission unit
SOCOM	satellite communications on-the-move
SOF	special operations forces
SOFA	status of forces agreement
SOP	standing operating procedure
SOTA	SIGINT operational tasking authority
SR	special reconnaissance
SRTD	signals research and target development
SSO	special security office
STANAG	standardization agreement (NATO)
STICEUR	Scientific and Technology Intelligence Center Europe
SUPR	supplemental imagery report
SWO	staff weather officer
TA	target acquisition
TACON	tactical control
***tactical questioning**	The expedient initial questioning for information of immediate tactical value. (FM 2-0) (¶6-6)
TAI	target area of interest
TCAE	technical control and analysis element
TCN	third-country national
***technical control**	The technical function to ensure adherence to existing policies or regulations and provide technical guidance for MI activities, particularly HUMINT, SIGINT, and CI operations. (FM 2-0) (¶3-23)
TECHINT	technical intelligence
TECOM	Test and Evaluation Command
TENCAP	Tactical Exploitation of National Capabilities
TEU	Technical Escort Unit

THREATCON	threat condition
TIB	Theater Intelligence Brigade
TIG	Theater Intelligence Group
TIM	toxic industrial material
TLP	troop-leading procedure
TOC	tactical operations center
TPFDD	Time Phased Force Deployment Data
TRADOC	US Army Training and Doctrine Command
TSA	target system assessment
TSCM	Technical Surveillance Countermeasures
TSE	theater support element
TTP	tactics, techniques, and procedures
TVA	threat vulnerability assessment
UAV	unmanned aerial vehicle
UN	United Nations
US	United States
USAF	United States Air Force
USAR	United States Army Reserve
USC	United States Code
USCENTCOM	United States Central Command
USCG	United States Coast Guard
USCS	United States Cryptologic System
USEUCOM	United States European Command
USJFCOM	United States Joint Forces Command
USMC	United States Marine Corps
USNORTHCOM	United States Northern Command
USPACECOM	United States Space Command
USPACOM	United States Pacific Command
USSID	United States Signal Intelligence Directive
USSOCOM	United States Special Operations Command
USSOUTHCOM	United States Southern Command
USSPACECOM	United States Space Command
USSTRATCOM	United States Strategic Command
USTRANSCOM	United States Transportation Command

VJ2 Vice J2

VTC video teleconference

WAN wide area network

WARNO warning order

WMD weapons of mass destruction

WO warrant officer

Bibliography

The bibliography lists field manuals published with new numbers followed by old number.

DOCUMENTS NEEDED

These documents must be available to the intended users of this publication.

JP 1-02. *Department of Defense Dictionary of Military and Associated Terms.* June 1999. [Online] Available http://www.dtic.mil/doctrine/jel/doddict/

FM 1-02/MCRP 5-2A. *Operational Terms and Symbols* (Drag Draft). 21 February 2003.

READINGS RECOMMENDED

These sources contain relevant supplemental information.

ARMY PUBLICATIONS

Most Army doctrinal publications are available online: http://155.217.58.58/atdls.htm

AR 381-10. *US Army Intelligence Activities.* 1 July 1984.

AR 381-12. *Subversion and Espionage Directed Against the U.S. Army (SAEDA).* 15 January 1993.

AR 381-20. *The Army Counterintelligence Program.* 15 November 1993.

AR 611-6. *Army Linguist Management.* 16 February 1996.

AR 715-9. *Contractors Accompanying the Force.* 29 October 1999.

FM 1 (100-1). *The Army.* 14 June 2001.

FM 2-19.402. *STRYKER Brigade Combat Team Intelligence Operations.* 1 March 2003.

FM 2-19.602. *Surveillance Troop.* 1 March 2003.

FM 2-33.5/ST. *Intelligence Reach Operations.* 1 June 2001.

FM 3-0. *Operations.* 14 June 2001.

FM 3-05.102. *Army Special Operations Forces Intelligence.* 31 August 2001 (formerly FM 34-36).

FM 3-05.30. *Psychological Operations.* 19 June 2000.

FM 3-06.11. *Combined Arms Operations in Urban Terrain.* 28 February 2002.

FM 3-13. *Information Operations.* 28 November 2003 (formerly FM 100-6).

FM 3-19. *NBC Reconnaissance.* 19 November 1993.

FM 3-19.1. *Military Police Operations.* 22 March 2001.

FM 3-19.4. *Military Police Leaders' Handbook.* 4 March 2002.

FM 3-19.40. *Military Police Internment/Resettlement Operations.* 30 September 1987.

FM 3-20.98. *Reconnaissance Platoon.* May 2001.

FM 34-2. *Collection Management and Synchronization Planning.* 8 March 1994.

FM 34-52. (FM 2-22-3) *Human Intelligence.* 28 September 1992.

FM 3-90. *Tactics.* 4 July 2001.

FM 3-90.3. *The Mounted Brigade Combat Team.* 1 November 2001.

FM 3-100.4. *Environmental Considerations in Military Operations.* 15 June 2000, Change 1, 11 May 2001.

FM 3-100.21. *Contractors on the Battlefield.* 26 March 2000.

FM 5-170. *Engineer Reconnaissance.* 5 May 1998.

FM 6-0. *Mission Command: Command and Control of Army Forces.* 11 August 2003.

FM 6-20-10. *Tactics. Techniques, and Procedures for the Targeting Process.* 8 May 1996.

FM 6-121. *Tactics, Techniques, and Procedures for Field Artillery Target Acquisition.* 25 September 1990.

FM 7-0. *Training the Force.* 22 October 2002.

FM 7-10. *The Infantry Rifle Company.* 14 December 1990.

FM 7-15. *Army Universal Task List.* 31 August 2003.

FM 7-92. *The Infantry Reconnaissance Platoon and Squad (Airborne, Air Assault, Light Infantry).* 23 December 1992.

FM 7-100. *Opposing Force Doctrinal Framework and Strategy.* 1 May 2003.

FM 17-97. *Cavalry Troop.* 3 October 1995.

FM 17-98. *Scout Platoon.* 10 April 1999.

FM 19-1. *Military Police Support for the Airland Battle.* 23 May 1988.

FM 27-10. *The Law of Land Warfare.* 18 July 1956.

FM 31-20-5. *Special Reconnaissance Tactics, Techniques, and Procedures for Special Forces.* 23 March 1993.

FM 34-2-1. *Tactics, Techniques, and Procedures for Reconnaissance and Surveillance and Intelligence to Counterreconnaissance.* 19 June 1991.

FM 34-3. *Intelligence Analysis.* 15 March 1990.

FM 34-8-2. *Intelligence Officer's Handbook.* 1 May 1998.

FM 34-10. *Division Intelligence and Electronic Warfare Operations.* 25 November 1986.

FM 34-25. *Corps Intelligence and Electronic Warfare Operations.* 30 September 1987.

FM 34-37. *Echelons Above Corps (EAC) Intelligence and Electronic Warfare (IEW) Operations.* 15 January 1991.

FM 34-54. *Technical Intelligence.* 30 January 1998.

FM 34-80. *Brigade and Battalion Intelligence and Electronic Warfare Operations.* 15 April 1986.

FM 34-130. *Intelligence Preparation of the Battlefield.* 8 July 1994.

FM 4-0. *Combat Service Support.* 29 August 2003.

FM 41-10. *Civil Affairs Operations.* 14 February 2000.

FM 100-25. *Doctrine for Army Special Operations Forces.* August 1999.

FM 101-5. *Staff Organization and Operations.* 31 May 1997.

FM 101-5-1. *Operational Terms and Graphics.* 30 September 1997.

FM 101-5-2. *US Army Report and Message Formats.* 29 June 1999.

ST 2-22.7. *Tactical Human Intelligence and Counterintelligence Operations.* April 2002.

ST 2-33.5. *Intelligence Reach Operations.* 1 June 2001.

ST 2-50. *Intelligence and Electronic Warfare Assets.* June 2002.

ST 2-50.4. *Combat Commanders' Handbook on Intelligence.* September 2001 (formerly FM 34-81, 28 September 1992).

JOINT PUBLICATIONS

Most joint publications are available online: http://www.dtic.mil/doctrine/jel/

JP 0-2. *Unified Action Armed Forces (UNAAF).* 10 July 2001.

JP 1. *Joint Warfare for the US Armed Forces.* 14 November 2000.

JP 1-06. *Joint Tactics, Techniques, and Procedures for Financial Management During Joint Operations.* 22 December 1999.

JP 2-0. *Joint Doctrine for Intelligence to Operations.* January 2000.

JP 2-01. *Joint Intelligence Support to Military Operations.* 20 November 1996.

JP 2-01.1. *Joint Tactics, Techniques, and Procedures for Intelligence Support to Targeting.* 9 January 2003.

JP 2-01.3. *Joint Tactics, Techniques, and Procedures for Joint Intelligence Preparation of the Battlespace,* 24 May 2000.

JP 2-02. *National Intelligence to Joint Operations.* 28 September 1998.

JP 2-03. *Joint Tactics, Techniques, and Procedures for Geospatial Information and Services Support to Joint Operations.* 31 March 1999.

JP 3-0. *Doctrine and Joint Operations.* 1 February 1995.

JP 3-07. *Joint Doctrine for Military Operations Other Than War.* 16 June 1995.

JP 3-07.2. *JTTP for Antiterrorism.* 17 March 1998.

JP 3-08. *Interagency Coordination During Joint Operations,* Volume II. 9 October 1996.

JP 3-09. *Doctrine for Joint Fire Support.* 12 May 1998.

JP 3-09.1. *Joint Tactics, Techniques and Procedures for Laser Designation Operations.* 28 May 1999.

JP 3-11. *Joint Doctrine for Operations in Nuclear, Biological, and Chemical (NBC) Environments.* 11 July 2000.

JP 3-13. *Joint Doctrine for Information Operations.* 9 October 1998.

JP 3-18. *Joint Doctrine for Forcible Entry Operations.* 16 July 2001.

JP 3-35. *Joint Deployment and Redeployment Operations.* 7 September 1999.

JP 3-51. *Joint Doctrine for Electronic Warfare.* 7 April 2000.

JP 3-55. *Doctrine for Reconnaissance, Surveillance, and Target Acquisition Support for Joint Operations (RSTA).* 14 April 1993.

JP 3-57. *Joint Doctrine for Civil-Military Operations.* 8 February 2001.

JP 3-60. *Joint Doctrine for Targeting.* 17 January 2002.

JP 4-0. *Doctrine for Logistic Support of Joint Operations.* 6 April 2000.

JP 5-0. *Doctrine for Planning Joint Operations.* 13 April 1995.

JP 5-00.2. *Joint Task Force Planning Guidance and Procedures.* 13 January 1999.

PUBLIC LAWS AND OTHER PUBLICATIONS

The United States Code is available online: http://uscode.house.gov/usc.htm

EO 12333. *United States Intelligence Activities.* 4 December 1981.

CJCS Instruction 1301.01, Policy and Procedures to Assign Individuals to Meet Combatant Command Mission-Related Temporary Duty Requirements. 1 July 2001.

CJCSM 3500.04C. *Universal Joint Task List.* 1 July 2002.

Military Standard 2525B, *DOD Interface Standard, Common Warfighting Symbology.* 30 January 1999.

PUBLICATIONS UNDER DEVELOPMENT

FM 2-01. *Intelligence Synchronization.* When published tentatively in July 2004, FM 2-01 will supersede FM 34-2, 8 March 1994, and FM 34-2-1, 19 June 1991.

FM 2-01.2. *Counterintelligence Analysis.* When published tentatively in April 2005, FM 2-01.2 will supersede FM 34-60, 3 October 1995.

FM 2-01.3. *Intelligence Preparation of the Battlefield.* When published tentatively December 2004, FM 2-01.3 will supersede FM 34-130, 8 July 1994.

Index

Entries are by paragraph number.

J-L

Foreword

We remain a nation at war in an era of persistent conflict, but we do not stand alone. Our nation has many multinational partners, equally committed to freedom, rule of law and stability. It is clear that we are stronger when we act with partners in today's operating environment. Therefore, security force assistance is no longer an "additional duty." It is now a core competency of our Army.

As distinctions separating war and peace blur and challenges to security increase, we must seek to help our multinational partners successfully confront their security challenges. Security force assistance builds our multinational partners' capability to defeat regular, irregular, and hybrid threats prevalent in an era of persistent conflict.

The two pillars of security force assistance are the modular brigade and Soldiers acting as advisors. Their tactical efforts build partner capability and improve the security situation. Over time, U.S. forces and partners foreign security forces collectively set the conditions to defeat common threats and ultimately achieve strategic success.

This manual addresses common characteristics and considerations for conducting security force assistance and clarifies what units and individual advisors must understand to work "by, with, and through" their counterparts. Since every situation and foreign security force is unique, units and individuals conducting security force assistance must carefully analyze the operational environment, especially the relationships of foreign security forces and their population. Conducting successful security force assistance requires adaptive units led by well-informed, culturally astute leaders. FM 3-07.1 provides commanders, staffs, and advisors with the common doctrine for achieving this goal.

Conducting foreign security assistance requires great skill in building relationships and "leading from behind." We must all increase our understanding of this important mission.

MARTIN E. DEMPSEY
General, U.S. Army
Commanding General
U.S. Army Training and Doctrine
Command

FM 3-07.1

Field Manual
No. 3-07.1

Headquarters
Department of the Army
Washington, DC, 1 May 2009

Security Force Assistance

Contents

Distribution Restriction: Approved for public release; distribution is unlimited.

Figures

Preface

This field manual (FM) is the Army's doctrinal publication for security force assistance (SFA). It provides doctrinal guidance and direction for how U.S. forces contribute to SFA. It focuses on the brigade combat team (BCT) conducting SFA and advising foreign security forces. It is based on lessons learned from previous advising efforts and recent combat operations with a view to the future. It supports the Army Education System instruction on the theory and conduct of SFA.

The two primary audiences for this manual are leaders in BCTs conducting SFA and Soldiers assigned as advisors. The BCT is the largest fixed tactical organization and the key formation of the Army's modular organization. Army modularity allows commanders to add selective units to assist the BCT as it conducts SFA. The BCT and higher echelons provide the framework for advisors to function and accomplish the mission—developing foreign security forces and, when appropriate, working by, with, and through foreign security forces to achieve the desired end state of the host nation's internal defense and development. Thus, Soldiers assigned as advisors are the key element of SFA.

This FM provides the conceptual framework for conventional forces to conduct SFA within the construct of full spectrum operations, across the spectrum of conflict. It addresses SFA at operational and tactical levels.

Army doctrine is consistent and compatible with joint doctrine. FM 3-07.1 expands on fundamental principles for SFA expressed in FM 3-07. FM 3-07.1 uses text and concepts developed with North Atlantic Treaty Organization partners and ABCA (America, Britain, Canada, Australia, and New Zealand program armies).

Terms that have joint or Army definitions are italicized and the number of the proponent manual follows the definition. FM 3-07.1 will be the proponent manual for foreign security forces as a term.

FM 3-07.1 applies to the Active Army, Army National Guard/U.S. Army National Guard, and U.S. Army Reserve unless otherwise stated.

Headquarters, US Army Training and Doctrine Command is the proponent for this publication. The preparing agency is the Combined Arms Doctrine Directorate, US Army Combined Arms Center. Send written comments and recommendations on Department of the Army (DA) Form 2028 (Recommended Changes to Publications and Blank Forms) directly to: Commander, US Army Combined Arms Center and Fort Leavenworth, ATTN: ATZL-CD (FM 3-07.1), 201 Reynolds Avenue (Building 285), Fort Leavenworth, KS 66027-1352. Send comments and recommendations by e-mail to leav-cadd-web-cadd@conus.army.mil. Follow the DA Form 2028 format or submit an electronic DA Form 2028.

Introduction

Our strategy emphasizes building the capacities of a broad spectrum of partners as the basis for long-term security. We must also seek to strengthen the resiliency of the international system to deal with conflict when it occurs.

National Defense Strategy, June 2008

In an era of persistent conflict, the United States supports the internal defense and development of international partners, regardless of whether those partners are highly developed and stable or less developed and emerging. While many of these partners are nations, they can also include alliances, coalitions, and regional organizations. U.S. support to these partners ranges from providing humanitarian assistance to major combat operations. U.S. support includes conducting conflict transformation, bolstering partner legitimacy, and building partner capacity. A vital part of these three aspects of U.S. support is assisting partner security forces.

Security force assistance (SFA) is not new for Army forces. In fact, General George Washington's Inspector General of the Army acted as an advisor for Army forces. Baron Friedrick Wilhelm von Steuben instilled discipline and professionalism into an army that previously lacked formalized training. His 1779 *Regulations for the Order and Discipline of the Troops of the United States*, adapted from the Prussian army, formed the doctrinal backbone of the Continental Army throughout the Revolutionary War. Additionally, the lineage of the Army's operations field manual, FM 3-0, can be traced to this document. As a benefactor of advisors such as von Steuben, the Army has since undertaken what is called SFA several times throughout its history.

Army doctrine defines *security force assistance* as the unified action to generate, employ, and sustain local, host-nation, or regional security forces in support of a legitimate authority (FM 3-07). SFA is part of the FM 3-0 construct of full spectrum operations. Similarly, it is conducted across the spectrum of conflict or in any of the operational themes. SFA is normally part of a larger security sector reform effort, while in other instances, SFA is not tied to reform but to building partner capacity.

Three general situations exist in which SFA may occur: an internally focused bilateral relationship, an externally focused bilateral relationship, and a multilateral relationship. Consequently, SFA supports the appropriate partner's plans. When SFA supports a host nation, it also supports that host nation's strategy. If SFA supports a host nation's externally focused efforts, it must support the host nation's national security strategy. SFA may support regional security forces, such as those of the African Union or the Organization of American States. In these cases, SFA supports that organization's plans.

Organization

This FM primarily focuses on the brigade combat team and advisors. It also provides a wider doctrinal framework for SFA. It is organized in the following manner:

- **Chapter 1 focuses above the brigade combat team.** It provides the overall doctrinal context for SFA.
- **Chapters 2 through 6 focus on the brigade combat team level.** These chapters provide a framework for SFA, its considerations for the brigade combat team operations process, considerations for augmenting the modular brigade for security force assistance, its considerations for the unit employment, and its sustainment considerations.
- **Chapters 7 through 10 focus on the individual advisor.** These chapters discuss the advisor, advisor cultural and communication considerations, the advisor working with counterparts, and cross-cultural influencing and communication.
- The appendices discuss Army special operations forces imperatives, legal considerations, and media considerations.

Chapter 1

The Strategic Context

This chapter places security force assistance in strategic context. First, it explains security force assistance and its relationship to full spectrum operations, spectrum of conflict, and operational themes (see FM 3-0). A discussion of U.S. national strategy and defense policies that impact security force assistance follows. This chapter then examines operations and programs related to security force assistance. Finally, it discusses the relationship of security force assistance and related operations and programs.

SECURITY FORCE ASSISTANCE AND THE STRATEGIC CONTEXT

1-1. *Security force assistance* is the unified action to generate, employ, and sustain local, host-nation or regional security forces in support of a legitimate authority (FM 3-07). Security force assistance (SFA) improves the capability and capacity of host-nation or regional security organization's security forces. These forces are collectively referred to as foreign security forces. ***Foreign security forces* are forces—including but not limited to military, paramilitary, police, and intelligence forces; border police, coast guard, and customs officials; and prison guards and correctional personnel—that provide security for a host nation and its relevant population or support a regional security organization's mission**.

1-2. SFA occurs within the framework of full spectrum operations (see FM 3-0). In most situations involving this assistance, there is relatively little weight on offensive and defensive operations from a U.S. perspective. However, when U.S. forces accompany foreign security forces (FSF) in combat, the weight of offensive and defense operations will change to address the situation and align with the foreign security force's efforts. SFA is not just a stability operation, although it is a key contributor to the primary stability tasks of establish civil security and establish civil control.

1-3. SFA can be conducted across the spectrum of conflict, from stable peace to general war. This assistance could focus on improving the security forces of a host nation that is currently under no immediate threat, on paramilitary forces to counter an insurgency, or on advising FSF in major combat operations against an external threat. Figure 1-1 illustrates how SFA spans the spectrum of conflict.

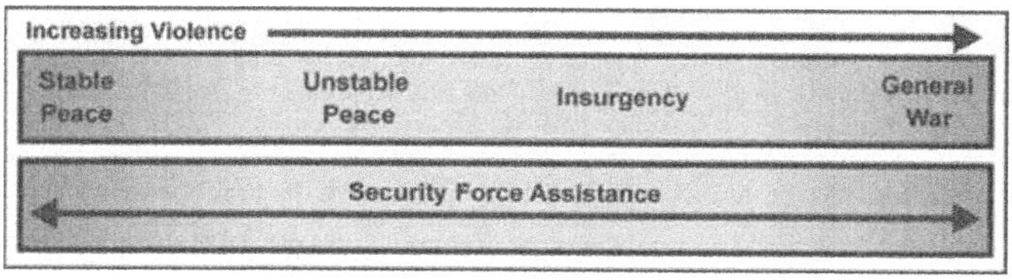

Figure 1-1. The spectrum of conflict and security force assistance

1-4. SFA can occur as part of any operational themes—peacetime limited engagement, limited intervention, peace operations, irregular warfare, and major combat operations. SFA often occurs during peacetime military engagement, peace operations, and irregular warfare. SFA involving paramilitary forces will often be limited to the irregular warfare operational theme, although paramilitary forces should eventually be absorbed, disbanded, or maintained in a long-term relationship with special operations forces.

1-5. While Army units are vital contributors, SFA is part of a comprehensive approach. This comprehensive approach should include close collaboration with military and civilian joint and multinational forces. The host-nation or regional security organization is the key actor within the comprehensive approach. It is vital that SFA is based on an objective, continuous assessment that examines the organization, training, equipment, rebuilding, and advising of the forces involved.

1-6. Law enforcement, military, intelligence, and border forces operate and cooperate within the security sector. U.S. forces understand how these units are intended to operate in the host-nation scheme, not necessarily using a U.S. model. U.S. forces should plan to help develop the respective capabilities so that these units can carry out their security functions.

1-7. Fully developed and institutionalized FSF plan, prepare, and execute at all the appropriate levels. Thus, U.S. forces must integrate SFA into the operations process at all levels. SFA should be designed to work seamlessly with the host-nation government at all levels, from ministries addressing the security sector to initial entry-level FSF training.

NATIONAL STRATEGY AND DEFENSE POLICIES

1-8. U.S. strategy and foreign policy affects all SFA. More importantly, SFA is critical to these policies as it is a primary tool for building partnership capacity with other nations. Key strategy documents for SFA include the National Security Strategy, the National Defense Strategy, the National Military Strategy, the Quadrennial Defense Review, and DODD 3000.05.

1-9. The National Security Strategy outlines the President's vision for providing enduring security for the American people in a volatile, uncertain, complex, and ambiguous strategic environment. The National Security Strategy of 2006 addresses stability operations within the broad engagement strategy for regional conflict. It supports SFA by "tailoring assistance and training of military forces to support civilian control of the military and military respect for human rights in a democratic society." This strategy also states that "[t]he world has found through bitter experience that success often depends on the early establishment of strong local institutions such as effective police forces and a functioning justice and penal system. This governance capacity is critical to establishing the rule of law and a free market economy, which provide long-term stability and prosperity."

1-10. The National Defense Strategy encompasses the Secretary of Defense's vision for the Department of Defense role in protecting the American people and U.S. interests. This strategy stems from the National Security Strategy. The 2008 National Defense Strategy highlights SFA. It states that "[o]ur strategy emphasizes building the capacities of a broad spectrum of partners as the basis for long-term security.... [l]ocal and regional conflicts in particular remain a serious and immediate problem...we will help build the internal capacities of countries at risk...[b]y helping others to police themselves and their regions, we will collectively address threats to the broader international system."

1-11. The National Military Strategy supports the National Security Strategy, National Defense Strategy, and Quadrennial Defense Review. The 2004 National Military Strategy states "[t]he Department will also work to strengthen alliances and partnerships by helping other nations increase their ability to defend themselves and protect common security interests." Working with other nations and strengthening partnerships appears throughout the National Military Strategy, which supports the importance of SFA.

1-12. The Quadrennial Defense Review is a congressionally mandated Department of Defense (DOD) review of strategy, programs, and resources. The 2006 Quadrennial Defense Review directed that joint ground forces "possess the ability to train, mentor and advise FSF and conduct counterinsurgency campaigns." This review also stated that conventional forces need to be capable of training, equipping, and advising local forces; deploying and engaging with partners; and supporting stability operations. Additionally, the review envisioned conventional ground forces continuing to perform more of the tasks traditionally performed by special operations forces. Conventional forces must understand foreign cultures and societies as well as be able to train, mentor, and advise FSF.

1-13. DODD 3000.05 provides guidance on stability operations. DODD 3000.05 establishes that stability operations are a core military mission and equal in importance to combat operations. DODD 3000.05 also reinforces the importance of SFA by stating "[t]he immediate goal often is to provide the local populace with security, restore essential services, and meet humanitarian needs. The long-term goal is to help develop host-nation capacity for securing essential services, a viable market economy, rule of law, democratic institutions, and a robust civil society."

RELATED OPERATIONS AND PROGRAMS

1-14. Several operations or programs directly or potentially relate to SFA. These operations and programs provide the context for SFA. They include security cooperation, security assistance, foreign internal defense, internal defense and development (IDAD), and security sector reform. For more information on operations related to SFA, see DOD 5105.38M, DOD security cooperation guidance, JP 3-07.1, FM 3-07, FM 3-05.137, and AR 11-31.

SECURITY COOPERATION

1-15. *Security cooperation* is all Department of Defense interactions with foreign defense establishments to build defense relationships that promote specific U.S. security interests, develop allied and friendly military capabilities for self-defense and multinational operations, and provide U.S. forces with peacetime and contingency access to a host nation (JP 3-07.1). Finally, security cooperation occurs across the spectrum of conflict and is not exclusively a peacetime activity. Security cooperation includes security assistance programs administered by the DOD as well as activities that enhance interoperability and the collective capability of combined forces (using Title 10 or exercise funding and authority). Security cooperation consists of a focused program of bilateral and multilateral defense activities conducted with foreign countries to serve mutual security interests and build defense partnerships. Security cooperation is governed by various sections of Title 10 and specific public laws addressing DOD interactions with other nations.

1-16. AR 11-31 establishes Department of the Army policy and prescribes responsibilities and procedures for the planning, integration, programming, budgeting, and execution of Army security cooperation activities. It is the policy of the U.S. Army to conduct security cooperation activities in compliance with higher-level guidance and in the execution of Army responsibilities under Title 10 and Title 22 which governs the transfer, exchange, conduct, and development of articles and services via a variety of U.S. Government programs.

1-17. Army security cooperation consists of official, cooperative, and noncombat interactions among any Army elements, Active Army, or Reserve Components. This includes any U.S. Government or nongovernment entity supporting the military and civilian joint and multinational organizations.

1-18. Commanders distinguish security cooperation from SFA based its emphasis in building relationships and capacities by using programmatic activities. Thus, security cooperation may support or be supported by SFA.

SECURITY ASSISTANCE

1-19. *Security assistance* is a group of programs authorized by the Foreign Assistance Act of 1961, as amended, and the Arms Export Control Act of 1976, as amended, or other related statutes by which the United States provides defense articles, military training, and other defense-related services by grant, loan, credit, or cash sales in furtherance of national policies and objectives (JP 3-57). Security assistance is a specific subset of security cooperation and may focus on external or internal threats.

1-20. DOD 5105.38-M describes the scope of security assistance programs in detail. Security assistance allows the transfer of military articles and services to friendly foreign governments. These transfers may be carried out via sales, grants, leases, or loans. If these transfers are essential to the security and economic well-being of allied governments and international organizations, they are equally vital to the security and economic well-being of the United States. U.S. forces can use security assistance to deliver defense weapon systems to foreign governments; to train international students; to advise other governments on improving internal defense capabilities; and to provide guidance and assistance in establishing infrastructures and economic bases for regional stability. Finally, security assistance cannot be conducted using Title 10 or exercise funds.

1-21. The military component of security assistance, implemented by DOD with policies established by the Department of State, has four principal components. They are international military education and training, foreign military sales, foreign military financing program, and peace operations. The foreign military financing program and international military education and training fall inside the military assistance budget process of the Department of State. Security assistance can also include funding peace operations.

International Military Education and Training

1-22. International military education and training contributes to internal and external security of a country by providing training to selected foreign militaries and related civilian personnel on a grant aid basis (see JP 3-07.1). These programs help to strengthen foreign militaries through training for the proper functioning of a civilian-controlled, apolitical, professional military. International military education and training serves as a foreign policy tool where the United States shapes doctrine; promotes self sufficiency in maintaining and operating U.S. acquired defense equipment; encourages the value of rule of law; and occasionally has a marked effect on the policies of the recipient governments. Foreign students, many of who occupy the middle and upper echelons of their country's military and political establishments, are taught U.S. doctrine and weapons systems employment resulting in greater cooperation and interoperability. All international military education and training activities are SFA.

Foreign Military Sales

1-23. Foreign military sales is a nonappropriated program through which foreign governments can purchase defense articles, services, and training from the United States (see JP 3-07.1). Eligible nations use this program to help build national security infrastructures. However, nations that require assistance are often unable to finance their needs. SFA supports foreign military sales when these sales include U.S. forces providing equipment and related training. It does not support foreign military sales when these sales provide equipment without related training.

Foreign Military Financing Program

1-24. The foreign military financing program provides funding to purchase defense articles and services, design and construction services, and training through foreign military sales or commercial channels (see JP 3-07.1). This program can assist nations with weak economies that would otherwise be unable to afford U.S. assistance. The foreign military financing program can fund foreign military sales. SFA and the foreign military financing program may be separate activities, although they can also be complementary.

Foreign Internal Defense

1-25. *Foreign internal defense* is the participation by civilian and military agencies of a government in any of the action programs taken by another government or other designated organization to free and protect its society from subversion, lawlessness, and insurgency (JP 3-05). U.S. foreign internal defense efforts involve all instruments of national power to support host-nation IDAD programs. The military instrument of foreign internal defense includes indirect support, direct support (not involving U.S. combat operations), and combat operations. Foreign internal defense can occur across the spectrum of conflict and within any operational theme. Foreign internal defense is often a part of irregular warfare or peacetime military engagement operational theme. SFA may support the military instrument of foreign internal defense. JP 3-07.1 discusses foreign internal defense in detail.

Note: It is a national policy decision for U.S. forces to accompany FSF on combat operations.

Military Instrument of Foreign Internal Defense

1-26. The military instrument of foreign internal defense supports the other instruments of national power through activities ranging from peacetime military engagement to major combat operations. Although foreign internal defense may include capacity building across the host nation, U.S. forces focus on combating internal threats to help the host nation maintain legitimacy and influence over the population. As such, foreign internal defense may be part of irregular warfare.

1-27. SFA and foreign internal defense support the host nation's IDAD. Consequently, forces conducting SFA must have proper mindset—they are part of a developmental and supporting effort. This applies even if SFA is part of working with a potential adversary or an existing enemy. SFA and foreign internal defense focus on working by, with, and through FSF and not conducting unilateral efforts to achieve objectives.

Indirect Support

1-28. Indirect support efforts are foreign internal defense operations that emphasize the principle of host-nation self-sufficiency. Indirect support focuses on building strong national infrastructures through economic and military capabilities that contribute to self-sufficiency. For example, indirect support could include U.S. assistance in improving a host-nation's intelligence infrastructure. Contributions from U.S. forces to indirect support come from security cooperation guidance through security assistance. They are supplemented by multinational exercises, exchange programs, and selected joint exercises. Indirect support that is focused on assisting security forces is SFA.

1-29. Although they indirectly assist those involved, military exchange programs are not SFA. Military exchange programs foster mutual understanding between forces; familiarize each force with the organization, administration, and operations of the other; and enhance cultural awareness. These exchange programs often have long-term implications for strengthening democratic ideals and respect for human rights among supported governments. However, such exchange programs (funded with Title 10 monies) do not themselves become vehicles for security assistance training, as they do not intend to assist security forces but to foster mutual understanding and enhance cultural awareness.

1-30. Joint and multinational exercises strengthen U.S. and host-nation relations and interoperability of forces, but they are not SFA. They complement security assistance and civil-military operations by validating host-nation needs and capabilities and by providing a vehicle for the conduct of humanitarian and civic assistance programs (see JP 3-07.1). Strict legal restrictions clarify the type of support that U.S. forces can provide and on the monetary limits of such support. These efforts are not SFA as they do not intend to directly assist security forces; instead, they intend to improve relations and interoperability. See appendix B.

Direct Support

1-31. Direct support does not involve combat operations but includes the use of U.S. forces providing direct assistance to the host-nation civilians or military. Direct support often focuses on civil-military operations (primarily, providing services to local civilians), psychological operations, intelligence and communications sharing, and logistic support. Sometimes direct support will be SFA, such as psychological operations, military training support, logistic support, and intelligence and communications sharing. SFA is not a civil affairs activity such as foreign humanitarian assistance, humanitarian and civic assistance, or military civic action.

1-32. Direct support differs from security assistance in that direct support is joint- or Service-funded. Direct support does not usually involve transferring arms and equipment and does not often include training local military forces. Direct support operations are normally conducted when the host nation has not attained or regained self-sufficiency and when it still faces social, economic, or military threats beyond its capability.

Combat Operations

1-33. The introduction of U.S. combat forces into foreign internal defense operations requires a Presidential decision. These operations are a temporary solution until host-nation forces can stabilize the situation and provide security for civilians. In all cases, U.S. combat operations support the host-nation IDAD program and remain strategically defensive in nature. When U.S. forces are advising, mentoring, partnering, or augmenting host-nation forces in foreign internal defense combat operations, this is SFA. When Army units conduct independent combat operations alongside FSF, this is not SFA.

1-34. U.S. tactical participation in host-nation internal conflicts requires judicious and prudent rules of engagement and guidelines for applying force. Inappropriate destruction and violence attributed to U.S. forces may easily reduce the legitimacy and sovereignty of the supported government. Adversaries may also use such incidents to fuel anti-American sentiments and assist their cause.

SECURITY RELATIONSHIPS

1-35. A complex relationship exists among security cooperation, security assistance, and the military instrument of foreign internal defense. The left side of figure 1-2 depicts this relationship, including how aspects of foreign internal defense and security assistance overlap. The right side of figure 1-2 illustrates how foreign internal defense focuses on internal threats to a host nation and how security assistance focuses on external threats. The column depicts how security (military, intelligence, and civilian), economic, and governance are considerations common to both foreign internal defense and security assistance.

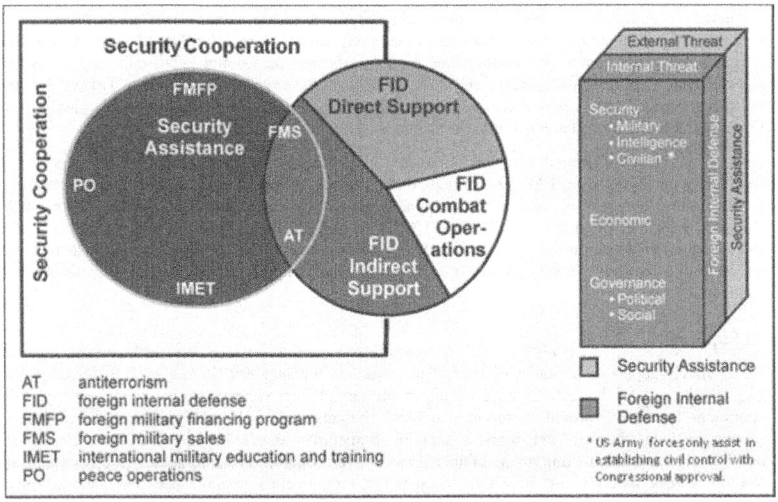

Figure 1-2. Security cooperation, security assistance, and foreign internal defense

INTERNAL DEFENSE AND DEVELOPMENT

1-36. *Internal defense and development* is the full range of measures taken by a nation to promote its growth and to protect itself from subversion, lawlessness, and insurgency. It focuses on building viable institutions (political, economic, social, and military) that respond to the needs of society (JP 3-07.1). IDAD aims to forestall or defeat the threat and to correct core grievances that prompt violence. Ideally it is a preemptive strategy. However, if an insurgency or other threat develops, it becomes an active strategy to combat that threat. To support the host nation effectively, U.S. forces, especially planners, consider the host-nation's IDAD strategy.

1-37. IDAD focuses on building viable institutions that respond to the needs of society. IDAD is a U.S. doctrinal term; it is not used universally by all countries. Internal defense and development blends four interdependent functions to prevent or counter internal threats: balanced development, security, neutralization, and mobilization. Finally, IDAD involves a cyclic interaction of execution, assessment, and adaptation. JP 3-07.1 discusses IDAD and its functions in detail.

Security Sector Reform

1-38. *Security sector reform* is the set of policies, plans, programs, and activities that a government undertakes to improve the way it provides safety, security, and justice (FM 3-07). Security sector reform aims to provide an effective and legitimate public service that is transparent, accountable to civil authority, and responsive to the needs of the public. It may include integrated activities to support defense and armed forces reform; civilian management and oversight; justice, police, corrections, and intelligence reform; national security planning and strategy support; border management; disarmament, demobilization, and reintegration; and concurrent reduction of armed violence. SFA is a vital component of security sector reform when security sector reform includes U.S. assistance to FSF. See chapter 6 of FM 3-07 for an additional discussion of security sector reform.

Security Force Assistance Relationship

1-39. Paragraph 1-35 explained and figure 1-2 supported the complex relationship among security cooperation, security assistance, and the military instrument of foreign internal defense. Figure 1-3 explains the relationship of SFA to security cooperation, security assistance, and the military instrument of foreign internal defense. SFA supports the military instrument of foreign internal defense, much of security assistance efforts, and some security cooperation efforts. The right side of figure 1-3 also shows that SFA addresses security issues, but not economic or governance issues.

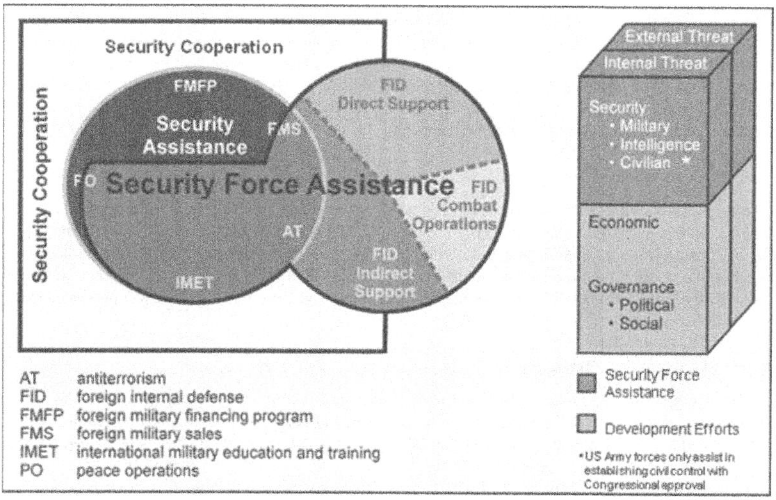

Figure 1-3. Relationship of security force assistance with security cooperation, security assistance, and foreign internal defense

COUNTRY TEAM

1-40. The country team will play a vital role with respect to SFA. The country team is the senior in-country U.S. coordinating and supervising body headed by the U.S. chief of mission. The chief of mission leads the country team and integrates U.S. efforts in support of the host nation. The chief of mission is the highest U.S. civil authority in a foreign country. As the senior U.S. Government official permanently assigned in the host nation, the chief of mission is responsible to the President for policy oversight of all U.S. Government programs. The Foreign Service Act assigns the chief of mission to a foreign country with responsibility for the direction, coordination, and supervision of all government executive branch employees in that country except for Service members and employees under the command of a U.S. area military commander.

1-41. The country team is composed of the senior member of each represented department or agency. As permanently established interagency organizations, country teams represent a priceless resource. They often provide deep reservoirs of local knowledge and interaction with the host-nation government and population. See FM 3-07 and JP 3-07.4 for more information on the country team.

THEATER COMMAND STRUCTURES

1-42. SFA is inherently multinational. Nations do not relinquish their national interests by participating in multinational operations. As in any multinational operation, commanders conducting SFA prepare to address issues related to legality, mission mandate, and prudence early in the planning process. In multinational operations, consensus often stems from compromise.

1-43. Political considerations heavily influence the ultimate shape of the command structure in which U.S. forces will conduct SFA. However, participating nations should strive to achieve unity of command for the operation to the maximum extent possible. All participants must understand missions, tasks, responsibilities, and authorities. While well defined in U.S. doctrine, command relationships are not necessarily part of the doctrinal lexicon of nations with which the United States may operate in coalition or combined operations.

1-44. Command and control relationships of higher echelons take precedence when conducting SFA. As a result, the brigade combat team (BCT) may be a supported command from a U.S. perspective as well as acting as a supporting command for FSF in the BCT's area of operations. For example, a BCT may support a joint interagency task force conducting international police training. Another example is a joint command supporting a BCT with joint fires; intelligence, surveillance, and reconnaissance; or other capabilities while the BCT supports FSF in training. The basic structures for multinational operations fall into one of three types: lead nation, integrated, or parallel command. Alliance and coalition commands use these structures.

Lead Nation Command Structure

1-45. A lead nation command structure exists when all member nations place their forces under the control of one nation. The lead nation command has a dominant lead nation command and staff arrangement with subordinate elements retaining strict national integrity. This command may also have an integrated staff and multinational subordinate forces. Integrating the staff allows the commander to draw upon the expertise of allied or coalition partners in areas where the lead nation may have less experience. This is the most desirable command structure for SFA.

Integrated Command Structure

1-46. Multinational commands organized under an integrated command structure provide unity of effort in a multinational setting. An integrated, combined command has several distinguishing features. Besides a designated single commander, the staff consists of representatives from all member nations. Subordinate commands and staffs are integrated into the lowest echelon necessary. Although effective for SFA, the integrated command structure works less effectively than the lead nation command structure.

Parallel Command Structure

1-47. Under a parallel command structure, no single force commander is designated. The coalition leadership develops a means for coordination among the participants to attain unity of effort. They can accomplish this by using coordination centers. Nonetheless, because of the absence of a single commander, the use of a parallel command structure should be avoided if at all possible. This is the least desirable command structure for SFA.

Alliance Command Structures

1-48. In combined commands, national political objectives are addressed and generally subsumed within multinational force objectives at the alliance treaty level. Combined command relationships often reflect either an integrated command structure or a lead nation command structure. In combined operations such as SFA, these structures should be used to the maximum practical extent. Combined command and force structures often mirror the degree of allied member participation. Senior military officers from member nations often lead subordinate commands. Effective operations within an alliance, including SFA, require senior political and military authorities to agree on the type of command relationships that will govern the operations of the forces. Potential political sensitivities associated with SFA can impact command relationships and operating procedures.

Coalition Command Structures

1-49. Coalitions often form in response to unforeseen crises that occur outside the area of an established alliance, which is often the case for SFA. Coalition command relationships routinely evolve as a coalition develops. Coalitions often have one of two basic structures: lead nation or parallel. For SFA, all participating coalition members should augment the host-nation headquarters staff with representatives. These representatives might include designated deputies or assistant commanders, planners, and logisticians. Such staffing provides the coalition commander with representative leadership and accessible expertise on the capabilities of the coalition members as well as facilitates the planning process. This integration of the multinational command elements into the coalition planning process should occur as early as practical.

Integration of Conventional and Special Operations Forces

1-50. Many criteria determine when conventional forces, special operations forces, or a combination are appropriate to conduct SFA. Both force levels and force characteristics suggest optimal, acceptable, and undesirable force package options in planning and resourcing SFA. Options for the deployment of a modular brigade augmented for SFA, a select number of conventional military transition teams, or special operations forces depend on conditions of the operational environment, priorities of the IDAD strategy, overall U.S. national policy, and forces available.

1-51. Rarely will U.S. forces conducting SFA be homogenous in terms of conventional or special operations forces, but rather contain both. FM 3-0 and other Army doctrine define command and control relationships and support relationships when both forces operate in the same area. Recent operational experience has shown that, in improving the effectiveness of FSF, the supported commander conducting SFA may be either a conventional force or special operations force commander. Subordinate commanders and staffs of both conventional and special operations forces may act in the supporting role to the U.S. ambassador. Typically this is done through the security assistance officer or the defense attaché at the American Embassy.

1-52. If U.S. forces are to train a large number of FSF in a short time, then the requirement may exceed the capacity of available special operations forces. Large SFA missions occur at the geographic command level or higher headquarters. Planners at these levels determine the number of conventional and special operations forces as part of their mission analysis aimed at training enough FSF in the given time frame. To best prepare for a large SFA, U.S. forces might deploy one or more augmented modular brigades supported by Army special operations forces. These forces conduct complementary foreign internal defense tasks under Title 22 authorities. Another option involves using special operations forces training FSF over several years while using a broad mix of funding and authorities. Another option uses a modular brigade to support a select number of company- or battalion-sized foreign units.

1-53. A force analysis of how best to integrate conventional and special operations forces to accomplish SFA considers host-nation and U.S. national policies. Skill sets among conventional and special operations forces trainers, advisors, and mentors avoid universally excluding or including final, integrated force packages for SFA. Planners leverage skill sets by integrating and coordinating conventional and special operations units with assigned tasks within SFA. Force analysis of the FSF to be trained often suggests a two-phase approach to training. This analysis also suggests a transition to the FSF by either conventional or special operations forces using basic and advanced training. The type of training depends on the nature of the foreign security force being trained. Additionally, integrating the advisor may be a phased process using either conventional or special operations forces as the initial advisor. This method lends itself to the classic crawl-walk-run approach to assisting FSF.

Chapter 2
Framework

This chapter discusses a doctrinal framework for the conduct of security force assistance. This security force assistance framework is composed of the mindset required of units and Soldiers, imperatives for success, inherent tasks and activities, and the three types of security force assistance.

MINDSET

2-1. Conducting successful security force assistance (SFA) requires a specific mindset. This mindset focuses on working by, with, and through foreign security forces (FSF) to support the host nation's internal defense and development (which includes local security requirements) or regional organization's charter. Soldiers conducting SFA must also understand that legitimacy is vital. The relevant population must perceive FSF as legitimate for long-term success. Those conducting SFA must understand that the military instrument of national power is only one part of a comprehensive approach. The imperatives of SFA provide the foundation for proper mindset.

Note: Host nation includes partner organization.

2-2. SFA focuses on how FSF organizations, institutions, systems, capabilities, and limitations—not U.S. organizations, systems, procedures, and equipment—can be enabled to address the host-nation or partner organization problems. Local forces have advantages over outsiders. They inherently understand the local culture and behavior that outsiders simply lack. To tap into those advantages, advisors must resist blatant military solutions. To overcome the temptation to do what they know and do best, whether relevant or not to the situation, advisors must accept that they are bound by unique situations.

IMPERATIVES

2-3. The imperatives of SFA come from the historical record and recent experience. These imperatives do not replace the principles of war or the principles of joint operations. Rather, they provide focus on how to successfully conduct SFA. The six imperatives apply to SFA at every level of war, for any echelon, and for any Soldier.

2-4. Since there is a close relationship between SFA and Army special forces conducting foreign internal defense, SFA planners should also consider the twelve Army special operations forces imperatives, especially at the tactical level (see appendix A).

UNDERSTAND THE OPERATIONAL ENVIRONMENT

2-5. An in-depth understanding of the operational environment—including available FSF, opposing threats, and civil considerations—is critical to planning and conducting effective SFA. Units and Soldiers conducting SFA must clearly understand the theater, population, and FSF with which they are working, especially FSF capabilities. Diplomatic, informational, military, economic, sociological, psychological, and geographic research and understanding are essential prerequisites for successful SFA. Tactically, successful SFA requires identifying the friendly and hostile decisionmakers, their objectives and strategies, and the ways they interact. Further, the conditions of the operational environment can rapidly change, and those conducting SFA must anticipate these changes and exploit potential and possibly fleeting opportunities.

PROVIDE EFFECTIVE LEADERSHIP

2-6. Leadership, a critical aspect of any application of combat power, proves important in the dynamic and complex environments associated with SFA. The operational environment in which SFA occurs places a premium on effective leadership at all levels, from the most junior to the most senior general officer and agency director. Leading FSF or a combined group of U.S. and foreign security forces is inherently more challenging due to differences in culture, language, training, and other aspects. Leaders focus on transitions; their decisions move the foreign security force toward independent operations.

BUILD LEGITIMACY

2-7. Ultimately, SFA aims to develop security forces that contribute to the legitimate governance of the local populace. Significant policy and legal considerations may apply to SFA activities. Legitimacy is the most crucial factor in developing and maintaining internal and international support. The United States cannot sustain long-term SFA without legitimacy. Legitimacy is a concept that goes beyond a strict legal definition; it includes the moral and political legitimacy of a host-nation government or partner organization. Local civilians and the international community determine the government's legitimacy based on collective perception of the credibility of its cause and methods.

MANAGE INFORMATION

2-8. Successful SFA disseminates timely and protected relevant information, integrates it during planning, and leverages it appropriately during execution. Effective and efficient information management supports decisionmaking throughout capacity building. Managing information encompasses the collection, analysis, management, application, and preparation of information. Maintaining continuity between rotating SFA elements is critical. For long-term SFA efforts, FSF expect more professional abilities from U.S. advisors. To meet expectations and success in general, units conducting SFA establish and integrate lessons learned. Capturing and incorporating evolving tactics, techniques, and procedures and aspects of the operational environment is vital; with each successive rotation, planners strive to maintain continuity and momentum. Finally, planners synchronize effective information engagement with the entire mission. See appendix C. SFA operations inherently affect the dynamic operational environment.

ENSURE UNITY OF EFFORT

2-9. SFA often includes many actors, making unity of effort essential for success. SFA will include U.S. and foreign security forces, including conventional forces, special forces, or a combination. Other civilian and military joint and military organizations are often involved in SFA. Planners integrate them into one cohesive effort.

SUSTAIN THE EFFORT

2-10. Sustainability consists of two major components: the ability to sustain SFA effort throughout the operation and the ability of the FSF to sustain their operations independently. While each situation will vary, Army personnel conducting SFA must avoid assisting FSF in techniques and procedures beyond the FSF's capability to sustain. U.S. tactics, techniques, and procedures must be modified to fit the culture, educational level, and technological capability of the FSF. Those involved in SFA must recognize the need for programs that are durable, consistent, and sustainable by both the U.S. and FSF. They must not begin programs that are beyond the economic, technological, or cultural capabilities of the host nation to maintain without U.S. assistance. Such efforts are counterproductive.

TASKS

2-11. SFA aims to establish conditions that support the partner's end state, which includes legitimate, credible, competent, capable, committed, and confident security forces. This requires a force capable of securing borders, protecting the population, holding individuals accountable for criminal activities, regulating the behavior of individuals or groups that pose a security risk, and setting conditions in the operational area that enable the success of other actors.

2-12. This section addresses the six SFA tasks: organize, train, equip, rebuild and build, and advise and assist. When supporting the development of FSF, commanders understand their role with the host-nation or regional security organization. These tasks facilitate SFA planners to assess and allocate resources based on conditions.

ORGANIZE

2-13. Organize is a SFA task that encompasses all measures taken to assist FSF in improving its organizational structure, processes, institutions, and infrastructure. U.S. forces must understand the existing security organizations of FSF to better assist them. Subsequently, SFA personnel may help the host nation organize its security forces to meet the needs of its security environment.

2-14. Organizing a foreign security force depends on the host nation's social and economic conditions, cultural and historical factors, and security threats. SFA aims to create an efficient organization with a command, intelligence, logistic, and operations structure viable for the host nation. Conventional forces with limited special purpose teams (such as explosive ordnance disposal) are preferred. Doctrine or standing operating procedures should apply across the force, as should unit structures. The organization must facilitate the collection, processing, and dissemination of intelligence across and throughout all security forces. As the foreign security force strengthens, U.S. leaders and trainers should expect more independent organizational decisions.

2-15. The host nation determines the structure of its military forces, to include approving all organizational designs. These may include changing the numbers of forces, types of units, and internal organizational designs.

2-16. Organization should address all FSF, from the ministerial level to the patrolling police officer and soldier. Building a competent FSF infrastructure—including civilian information systems—is critical for success. Commanders work with ministries responsible for national and internal security, including the ministry of national defense, the interior ministry, and the justice ministry. Commanders assess strengths and weaknesses of the ministerial organization as well as training requirements of their employees. The joint and multinational advisor team at the ministry level helps the host nation develop a procurement and management system to meet its requirements.

2-17. Organizing a foreign security force requires resolving issues related to—

- Recruiting.
- Promotion screening and selection.
- Pay and benefits.
- Leader recruiting and selection.
- Personnel accountability.
- Demobilization of security force personnel.

Recruiting

2-18. Recruiting is critical when establishing FSF. The recruiting program assimilates local culture and themes that resonate with the populace. The program ensures that FSF have members from all major demographic groups. Joint and multinational partners should encourage and support efforts to recruit from minority populations. Mobile recruiting can target specific areas, ethnic groups, or tribes. Forces should encourage moderate groups and factions within hostile or potentially hostile ethnic groups to join the foreign security force. Often host-nation governments resist recruiting disaffected ethnic groups. However, even moderate success in recruiting from these groups provides enormous payoffs. It builds the security forces' legitimacy and often quiets legitimate fears of such groups regarding their relationship to the host-nation government. Effectively disarming, demobilizing, and reintegrating former enemies or other armed groups should be part of the overall plan. These programs should be mentioned in recruiting efforts.

2-19. A proper recruiting program specifies appropriate behavioral, physical, and moral standards. Ideally, recruits are centrally screened and inducted. Recruiting centers reside in safe areas safe. All recruits undergo a basic security check and are vetted against lists of suspected enemies. Foreign agencies and personnel perform this screening. Membership in illegal organizations needs to be carefully monitored. Past membership need not preclude a person from joining the foreign security force; however, any ongoing relationship with an illegal organization requires constant monitoring. FSF personnel limits the members of an illegal unit, tribal militia, or other militant faction to a single military or police unit.

Promotion Screening and Selection

2-20. Selection for promotion must stem from proven performance and aptitude for increased responsibility. Objective evaluations ensure promotion is by merit and not through influence or family ties. One method of selection identifies the most competent performers, trains them, and recommends them for promotion. The second method identifies those with social or professional status within the training group, then trains and recommends them for promotion. The first method may lead to more competent leaders but could be resisted for cultural reasons. The second method ensures the new leader will be accepted culturally but may sacrifice competence. The most effective solution comes from combining the two methods.

Pay and Benefits

2-21. Appropriate compensation precludes a culture of corruption in FSF. Spending the money needed for adequate wages and producing quality security forces is less costly than ending up with corrupt and abusive forces that alienate the populace. Paying the police adequately is especially important; their duties and contact with civilians often expose them to opportunities for corruption.

2-22. Payroll procedures and systems are also vitally important. They must be transparent and accountable, so FSF members receive their full pay and entitlements. Centralized administration of compensation through secure, automated systems minimizes the risk for corruption and manipulation; however, the supporting infrastructure will have to be available or developed over time.

2-23. Effective FSF can help improve host-nation social and economic development through the benefits each member receives. Every recruit should receive a basic education, job training, and morals and values inculcation.

Leader Recruiting and Selection

2-24. Officer candidate standards should be high. Candidates should have good health and pass an academic test with higher standards than the test for enlisted recruits. Recruiters carefully vet officer candidates to ensure that they do not have close ties to any radical or enemy organization.

2-25. Noncommissioned officers should be selected from the best enlisted FSF members. Objective standards, including proficiency tests, should be established and enforced to ensure that promotion to their ranks comes from merit. Many armies lack a professional noncommissioned officer corps; establishing one for a foreign security force may prove difficult. In the meantime, adjustments will have to be made, placing more responsibility on commissioned officers.

Personnel Accountability

2-26. Host-nation leaders carefully track and account for FSF. Proper personnel accountability reduces corruption, particularly in countries with manual banking systems where soldiers are paid in cash. In addition, many personnel failing to report for duty can indicate possible attacks, low unit morale, or enemy and militia influences on the foreign security force.

Demobilization of Foreign Security Force Personnel

2-27. Host nations develop programs to keep a class of impoverished and disgruntled former officers and soldiers from forming. As the foreign security force mature, officers who perform poorly or fail to meet the standards will need to be removed. Providing some form of government-provided education grants or low-interest business loans enables discharged personnel to earn a living outside the military. All soldiers who serve for several years and are then removed should receive a lump-sum payment or pension to ease their transition to civilian life. These programs should not apply to those guilty of major human rights abuses or corruption. Transition or demobilization planning should start as soon as commanders anticipate the need (it may not be required in all cases). Similar programs may be required when demobilizing nongovernment militias.

TRAIN

2-28. Train is a SFA task to assist FSF by developing programs and institutions to train and educate. These efforts must fit the nature and requirements of their security environment. The imperatives of Army special operations forces for facilitating a comprehensive approach and engaging the threat discriminately apply to the train task (see appendix A).

Training the U.S. Trainers

2-29. Soldiers assigned training missions should receive training on the requirements of developing FSF. The training should emphasize the host nation's cultural background, introduce its language, and provide cultural tips for developing a good rapport. Training should also include protection for troops working with FSF. U.S. trainees must become familiar with FSF organization and equipment, especially weapons not in the U.S. inventory. This training must emphasize the following:

- Sustaining training and reinforcing individual and team skills.
- Using the smallest possible student-to-instructor ratio.
- Developing host-nation trainers.
- Training to standards—not to time.
- Providing immediate feedback; using after action reviews.
- Respecting the culture but learning to distinguish between cultural practices and excuses.
- Learning the language.
- Working with interpreters.

2-30. U.S. forces should show respect for local religions and traditions. They should willingly accept many aspects of the local and national culture, including food (if sanitation standards permit). U.S. forces must stress that they do not intend to undermine or change the local religion or traditions. However, they have to strive to reduce how dysfunctional social practices affect the foreign security force's ability to conduct operations. U.S. trainers and advisors must have enough awareness to identify and stop inappropriate behavior, or at least report it to the appropriate chains of command.

Establishing Training Standards

2-31. Training in SFA involves many of the individual and collective skills performed in conventional military operations. All levels of training for all components should include values training. Metrics for evaluating units should include subjective measures, such as loyalty to the host-nation government, as well as competence in military tasks. Soldiers know how to evaluate military training. However, the acceptance of values, such as ethnic equality or the rejection of corruption, may be a better measure of training effectiveness. Gauging this acceptance is far more difficult than evaluating task performance. While the operational environment varies widely, FSF and trainers can still establish clear measures to evaluate the training of individuals, leaders, and units.

2-32. Effective training programs require clear, detailed individual, leader, and unit performance standards. These standards take into account cultural factors that directly affect the ability of the individual or unit to operate. For example, training a staff or unit to conduct effective operations requires more time in countries where the average soldier is illiterate. Building a security force from the ground up takes far more time than creating one around a trained cadre. Thus, using existing military personnel to form units and cadres for units often proves better than creating novice security forces.

2-33. Often, poorly trained leaders and units commit more human rights violations than well-trained, well-led units. Leaders and units unprepared for the pressure of active operations tend to use indiscriminate force, target civilians, and abuse prisoners. These actions can threaten the popular support and government legitimacy essential for SFA success. Badly disciplined and poorly led FSF can facilitate insurgent recruiting and propaganda efforts.

2-34. Setting realistic measures for FSF and following through on training plans consume time. The pressure is strong to find training shortcuts, employ quick fixes, or train personnel on the job. Trainers should resist such approaches. Such approaches often create more problems than they solve. However, trainers should also avoid the temptation to create long, complex training programs based on unrealistic standards. Effective programs account for the host nation's culture, resources, and short-term security needs. No firm rules exist on how long particular training programs should take. Trainers can use existing and historical training programs to determine how long training should take. To a certain extent, the enemy threat dictates the length of training. As security improves, training programs can expand to facilitate achievement of the long-term end state.

Training Foreign Security Forces

2-35. Members of FSF develop through a systematic training program (individual training and education as well as collective training). The program builds basic skills, then teaches them to work as a team, and finally allows them to function as a unit. Basic military, intelligence, or law enforcement training focuses first on basic skills such as legal considerations, first aid, marksmanship, and fire discipline. Leaders have training in tactics, including patrolling, urban operations, and legal evidence collection. Everyone must master rules of engagement and the law of armed conflict. FSF units should train to standard for conducting the major operations they will face.

Training Leaders

2-36. The effectiveness of FSF directly relates to the quality of their leadership. Building effective leaders requires a comprehensive program of officer, staff, and specialized training. The ultimate success of any SFA effort depends on creating viable FSF leaders able to carry on the fight at all levels and build their nation on their own.

2-37. The standards of leader training reinforce different levels of authority within the foreign security force. Clearly established responsibilities for commissioned and noncommissioned officers specify what is expected of recruits and leaders. To ensure civilian control, subordinate relationships to civilian authorities also are reinforced. In addition, training should establish team dynamics. In some cultures, security forces may need training to understand the vital role of members not in primary leadership positions. For example, noncommissioned officers may be a new or different concept for some FSF.

Employing Newly Trained Forces

2-38. SFA must build the morale and confidence of FSF. Committing poorly trained and badly led forces results in high casualties and invites tactical defeats. While defeat in a small operation may have little strategic consequence in a conventional war, often a small tactical defeat of FSF has serious strategic consequences. If the foreign security force fails, the local populace may begin to lose confidence in the host-nation government's ability to protect them. FSF must prepare for operations so they have every possible advantage. As much as possible, FSF should begin with simpler missions. As their confidence and competence grows, these forces can assume more complex assignments. Collaborating with joint or multinational units can help new forces to adjust to combat stress.

2-39. Newly trained units should enter their first combat operation in support of more experienced foreign units. Operational performance of such inexperienced organizations should be carefully monitored and evaluated to correct weaknesses quickly. The employment plan for FSF should allow enough time for additional training after each operation. Gradually introducing units into combat allows the command to identify poor leaders for retraining or other actions. Competent leaders are also identified and given greater authority and responsibility.

EQUIP

2-40. Equip is a SFA task encompassing all efforts to assess and assist FSF with the procurement, fielding, and sustainment of equipment. All equipment must fit the nature of the operational environment. The SFA principle of ensuring long-term sustainment is a vital consideration for the equip task.

2-41. The strategic plan for security force development should outline equipment requirements. Appropriate equipment is affordable and suitable against the threat. Forces must be able to train on the equipment. Interoperability may be desired in some cases. A central consideration includes the host nation's long-term ability to support and maintain the equipment.

2-42. The initial FSF development plan should use a comprehensive approach with local, regional, national, and other appropriate actors. Fundamentally, it should support the internal defense and development strategy. It should include phases with objectives for FSF to meet over three to four years. Since potential enemies adapt rapidly and situations change, commanders must continually assess the direction and progress of developing FSF.

2-43. The requirement to provide equipment may be as simple as assisting with maintenance of existing formations or as extensive as providing everything from shoes to communications and investigation kits. If the enemy uses heavy machine guns and rocket-propelled grenades, then FSF need comparable or better equipment. This especially applies to police forces, which are often lightly armed and vulnerable to well-armed enemies.

2-44. Primary considerations should include maintainability, ease of operation, and long-term sustainment costs. Few developing nations can support highly complex equipment. In most operations, having many versatile vehicles that require simple maintenance is often better than having a few highly capable armored vehicles or combat systems that require extensive maintenance. Effective maintenance systems for FSF often begin with major maintenance performed by contractors. The program then progresses to arrangements with U.S. forces as they train foreign personnel to conduct the support mission.

2-45. Sources of materials for FSF include U.S. foreign military sales (not technically SFA), multinational or third-nation resale of property, contracts with internal suppliers, or purchases on the international market. The organizations responsible for equipping FSF require flexibility to obtain equipment that meets their needs for quality, timeliness, and cost. As part of their training, FSF also need to learn property accountability to reduce corruption and ensure proper equipment usage. FSF provide equipment the same level of control and protection that U.S. forces provide for similar equipment. (See AR 12-1 and DOD 5105.38M.)

REBUILD AND BUILD

2-46. Rebuild and build is a SFA task to assess, rebuild, and build the existing capabilities and capacities of FSF and their supporting infrastructure. This task requires an in-depth analysis of the capability, capacity, and structures required to meet the desired end state and operational environment. Some FSF may require assistance in building and rebuilding, while other FSF may only need assistance in building.

2-47. FSF need infrastructure support. People need buildings for storage, training, and shelter. Often requirements include barracks, ranges, motor pools, and other military facilities. Construction takes time; the host nation needs to invest early in such facilities if they are to be available when needed. Forces must plan to protect any infrastructures, including headquarters facilities, since these building are attractive targets for insurgents. (See FM 3-34.400 for information on hardening measures to increase infrastructure survivability and improve protection.)

2-48. During some operations, such as counterinsurgency, foreign military and police forces often operate from local bases. Building training centers and unit garrisons requires a long-term force-basing plan. If possible, garrisons should include housing for the host-nation soldiers and their families; government-provided healthcare for the families; and other attractive benefits.

2-49. The host nation may need to make large investments in time and resources to restore or create the nationwide infrastructure necessary to effectively command and control FSF. The host nation will build functional regional and national headquarters and ministries as well as local bases and police stations.

ADVISE AND ASSIST

2-50. Advise and assist is a SFA task in which U.S. personnel work with FSF to improve their capability and capacity. Advising establishes a personal and a professional relationship where trust and confidence define how well the advisor will be able to influence the foreign security force. Assisting is providing the required supporting or sustaining capabilities so FSF can meet objectives and the end state. The level of advice and assistance is based on conditions and should continue until FSF can establish required systems or until conditions no longer require it. Leaders cannot permit FSF to fail critically at a point that would undermine the overall effort. See chapter 6 for a discussion on advisors.

ACTIVITIES

2-51. SFA is inherently a developmental effort. Success is measured by the increase in the foreign security force's capability, capacity, competency, commitment, and confidence in areas in which U.S. forces are assisting them and as a whole. SFA activities include plan and resource, generate, employ, transition, and sustain. Figure 2-1 depicts the SFA activities. These activities blend over time based on the capability and capacity of FSF.

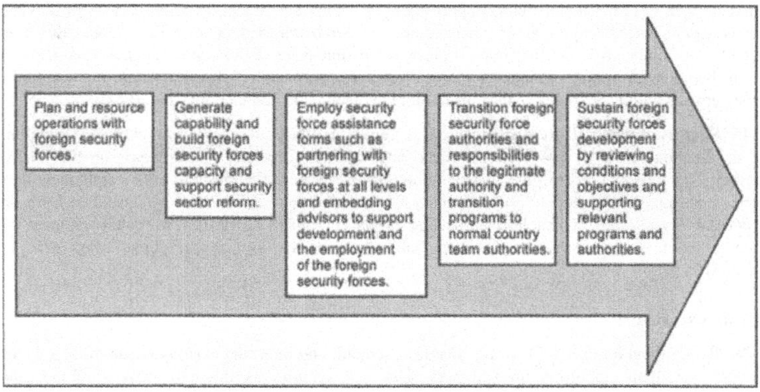

Figure 2-1. Security force assistance activities

PLAN AND RESOURCE

2-52. The plan and resource SFA activity begins as commanders understand the operational environment and determine the requirements of FSF. It also ensures that the United States provides SFA that achieves the objectives and end state of both the host nation and United States. The host nation and United States should then analyze the resource requirements and efforts so that developing FSF have sufficient and appropriate resources. Chapter 4 provides additional detail on planning for facilities and infrastructure.

GENERATE

2-53. The generate SFA activity includes generation of FSF, including the forces, leaders, and warfighting function capabilities (movement and maneuver, intelligence, fires, sustainment, command and control, and protection) based on the desired end state.

Employ

2-54. The employ SFA activity involves FSF transitioning from force generation to mission employment. It does not rely on the maturity of the force or supporting institutions as a whole but is focused on the foreign element in question. Conditions determine when to use an element of FSF. Leaders of U.S. and foreign security forces assess the risk associated with employment and mitigate that risk as much as possible.

Transition

2-55. The transition SFA activity defines the transition between two security forces, when applicable. This could be a transition of authority between U.S. forces to a new foreign security force. Another example could be from a regional foreign security force to a foreign security force with U.S. advisors. It could also be from a host-nation military force to a host-nation police force, with or without U.S. advisors.

Sustain

2-56. The sustain SFA activity occurs when the institutional capacity of the foreign security force has been developed to a point where it is self sustaining. It may continue to have SFA contact through combined exercises, educational opportunity exchange, intelligence sharing, and foreign military sales.

TYPES

2-57. The U.S. or other actors use three types of SFA—augmenting, partnering, and advising—to develop confident, competent, capable, committed, and credible FSF. These types of SFA define the relationship between outside actors and the foreign security force. They may be employed simultaneously, sequentially, or in combination. The progression and types of SFA are determined by the operational environment, the assessment of the FSF, and by resources available. Each of these types requires decidedly different requirements, objectives, and legal considerations.

Advising

2-58. Advising is the primary type of security force assistance. Advising is the use of influence to teach, coach, and advise (see chapter 6) while working by, with, and through FSF. This type of SFA relies on the ability of the advisor to provide relevant and timely advice to FSF. Advising helps FSF conduct independent decisionmaking and operations. Advisors may also provide FSF with direct access to joint and multinational capabilities such as air support, artillery, medical evacuation, and intelligence. However, the advisor and advisor team require proper manning and equipment to perform these secondary support functions while staying focused on advising. Multiple sources will pressure advisors as they perform their missions. For example, nearby maneuver units may have other priorities and may focus on accomplishing their missions with or without FSF.

2-59. Advisor teams require a clearly defined and structured chain of command under which to operate. This is not only for logistics and support, but also keeps the advisor focused on developing FSF. Advisor teams will often find themselves answering to their higher military assistance group, the brigade combat team they are attached to, and the host-nation unit with which they are embedded. A well-defined chain of command alleviates confusion concerning who tasks or monitors the team's progress as well as ensures constant and adequate team sustainment.

2-60. Advisors are not partners; U.S. forces act as partners. Advising and partnering are complementary but inherently different activities. Advising requires relationship building and candid discourse to influence development of a professional security force. Partnering incorporates training with combined operations to achieve the same SFA goals. Advisors perform partnership shaping functions, shape discussions with their counterparts, and create opportunities for the partner units.

Partnering

2-61. Partnering attaches units at various levels to leverage the strengths of both U.S. and foreign security forces. As a foreign security force's capabilities mature, the echelon and degree of partnering decrease. As the foreign security force conducts more autonomous operations, U.S. forces still provide quick reaction forces and other assistance as appropriate.

2-62. Partner units should establish combined cells for intelligence, operations, planning, and sustainment. While effective coordination is always required and initial efforts may require completely fused efforts, FSF should eventually build the capability and capacity to conduct all efforts autonomously. These combined cells have several functions. They support transparent operations and a comprehensive approach. They also enhance the relationships among U.S. and foreign security forces by demonstrating trust. Finally, they develop the capacity of FSF in key staff areas. However, combined cells are not without risk. Operations and plans always at risk of compromise and prudent precautions should be taken.

2-63. Another partnering technique establishes relationships among command and staff elements of U.S. and foreign security forces. It requires less reorganization. It also allows counterparts to understand each other since foreign personnel can observe U.S. personnel performing their duties. This construct works well when the foreign culture is extremely sensitive to maintaining honor. In this case, foreign personnel can observe without having to expose their lack of knowledge or appear subordinate.

2-64. Unit partnerships do not replace advisor roles or functions. If partnering and advising are used in combination, it forms a three-part relationship amongst FSF, advisors, and the partner units. Partner units should look to the advisor to identify, shape, and facilitate operational partnering opportunities and training events. Advisors support U.S., coalition, and partner unit objectives, but, depending on the operational phase, the partner unit may support advisors or advisors may support the partner unit. Therefore, some level of advisor skills training should be included in the partner unit training program if those units will be conducting SFA activities.

Augmenting

2-65. Augmenting is an arrangement where FSF provide individuals or elements to combine with U.S. units, or U.S. individuals or elements combine with FSF. Augmentation improves the interdependence and interoperability of U.S. and foreign security forces. Augmentation can occur at many levels and in many different forms. For example, a U.S. squad can be augmented with host-nation individuals, a U.S. company can be augmented with a host-nation platoon, or a U.S. battalion can be augmented with a company from a foreign security force. Similarly, augmentation can be of short duration for a specific operation or of a longer duration for an enduring mission. Augmenting immerses FSF in a U.S. environment to provide language and cultural awareness to the U.S. unit. U.S. forces can also augment FSF.

Combining Tasks, Activities, and Assessment

2-66. Execution of successful SFA links its activities, tasks, and assessment. Figure 2-2 depicts how to arrange SFA activities, tasks, and assessment in time and purpose. The tasks that occur in one SFA activity will normally continue throughout operations. For example, organize, train, equip, rebuild and build, and advise and assist tasks will continue throughout the SFA activities of employment, transition, and sustain.

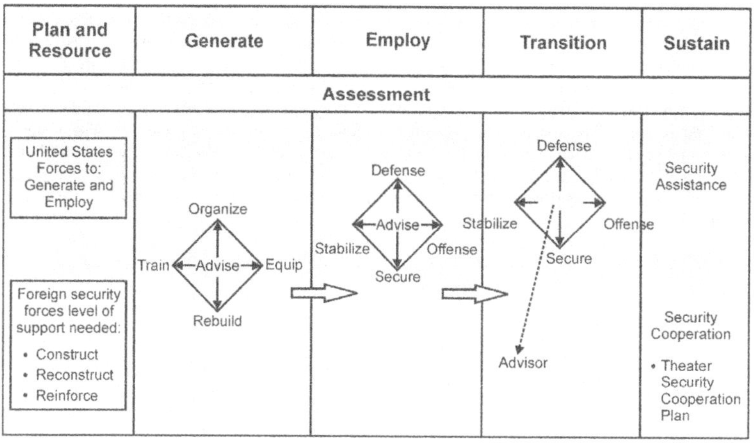

Figure 2-2. Security force assistance tasks, activities, and assessment

Plan and Resource

2-67. The plan and resource SFA activity determines the level of support and number of U.S. forces needed. As all efforts in SFA this activity is part of a comprehensive approach. Planners assess whether they will be constructing a foreign security force from the ground up, reconstructing a foreign security force based on existing capabilities and structure, or merely reinforcing an existing foreign security force. Understanding what capability and capacity the foreign security force should have and what capability and capacity they currently have is vital. Commanders and planners can then determine the size and structure, resources, and funding for the U.S. unit or units that will facilitate or assist the generation of FSF. Chapters 3 and 4 discuss planning and execution considerations for SFA.

Generate

2-68. In the generate SFA activity, U.S. forces focus on organizing, training, equipping, and advising FSF. During generation of FSF, the condition of the existing foreign security force's infrastructure and capabilities determine the assistance needed from U.S. forces. The generate activity adapts to the needs and requirements of the foreign security force. It can be designed to take the individual recruit from initial entry through to an employable unit, or it can augment an existing foreign unit with additional or enhanced capabilities. Like all SFA efforts, generate focuses on bringing a foreign security force's capabilities to a level where it can conduct its mission. With help from advisors in the next phase, FSF can work toward independent and sustained employment. In large SFA efforts, an organization may be required to command and control this activity.

2-69. Once elements of FSF are trained, the focus shifts from organizing, training, equipping, and rebuilding to employing the foreign security force. However, not all FSF units and organizations are ready for employment at the same time. As one unit is employed, others may be part of the generation activity, while still others may be completing the transition activity. FSF may be employed for offensive, defensive, or stability operations while other FSF may be employed against internal or external threats. The primary SFA task for U.S. forces is to advise and assist to facilitate the foreign security force operations.

Transition

2-70. The transition activity focuses on removing the U.S. elements and advisors once FSF are ready to conduct independent operations. All involved actors work in a comprehensive approach to assess when this can be achieved and to coordinate the transition itself. As FSF gradually progress toward the transfer of authority, close relationships forged between the FSF and their partners prove essential to sustainable development and successful transition. Genuine relationships engender trust and confidence, enabling increased responsibility and a well-executed transition process. These relationships also foster a clear understanding of command responsibilities and authorities. Such an understanding ensures FSF approach transition prepared to assume the full weight of their future role in the security sector. Success in developing FSF often depends more on relationships and personalities at the unit level than any other factor.

Chapter 3

Brigade Operations Process

This chapter examines operational considerations for the modular brigade augmented for security force assistance (SFA). It uses battle command and the operations process to examine these considerations.

BATTLE COMMAND

3-1. *Battle command* is the art and science of understanding, visualizing, describing, directing, leading, and assessing forces to impose the commander's will on a hostile, thinking, and adaptive enemy. Battle command applies leadership to translate decisions into actions—by synchronizing forces and warfighting functions in time, space, and purpose—to accomplish missions (FM 3-0). Commanders are central to battle command for brigade combat team (BCT) operations just as they are for any other operation. However, there are key considerations for successful battle command of a BCT conducting SFA. Most importantly, BCT commanders must work with their foreign counterparts, which may require facilitating operations when the foreign security forces (FSF) commander has the lead for the operations process.

UNDERSTAND

3-2. Successful SFA requires that BCT commanders thoroughly understand the operational environment. Conditions of the operational environment affect what SFA commanders provide. Commanders must understand the conditions that make up the current situation, including the relevant population, other actors, and FSF. From this understanding, commanders visualize desired conditions that represent a desired end state. Commanders must understand the capacity and capabilities of foreign security. The host-nation or regional security organization helps to determine desired conditions associated with the foreign security forces. After envisioning a desired end state, commanders then conceptualize how to change the current situation to the desired end state.

3-3. BCT commanders must understand the host-nation's security and political concerns. Security concerns focus on internal and external threats, their capability to affect the operational environment, and the foreign security force's strength and capability to counter those threats. Politically, the considerations associated with the security sector are the legitimacy of the foreign security force, FSF capability and capacity, host-nation or partner organization policy, and the ability to support the host-nation and U.S. Government objectives.

3-4. Commanders use the information obtained from intelligence preparation of the battlefield and intelligence, surveillance, and reconnaissance (ISR) to determine objectives and requirements for assessing and mitigating risk. Risk applies to how well FSF, U.S. Government, host-nation, partner organization, and international organizations can tolerate changes in the operational environment, as well as the challenges and conditions inherent to the operation.

VISUALIZE

3-5. *Commander's visualization* is the mental process of developing situational understanding, determining a desired end state, and envisioning the broad sequence of events by which the force will achieve that end state (FM 3-0). Visualization begins with assignment of the mission, which provides the focus for developing the commander's visualization. However, visualization must be continuous as SFA is dynamic. During planning, commander's visualization provides the basis for developing plans and orders.

3-6. The nature and extent of assistance provided to FSF are determined by this desired end state. SFA is often a slow process that does not lend itself to quick solutions. To ensure long-term success, commanders and host-nation or regional security organizations clarify early what conditions they desire. These conditions include the desired end state for FSF, which often include being—

- **Flexible**. FSF can accomplish broad missions. Flexibility requires an effective command and control organization appropriate for the partner or host-nation. This can include both internal and external threats.
- **Proficient**. The level of proficiency of FSF should include—
  - The capability of working effectively within a comprehensive approach.
  - The capability to integrate their operations with those of multinational partners.
  - Nonmilitary security forces competent in maintaining civil order, enforcing laws, controlling borders, securing key infrastructure, and detaining criminal suspects.
  - Nonmilitary security forces thoroughly trained in modern police ethos and procedures and which understand the basics of investigation, evidence collection, and proper court and legal procedures.
- **Self-sustained**. FSF can manage their own equipment throughout its lifecycle and perform administrative support.
- **Well led**. Leaders at all levels possess sound professional standards and appropriate military values; they are selected and promoted based on competence and merit.
- **Professional**. FSF are—
  - Honest, impartial, and committed to protecting and serving the entire population; they operate under the rule of law and respect human rights.
  - Loyal to the partner organization or host-nation government and serving partner or national interests; they recognize their role as the servants and not masters.
- **Integrated into society**. In the case of a host nation, FSF represent the host nation's major social groups; they are not seen as elements of just one faction.

3-7. These conditions may vary in different countries, but well-trained FSF should—

- Provide reasonable levels of security from external threats without threatening regional security.
- Provide reasonable levels of internal security without infringing on the populace's civil liberties or posing a coup threat to the host-nation government.
- Be founded upon the rule of law, including international principles of the rule of law.
- Be sustainable after joint and multinational forces depart.

3-8. When addressing internal threats, foreign military and police forces often perform unconventional functions. The military may have to fill an internal security role usually reserved for the police, or police may have forces so heavily armed that they would normally be part of the military. In the near term, FSF should—

- Integrate military capabilities with those of local, regional, and national police.
- Maintain the flexibility to transition to more conventional roles of external and internal defense based on long-term requirements.

3-9. To achieve this end state and intermediate objectives, the host nation should develop an internal defense and development (IDAD) strategy (with multinational assistance if required). This strategy should have a section on IDAD that addresses all aspects of SFA.

3-10. U.S. doctrine divides force development into domains: doctrine, organization, training, materiel, leadership and education, personnel, and facilities (DOTMLPF), which may be useful to analyze the IDAD. DOTMLPF elements are tightly linked, simultaneously pursued, and difficult to prioritize. Commanders monitor progress in all domains. Often military forces involved in such programs are tempted to impose their own doctrine and judgment on the host nation. Foreign security force doctrine, like the remaining DOTMLPF domains, must be appropriate to host-nation capabilities and requirements. Paragraphs 3-59 through 3-70 use DOTMLPF as a tool to assess FSF.

DESCRIBE

3-11. After BCT commanders visualize how to conduct SFA, they describe it to their staffs and subordinates to facilitate shared understanding of the mission and commander's intent. Commanders ensure subordinates understand the visualization well enough to begin planning. Commanders express their initial visualization as—

- Initial commander's intent.
- Planning guidance, including an initial concept of operations.
- Commander's critical information requirements.
- Essential elements of friendly information that must be protected.

Commander's Intent

3-12. Like any operation, SFA requires a clear, concise statement of what the force must do and the conditions the force must meet to succeed with respect to the enemy, terrain, and desired end state. In such an operation, the end state should include a description of desired conditions with respect to FSF capability and capacity. The commander's intent links the mission and concept of operations. It describes the key tasks and end state that, along with the mission, form the basis for subordinates' initiative.

3-13. A sample end state could discuss—

- How civilians perceive FSF as credible and legitimate and representing a legitimate authority.
- How FSF have the capability to support stability and security, retain the initiative and freedom of movement, disrupt enemy activities, influence the population, operate with other friendly forces, and sustain themselves.

3-14. Like any operation, how the BCT supports the higher organization's mission is vital. Additionally, the BCT supports FSF. Some sample key tasks are:

- Support the organization of a division of foreign infantry and two battalions of foreign police.
- Train leaders of FSF to create a cadre that can sustain their own training programs.
- Oversee the equipping of FSF and provide new equipment training.
- Support the rebuilding or building of infrastructure required to develop FSF.
- Advise FSF to develop the required capabilities and competencies associated with shared objectives.
- Assist the partner's security forces in meeting their operational requirements.

Planning Guidance

3-15. Commanders develop planning guidance for the staff from the commander's visualization. Planning guidance may be as broad or detailed as circumstances require; however, it must convey to the staff the essence of the commander's visualization. Due to the complex and dynamic nature of FSF, SFA often requires detailed planning guidance. Commanders ensure the staff understands their visualization while still permitting the staff to explore different options. Some considerations are:

- Friendly courses of action (COAs) with differing SFA forms and differing methods on how to augment, partner, or advise.
- Command and control architecture, intelligence architecture, and communications architecture to support dispersed SFA.
- Enemy COAs as they seek to defeat friendly efforts with respect to SFA activities.
- Enemy adaptation given the development of the foreign security force.
- Guidance for security during the conduct of SFA, including the BCT securing its advisor teams. The BCT determines the staging, needs, and ad hoc arrangements of advisor teams.
- ISR guidance to include identifying the focus of effort as it relates to SFA.

Commander's Critical Information Requirements

3-16. Commander's critical information requirements (CCIRs) will be developed for SFA just as for any other operation. These requirements are what commanders need to make decisions or the information they think is important for understanding the operational environment. These CCIRs should be compatible with the FSF commander's information requirements. CCIRs include priority intelligence requirements and friendly force information requirements.

3-17. Priority intelligence requirements are vital to support decisionmaking. During SFA, priority intelligence requirements must be focused on the enemy and the operational environment, not on FSF. U.S. forces do not focus intelligence efforts, such as collection, on the foreign forces they assist.

3-18. Friendly force information requirements provide the information that commanders need to understand the conditions of their own forces, as well as the conditions of the foreign security forces. For example, a BCT commander may direct the staff to better define a foreign security force's critical equipment shortages or maintenance requirements. During execution, commanders may focus on when FSF meet certain objectives or conditions. This information allows the commander and FSF commander to make adjustments, such as changes to the task organization, support requirements, or even the type of SFA.

Essential Elements of Friendly Information

3-19. An *essential element of friendly information* is a critical aspect of a friendly operation that, if known by the enemy, would subsequently compromise, lead to failure, or limit success of the operation, and therefore should be protected from enemy detection (FM 3-0). Essential elements of friendly information should be determined through a comprehensive approach. Information sharing and trust are vital elements of SFA. U.S. commanders must be aware of the inherent risks in this context and employ prudent risk mitigation efforts. Thus, successful SFA requires a balance between sharing information and risk mitigation efforts. Although essential elements of friendly information are not CCIRs, they have the same priority.

Intelligence, Surveillance, and Reconnaissance

3-20. *Intelligence, surveillance, and reconnaissance* is an activity that synchronizes and integrates the planning and operation of sensors, assets, and processing, exploitation, and dissemination systems in direct support of current and future operations. This is an integrated intelligence and operations function. For Army forces, this activity is a combined arms operation that focuses on priority intelligence requirements while answering the commander's critical information requirements (FM 3-0). For the brigade, these efforts will often support FSF. Moreover, the BCT must have means that allow effective intelligence collaboration, such as an intelligence fusion cell with the participation of all appropriate actors.

3-21. ISR efforts focus on the enemy and operational environment within the context of SFA. ISR efforts in SFA are not used on the foreign security force. Rather, U.S. forces assess the foreign security force within a comprehensive approach.

DIRECT

3-22. BCT commanders direct all aspects of operations, although a comprehensive approach, including FSF and other appropriate actors is fundamental to successful SFA. The commander's direction takes different forms during planning, preparation, and execution. Commanders make decisions and direct actions based on their situational understanding, which they maintain by continuous assessment and coordination with FSF. SFA is directed by—

- Preparing and approving plans and orders with FSF.
- Assigning and adjusting missions, tasks, task organization, and control.
- Positioning units to maximize their support to FSF, anticipate threat actions, or create or preserve options.
- Positioning key leaders to ensure observation and supervision of FSF at critical times and places.

- Adjusting SFA and other support priorities and allocating resources based on FSF priorities, as well as opportunities and threats.
- Changing support arrangements, either internally or with respect to FSF.

ASSESS AND LEAD

3-23. Effective battle command requires commanders to continuously assess and lead. Assessment helps commanders to better understand current conditions and determine how the operation is progressing and how best to support FSF. Based on an assessment that is the product of a comprehensive approach, commanders modify plans and orders to better accomplish the mission. This may include changing the type or combination of SFA, support, sustainment, or task organization. If their assessment reveals a significant variance from the original shared commander's visualization, a comprehensive approach must be used to reframe the visualization and develop a new plan.

3-24. Commanders in a BCT conducting SFA lead by example and presence, as Army commanders do in any operation. However, commanders have the additional challenge of leading by, with, and through FSF. Command posts for BCTs conducting SFA have similar challenges in that they should be integrated headquarters as much as practical.

OPERATIONS PROCESS

3-25. As they would in any other mission, BCT commanders conducting SFA balance their time and the staff's time and resources among four major activities in a continuous learning and adaptive cycle called the operations process. The *operations process* consists of the major command and control activities performed during operations: planning, preparing, executing, and continuously assessing the operation. The commander drives the operations process (FM 3-0). While simple in concept, the operations process is complex in execution. The operations process activities may be sequential or simultaneous.

3-26. Figure 3-1 depicts the operations process activities as a cycle that is driven by battle command. This cyclic process applies equally to a BCT conducting major combat operational theme or operations within an irregular warfare operational theme. During the operations process involving SFA, commanders consider the presence and inclusion of other actors, including FSF. Policy, the operational environment, and the forms of SFA determine how U.S. forces, FSF, and other actors interact in the operations process.

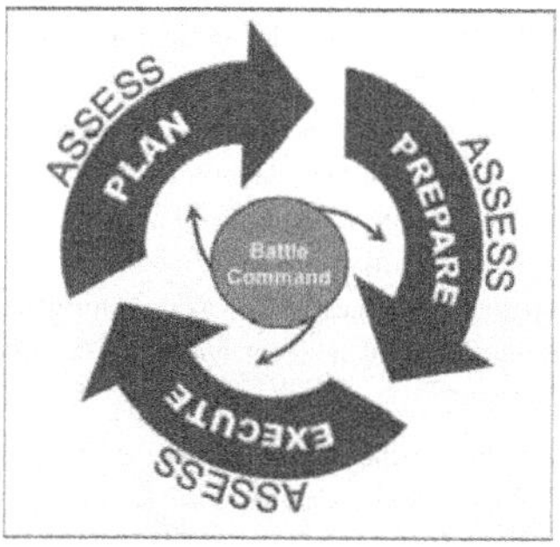

Figure 3-1. The operations process

PLAN

3-27. Planning for SFA, like any other operation, begins either with the anticipation of a new mission or the receipt of mission as part of the military decisionmaking process. *Planning* is the process by which commanders (and the staff, if available) translate the commander's visualization into a specific course of action for preparation and execution, focusing on the expected results (FM 3-0). SFA requires a comprehensive approach to ensure the U.S. commander's visualization synchronizes with policy and the foreign commander's visualization. SFA is the major requirement for a BCT conducting SFA, but the mission may require the brigade to perform other tasks.

3-28. Continuous and open to change, planning for SFA includes identifying how to best assist the FSF and developing a sequence of actions to change the situation. This planning must occur within a comprehensive approach that includes FSF, U.S. forces, and other actors. Similarly, this plan must be nested within policy, IDAD strategy, the campaign plan, and any other higher-echelon plans. It also involves anticipating consequences of actions and developing ways to mitigate them.

3-29. The plan, which includes the commander's intent, provides a common understanding to U.S. and foreign security forces on the actions to take. Plans and orders also provide decision points and branches that anticipate options that enable the force to adapt as the operation unfolds. This is especially important for SFA, as these operations tend to be prolonged efforts. BCTs conducting SFA often rotate before achieving all objectives. As a result, the planning team should establish objectives and milestones that can be achieved during the brigade's mission. These BCT objectives and milestones must support higher-echelon plans, including the campaign plan and IDAD strategy.

Functional Requirements

3-30. The composition of advisor teams is subject to objectives and conditions. For example, security conditions may require three armored vehicles, a crew-served weapon, and nine personnel to conduct independent movement. Another example involves the difference in the number of advisors. Newly generated FSF may require many advisors whereas experienced FSF may only require assistance in a few warfighting functions and consequently need fewer advisors. Understanding the functional requirements provides flexibility to the units assisting FSF. Three examples of determining functional requirements based on the understanding the operational environment include the BCT, a partnering or augmenting unit, and an advisor team.

Brigade Combat Team

3-31. A BCT headquarters has specific functions and requirements. These can include, but are not limited to—

- The functions of—
  - Carrying out SFA tasks and assisting FSF in support of objectives and the end state.
  - Facilitating partnerships between U.S. and foreign security forces.
  - Requesting, coordinating, and synchronizing resources.
  - Providing command and control of subordinate units and assigned advisor teams.
  - Coordinating with other commands and adjacent units.
  - Coordinating with other participants integrated within the comprehensive approach.
  - Developing, implementing, and synchronizing information engagement in assigned areas.
  - Developing relationships and an understanding with appropriate actors involved in SFA.

3-32. The requirements of—

- Having enough staff to handle the communications traffic generated by conditions.
- Being able to analyze information to aid commander and subordinate elements.
- Being able to articulate resource requirements in support of subordinate elements and efforts to conduct SFA tasks.
- Being organized and equipped to conduct operations and support and sustain FSF.
- Including regionally oriented host-nation officers to provide regional expertise.

Partnering or Augmenting Unit

3-33. Functions and requirements depend on the degree and level of partnering or augmenting. No prescriptive way to partner or augment exists; it is based on specific conditions and subject to the needs of policy, higher headquarters guidance, and needs of the FSF. A BCT may designate one battalion to partner or augment while designating other battalions to advise. A partnering or augmenting unit conducting SFA has specific functions and requirements. These can include, but are not limited to—

- The functions of—
  - Partnering with foreign security under the BCT level.
  - Increasing capability, capacity, competency, confidence, and commitment of FSF by conducting combined tactical operations.
  - Providing feedback on performance of FSF and certain shortfalls to embedded advisors and BCT headquarters.
  - Conducting sustainment and medical training with FSF at their home station or on operations.
  - Reporting conditions in the operational environment.
- The requirements of—
  - Having U.S. forces that are capable, competent, committed, and confident.
  - Being interoperable with FSF.
  - Being available for an appropriate period.

Military Advisor Team

3-34. The functions and requirements of a military advisor team are based on conditions—requirements, availability, and footprint. A military advisor team conducting SFA has specific functions and requirements. These can include, but are not limited to—

- The functions of—
  - Under the BCT headquarters, advising, training, and assisting FSF to which it is assigned.
  - Increasing capability, capacity, competency, confidence, and commitment of FSF by providing advice and support during battalion level and higher operations.
  - Assessing partner leaders, staff, and certain shortfalls to BCT headquarters.
  - Conducting sustainment training with FSF at their home or on operations.
  - Reporting on conditions in the operational environment.
- The requirements of—
  - Being manned with trained personnel who are properly equipped so as to be capable of executing their mission in current conditions.
  - Providing personnel with rank, education, and experience corresponding to the echelon and type of unit being advised.

Support and Sustainment Requirements

3-35. Support and sustainment requirements play a vital role in COA development (see chapter 6). The BCT may operate across a broad geographical area when conducting or supporting SFA. Partnering may be on certain levels with FSF and may support numerous embedded advisor teams or other actors. Since FSF often have varied relationships with higher echelons and institutions, the BCT may be involved vertically and spatially. These considerations may require more support and sustainment requirements than an initial analysis might indicate. Support and sustainment considerations include, but are not limited to—

- Distribution of FSF to which U.S. forces are providing SFA.
- The capability of FSF to provide support, sustainment, and medical functions.
- The technical capability and interoperability of U.S. and foreign security forces.
- The infrastructure that supports SFA.
- Overall security conditions where SFA occurs.

- The available contracting support.
- The ability of the environment to accept and absorb the U.S. forces' footprint.

Determining a Course of Action

3-36. COAs for SFA require an understanding of the breadth and width of the security sector. They may have to account for the roles ranging from the individual soldier or policeman to the ministerial level. COA development examines the roles other actors, for example, a gendarme or paramilitary force, in the security sector. Because the United States does not have a gendarme, U.S. forces may not fully appreciate its importance.

3-37. COAs for SFA also need to be flexible and inclusive. Commanders consider the following when examining a partner's security sector:

- The status and challenges of the host-nation security sector or regional security organization.
- The composition and disposition of the partner's security sector.
- Supporting institutions.
- Links among ministries, institutions, and components of the security sector.
- Partner capability and capacity shortfalls and impediments to their development.
- Environmental requirements to enable the building of security capability and capacity.
- Functional requirements to assist building capability and capacity.
- The impact of other actors within the comprehensive approach.

3-38. Once planners analyze the security sector, they can develop COAs that best combine SFA tasks, activities, and types (see chapter 2). While the advising type of SFA can be of any duration, partnering and augmenting tend to be enduring if used.

3-39. All SFA types tend to be more effective if they are long-term efforts. Advising fosters a more personal relationship and infers a stronger commitment to see the partner succeed. Partnering and augmenting include U.S. and foreign security forces conducting operations together. There are levels of partnering. Different advisory relationships use advisor teams embedded within various partner echelons and units, use one advisor team to advise multiple entities, or have a senior leader visit a partner unit to give advice.

3-40. Using any form or combination of forms depends on conditions (such as the level of competency), capability of FSF, and resources available. Policy, guidance, or formal agreements can also help commanders determine which form to use. Historically, a combination of forms appears to be the most effective. However, combining forms often proves more resource intensive and affected by time, geography, and the number of forces. COAs should remain flexible with respect to the form of SFA. Some example changes in conditions are the increased capability of one element in the security sector over another, the desire to increase one element over another based on partner objectives, or the increased targeting of one element over another by the enemy.

3-41. If a desired COA is not feasible given the scope of the security sector or due to other considerations, it may be useful to sequence SFA forms with portions of FSF. For example, a COA that affects smaller portions of the FSF, but in the right area and at the right time, may do more for achieving an objective or end state than a COA that attempts to support all FSF at one time.

3-42. BCTs are task-organized and augmented to accomplish a specific mission. How BCTs are specifically task-organized and augmented is based on the phase and types of SFA that policy and higher headquarters have directed. Chapter 4 describes augmentation in detail.

3-43. Attached units or units under operational control are task-organized under different headquarters as necessary to accomplish assigned and implied SFA missions. Deciding to task-organize specific formations for SFA may reflect previously approved contingency plans or result from current planning efforts. As these missions end, units return to their parent headquarters or are further task-organized by the controlling headquarters.

3-44. As with other operations, gaining commanders task-organize for SFA by designating operational control, attached, or support relationships to another unit or headquarters. Within each BCT, the brigade commander task-organizes all organic, assigned, attached, or operational control units to best suit the operational environment and types of SFA. Commanders may choose to use a support relationship (direct support, reinforcing, or general support reinforcing) among units of the BCT instead of a command relationship. This can also apply to relationships with foreign units.

3-45. SFA is often highly dispersed, as BCTs may have individuals or units assisting FSF over a wide area. This dispersion increases communications and sustainment requirements for the BCT. The nature of SFA also requires BCTs to help FSF meet requests for assets. To do so, the BCT and other units may need to task-organize to provide tactical support or sustainment. Dispersion and the nature of SFA often necessitate a decentralized leadership approach to facilitate effective and rapid support of FSF.

3-46. BCT commanders use the three types of SFA—advising, partnering, or augmenting—to accomplish the mission. The exact nature of a unit's assigned mission, the operational variables, and the mission variables drive modifications to headquarters and maneuver units. For example, a commander may need a unit tasked with advising, a unit tasked with partnering, and a unit tasked with augmenting. In this case, the unit tasked to advise would provide teams for the appropriate FSF while other foreign units continue their traditional missions. The unit that is partnering may provide the host-nation staff to assist the FSF headquarters. The third unit may provide squads to augment foreign platoons. Units may request additional support, sustainment, and medical support. However, BCTs conducting SFA must retain a reserve to conduct full spectrum operations.

3-47. Based on the overall mission analysis, including an analysis of functions and requirements, commanders and staffs begin to determine how to task organize. Due to the level of detail that may be involved, task organization should begin early in the military decisionmaking process, especially during COA development. No standard task organization exists since conditions vary. For example, a squad of infantry and three additional leaders may work as an advisor team for one mission, where another mission may require an advisor team with specialists for different warfighting functions. The art in SFA involves determining where, when, and if to allocate resources; it involves either breaking up or simply moving personnel and units from one foreign unit to another.

3-48. The techniques used by BCT commanders and staffs to allocate other resources and support are just as valid in SFA as in other types of operations. For example, if it is determined that the most important effort is advising, then the BCT commander may weight that effort with additional resources, such as communications equipment or experienced personnel.

3-49. BCTs emphasize unity of effort with FSF. Ideally, interoperability and shared resources extend to the lowest level. This means having one tactical operations center, dining facility, clinic, motor pool, and so on. Not only does this use resources more efficiently, but it also facilitates understanding, trust, and rapport between U.S. and foreign security forces. For Soldiers to effectively conduct SFA work by, with, and through FSF requires a close rapport and high level of trust.

3-50. Part of COA development for SFA includes an analysis of where to focus the brigade's subordinate units. This analysis includes determining which subordinate units support key elements of FSF. The brigade's subordinate units may support provincial reconstruction teams or transition teams that are not organic to the brigade. Another method is for the brigade may provide organic forces to make up transition teams or the military portion of provincial reconstruction teams or transition teams. Transition teams may include, but are not limited to police, border, point of entry, or military transition teams. Figure 3-2 depicts an example for a brigade providing forces as part of a provincial reconstruction team as well as providing the other transition teams. Figure 3-2 is an example of how a brigade may focus its subordinate units. Additionally, the teams are only examples. All of these teams may not be necessary for all brigades, and some brigades may require other types of advisor teams.

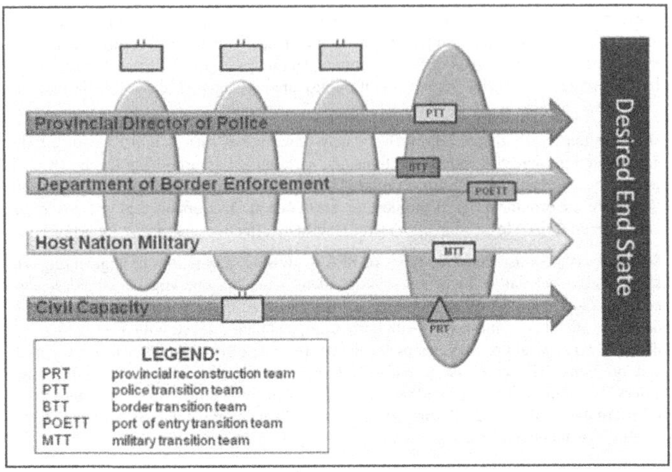

Figure 3-2. Example of brigade focus for subordinate units

PREPARE

3-51. *Preparation* consists of activities performed by units to improve their ability to execute an operation. Preparation includes, but is not limited to, plan refinement; rehearsals; intelligence, surveillance, and reconnaissance; coordination; inspections; and movement (FM 3-0). Like any operation, preparation for SFA creates conditions that improve friendly forces' opportunities for success. It facilitates and sustains transitions, including those to branches and sequels, which are of vital importance for the often dynamic operational environment for SFA. Chapter 5 covers preparation for units conducting SFA, including the BCT.

EXECUTE

3-52. *Execution* is putting a plan into action by applying combat power to accomplish the mission and using situational understanding to assess progress and make execution and adjustment decisions (FM 3-0). Execution of SFA focuses on supporting the needs of the FSF and, like many other operations, is often characterized by phases and transitions between phases. Chapter 4 examines BCT phases for SFA.

Acceptance of Prudent Risk

3-53. SFA, like any other operation, requires prudent, disciplined risk-taking focused on winning rather than preventing defeat. Because uncertainty exists in all operations, every decision involves risk. Among key elements of the art of command are deciding how much risk to accept and minimizing the effects of accepted risk (see FM 6-0). Due to SFA's inherent collaborative nature, determining prudent risk is complex and requires a comprehensive approach.

3-54. Risk reduction measures identified in planning add to the plan's flexibility during execution. A flexible plan can mitigate risk by partially compensating for a lack of intelligence. Again, SFA requires a thorough comprehensive approach to analyzing and agreeing upon risk reduction measures.

Execution Keys to Success

3-55. History and recent operations point to several keys to success for BCTs executing SFA:

- Communication is vital, especially when the BCT shares its operational area with other entities that have cultural differences and lack of communication. It can include FSF or multinational forces, as well as U.S. forces with another mission. For example, an advisor team can develop effective relationships with its foreign security force and local authorities only to have another unit conduct an operation that unhinges those relationships.
- BCTs establish close and continuing relationships with all advisor teams, other actors operating in their operational area, and foreign area officers with local or regional expertise. In addition to advisor teams assigned to military units in direct contact with the BCT, advisors may embed with local police, border, and other forces.
- BCTs establish close and continuing relationships with all foreign units (military, police, and others) operating in their operational areas.
- BCTs establish close and continuing relationships with all political entities and actors in their operational areas.
- Before each operation, as part of the military decisionmaking process, the BCT considers second- and third-order effects of the operation. The team coordinates the commander's intent with advisor teams of FSF and other units in the operational area.
- Unit leaders develop cultural awareness and use this awareness so that operations and relationships achieve the desired end state. In short, the closer a maneuver commander works with advisor teams and the more they interact with local political and cultural leaders, the better the overall chances of mission success.

ASSESS

3-56. The operations process and effective battle command require commanders to continuously assess and lead. Assessment helps commanders to better understand current situations and how SFA is progressing. *Assessment* is the continuous monitoring and evaluation of the current situation, particularly the enemy, and progress of an operation (FM 3-0). Assessment is conducted during planning, preparation, and execution and is evaluated against measures of effectiveness and performances to make decisions and adjustments (see FM 6-0).

3-57. Assessment of SFA requires a comprehensive approach that analyzes the overall operational environment and focuses on FSF and their impact. SFA assessment is performed in close collaboration with the U.S. country team, the host-nation government, and other multinational partners. These partners continually assess the security situation and its influence on other operations. From the assessment, planners develop short-, mid-, and long-range plans. As circumstances change, so do the plans. An insurgency might require employing host-nation forces at earlier stages of development. Some existing security forces may be so dysfunctional or corrupt that the organizations ideally should be disbanded rather than rehabilitated. In some cases, commanders will need to be replaced before their units will become functional. Political realities, however, may make both prospects undesirable. Minimizing and marginalizing particularly corrupt units or leaders while focusing on other units can be effective in this case.

3-58. While every situation differs, BCT leaders assess the following factors throughout planning, preparation, and execution of the operation:

- The operational environment.
- The civil considerations of the operational environment (see appendix B of FM 6-0).
- The core grievances and the threat (possibly including the prerequisites of insurgency).
- FSF elements.
- Methods, successes, and failures of FSF security efforts.
- State of training at all levels, and the specialties and education of leaders.

- Equipment and priority placed on maintenance.
- Logistic and support structure and their ability to meet the force's requirements.
- Extent of acceptance of ethnic and religious minorities.
- Laws and regulations governing the security forces and their relationships to national leaders.

Initial Assessment

3-59. An initial comprehensive assessment offers the BCT a baseline for establishing the scope of effort required. This assessment includes a troop-to-task analysis that determines the type and size of forces needed. FSF may require complete reestablishment or only assistance to increase capacity. They may completely lack a capability (such as internal affairs, corrections, or formal schools for leaders) or they may require temporary reinforcement. As with other military operations, efforts to assist FSF should reinforce success, not failure. For example, instead of building new police stations in every town, improve the good stations and use them as models for other organizations.

3-60. BCT leaders need to make decisions on what shortfalls to address first. The extent of the threats combined with resource limitations inevitably forces commanders to set priorities. However, making efforts to legitimize the host-nation leaders by encouraging them to make the decisions often proves more successful.

3-61. A key effort for SFA is the initial assessment of the DOTMLPF and policy capabilities of a foreign security force. During the early phases of a campaign, a geographic combatant command receives an initial assessment or a unit operating in the area has an assessment of FSF. The incoming BCT updates this assessment as it begins SFA.

3-62. Assessments require objective and subjective reporting, which may require a classification level. At a minimum, a BCT should assess—

- Doctrine.
- Organization.
- Training.
- Materiel and equipment.
- Leadership.
- Personnel.
- Command, control, communications, and intelligence.
- Operational effectiveness.

3-63. The doctrine assessment provides the BCT with an analysis on how the FSF operate. This assessment could focus on formal doctrine or more localized practices.

3-64. The organization portion of the overall assessment provides a perspective on how the FSF are organized. This includes the array of foreign institutions and units.

3-65. The level of training for FSF is a vital consideration. While this assessment is partly subjective, it should objectively examine individual and collective training strategies and mission-essential task lists. In the absence of mission-essential task lists, the BCT should compare FSF to U.S. equivalents.

3-66. The materiel and equipment assessment includes, but is not limited to, weapons, vehicles, equipment, and communications. It should be divided into equipment on hand, equipment readiness, and sustainment. Equipment on hand is quantitative and based on the type of equipment available to security operations. Equipment readiness reflects the operational readiness condition or serviceability of on-hand equipment. Sustainment readiness provides a qualitative and quantitative assessment that reflects how well FSF perform supply, maintenance, transportation, and medical functions. This includes facilities, systems, and structures.

3-67. The leadership assessment analyzes FSF commanders, officers, noncommissioned officers, and other leaders. This includes an assessment of staff members.

3-68. The foreign personnel assessment is quantitative and examines manning, staff manning, and administrative capabilities. It also examines officers, noncommissioned officers, and other personnel.

3-69. The command, control, communications, and intelligence assessment is qualitative and should reflect how well FSF conduct these functions.

3-70. The operational effectiveness assessment examines the foreign security force's ability to execute operations. This portion of the assessment should not be confused with the examination of training, as relatively high levels of training do not necessarily mean that FSF will perform in operations. This assessment is highly subjective.

Assessment During Operations

3-71. Assessment for the BCT conducting SFA should be part of a comprehensive approach. Assessment is directly tied to the commander's decisions throughout planning, preparation, and execution. Critical to effective assessment is developing criteria to evaluate progress toward task accomplishment, achievement of objectives, and attainment of the end state conditions. Assessment criteria are expressed as measures of effectiveness or measures of performance.

3-72. During operations, BCT and FSF commanders and staffs periodically assess the underlying framework of the plan itself. This involves reexamining the logic and assumptions used to develop the original plan, and it requires a deliberately conceived effort to learn as a part of the operation. BCT and FSF collaboration among higher, lower, and adjacent commanders, staffs, and other actors, backed up by qualitative and quantitative assessments, contributes to this learning. Based on this reexamination, commanders update their visualization, direct changes to the order, or develop an entirely new plan to adapt the force to better accomplish the mission.

Chapter 4

The Modular Brigade Augmented for Security Force Assistance

This chapter examines the typical phases for the brigade combat team conducting security force assistance, including a discussion of baseline augmentation by phase. The final section examines some of the subordinate units and elements of the modular brigade augmented for security force assistance.

THE BRIGADE COMBAT TEAM

4-1. The brigade combat team (BCT) is the cornerstone for Army modularity and is designed to operate at the tactical level across the spectrum of conflict. The modular brigade can be augmented, based on the requirements of the operational environment, with enabling assets and capabilities to support distributed security force assistance (SFA). However, the BCT augmented for SFA retains the capability to conduct full spectrum operations—offense, defense, and stability. Any of the three modular BCTs—heavy, infantry, or Stryker—can support SFA. Battalion-sized maneuver, fires, reconnaissance, and sustainment subordinate elements of the BCT support foreign security forces (FSF).

4-2. The gaining theater army force tailors all modular Army forces for use in major combat operations and contingencies. The theater army commander, working for the geographic combatant commander, determines the mix of forces and capabilities required for the BCT. This force tailoring can be done as part of the joint deliberate planning process or as a result of crisis action planning. Force tailoring includes providing BCTs with additional forces, personnel, or capabilities, including the possibility of an embedded provincial reconstruction team. Additional assets and capabilities can include, but are not limited to, command and control, communications, sustainment, engineering, military police, and intelligence.

4-3. In recent operations, the BCT has proven its versatility and capability to learn at the individual and collective levels. It supported other U.S. units conducting SFA and directly increased the effectiveness of FSF. The BCT has operated under varying levels of enemy activity, host-nation capabilities, objectives, and other conditions.

4-4. Because the BCT can operate in nonpermissive and permissive environments, it can conduct SFA across the spectrum of conflict. Thus, it can support civilian and military joint and multinational actors. The support can range from movement security and sustainment to augmentation of reconstruction teams and security for elections.

4-5. BCTs conducting SFA may support FSF development, assist FSF operations, and support and assist the development of host-nation institutions and infrastructure. While providing this support to the host nation, the BCT remains capable of conducting full spectrum operations independently.

BRIGADE COMBAT TEAM PHASES

4-6. If a force lacks the means to accomplish its mission in a single effort, then commanders normally phase the operation. A *phase* is a planning and execution tool used to divide an operation in duration or activity. A change in phase usually involves a change of mission, task organization, or rules of engagement. Phasing helps in planning and controlling and may be indicated by time, distance, terrain, or an event (FM 3-0). Phasing SFA assists in planning and controlling these dispersed and complicated operations. These phases often involve key changes in supporting SFA tasks and types employed.

4-7. BCT phases for SFA are initial response, transformation, and fostering sustainability. These phases mirror the stability operations framework described in FM 3-07 and are based on the operational environment. SFA can start in any phase or may even move to a previous phase due to changes in the conditions of the operational environment.

4-8. Figure 4-1 illustrates an SFA operation that includes all three SFA phases. The horizontal axis depicts an improving security situation, and the vertical axis depicts security capability. The example depicts an ideal situation, although this may not be the case in actual execution. Issues that arise require proper assessment and potentially the use of branches or sequels.

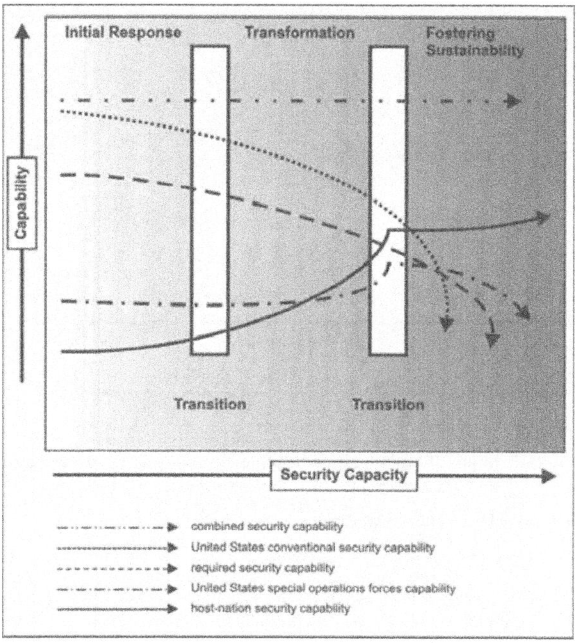

Figure 4-1. Example of phases of security force assistance

4-9. In figure 4-1, the combined security capability represents the overall security capacity of all security forces involved; it is the sum of U.S. conventional security, host-nation security, and the U.S. special operations forces security capabilities. This example assumes that the combined security capability exceeds the required security capability to effectively maintain a secure environment and counter any threat. The U.S. conventional security capability refers to tactical U.S. elements, such as the BCT. It does not refer to higher tactical or operational headquarters or support, sustainment, or medical assets. The U.S. special operations forces capability is depicted as declining in the end of the sustainment phase; as the FSF capability improves, the need for special operations forces declines. The host-nation or regional security organization and U.S. policy determine the end state for special operations.

4-10. For figure 4-1, a major change to enemy resources may require more FSF or additional. Drastic changes to the situation may even require a new operation plan. The figure focuses on capability, not forces, so host-nation security capability may increase without increasing the number or size of host-nation forces.

4-11. Figure 4-2 illustrates how the area of operations for a modular brigade augmented for SFA may change throughout the SFA phases. The differences between the initial response phase and the transformation phase may not change on the surface, but the BCT's relationship with FSF changes drastically. For example, FSF in the north met required conditions so that one of the U.S. brigades conducting SFA was no longer required. The responsibility of the other U.S. brigade thus expanded from providing assistance to one division to two divisions. This brigade's span of control and area of operations similarly expanded. The latter stage of the transformation phase can differ greatly. Areas of operation generally increase in size as they provide SFA to more FSF.

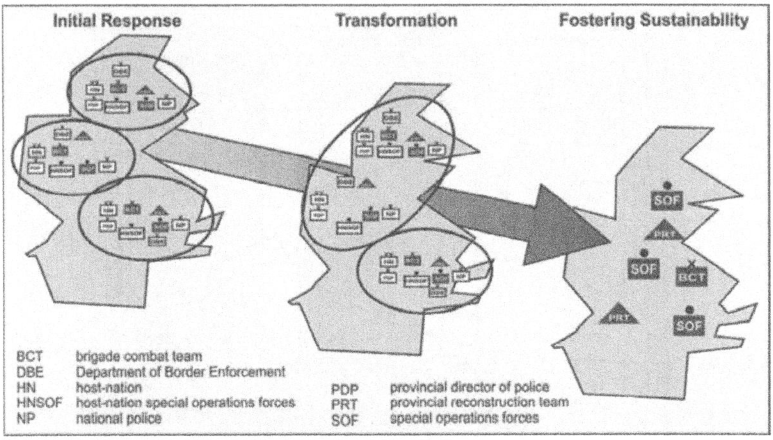

Figure 4-2. Example of changes to a brigade combat team's area of operations

AUGMENTATION BY PHASE SUMMARY

4-12. BCT augmentation is based on the operational environment and mission. As the three phases are based on the operational environment, they provide a baseline for augmentation. Potential augmentation may require military police, legal, public affairs, civil affairs, psychological operations, engineering, sociocultural experts, sustainment, and military transition team personnel. Table 4-1 illustrates a baseline for augmentation by phase for BCTs.

4-13. An embedded provincial reconstruction team is a key element of a BCT conducting security for assistance. The BCT has the lead on the following tasks: establish civil security, establish civil control (when approved by Congress), and develop and enable foreign security forces. An embedded provincial reconstruction team has the lead for the following tasks: support to economic and infrastructure development, restore essential services, and support to governance. Thus, the BCT and an embedded provincial reconstruction team are able to effectively support FSF and execute all five stability tasks.

Table 4-1. Baseline augmentation for modular brigade

CAPABILITY	INITIAL PHASE (1 HN DIV)	TRANSFORMATION PHASE (2 HN DIV)	TRANSFORMATION PHASE (3 HN DIV)
Police Provincial District	171 - MP Company 2 - MP Team (LTC/SGM) 3 - MP MAJ	171 - MP Company 4 – 2 x MP Team (LTC/SGM) 6 - MP MAJ	171 - MP Company 6 – 3 x MP Team (LTC/SGM) 9 - MP MAJ
Legal HN BDE HN DIV	2 - JAG Team (MAJ/SFC) 2 - JAG Team (CPT/SSG)	4 – 2 x JAG Team (MAJ/SFC) 4 – 2 x JAG Team (CPT/SSG)	6 – 3 x JAG Team (MAJ/SFC) 6 – 3 x JAG Team (CPT/SSG)
Public Affairs	2 - PA (Officer/NCO)	2 - PA (Officer/NCO)	2 - PA (Officer/NCO)
Civil Affairs & Psychological Operations	32 - CA Company = 5 CA Teams 70 – Tactical PSYOP Company = 9 TPTs	32 - CA Company = 5 CA Teams 70 – Tactical PSYOP Company = 9 TPTs	197 - CA Battalion (4 CA Companies) = 20 CA Teams 70 – Tactical PSYOP Company = 9 TPTs
Engineer	8 - FEST-A Team 2 - EOD (CPT/SFC)	8 - FEST-A Team 2 - EOD (CPT/SFC)	8 - FEST-A Team 2 - EOD (CPT/SFC)
Sociocultural Experts	8	8	8
Military Transition Team	56- MTT Augment	4 - MTT Augment	8 - MTT Augment
TOTAL	358	315	493
EPRT	~15-20	~15-20	~15-20
Sustainment Truck Squad Supply Platoon CCT	 21 personnel 32 personnel 4 personnel	 4 personnel	 4 personnel
FMST	4 personnel	4 personnel	4 personnel

CA	civil affairs	LTC	lieutenant colonel
CPT	captain	MAJ	major
CCT	contingency contracting team	MP	military police
DIV	division	MTT	military transition team
JAG	Judge Advocate General	NCO	noncommissioned officer
EOD	explosive ordnance disposal	PA	public affairs
EPRT	embedded provincial reconstruction team	SFC	sergeant first class
FEST-A	forward engineering support team – augmenting	PSYOP	psychological operations team
FMST	financial management support team	SGM	sergeant major
HN	host-nation	SSG	staff sergeant
HTT	human terrain team	TPT	tactical psychological operations team

Initial Response Phase

4-14. The initial response phase seeks to assist FSF stabilize the operational environment in a crisis state. This phase occurs during or immediately after a conflict where the operational environment prevents civilian personnel from operating effectively. The operational environment is typified as nonpermissive.

Thus, the objective of this phase is to improve the security situation, reducing the threat to the populace and creating the conditions that allow civilian personnel to safely operate.

4-15. SFA in the initial response phase is normally required when FSF lack the capability or capacity to provide the required level of security. This phase often requires SFA efforts to help generate and train or assist new and existing FSF. This phase may require a combination of the types of SFA and considerable support, sustainment, and medical resources. BCT activities during the initial response may have to be conducted with multinational major combat operations, to include providing a safe, secure environment for the local populace.

4-16. The BCT's SFA efforts during this phase focus on improving the foreign security force's capability and capacity so all security forces—U.S., other, and foreign security forces—provide a secure environment and reduce the threat. Figure 4-3 illustrates the BCT's span of control. The figure also depicts three BCTs' operational areas in the initial response phase. As security conditions improve, transition to the transformation phase begins.

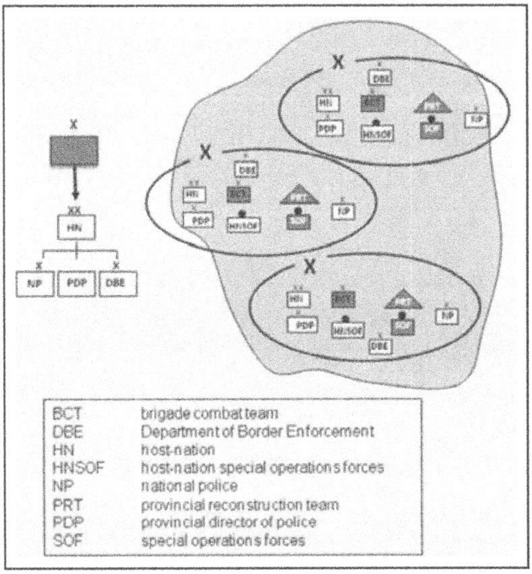

Figure 4-3. Example of initial response span of control and area of operations

4-17. The BCT requires augmentation for the initial response phase. These embedded units provide key enabling capabilities for the brigade, supported FSF, and other supported actors. The embedded provincial reconstruction team is one of the most important of the brigade's augmentation, as it provides interim civilian expertise in developing local institutions to take the lead in national governance, the provision of basic services, fostering economic development, and enforcement of the rule of law. Column two of table 4-1 summarizes augmentation during the initial phase. Although each brigade's mission and conditions in its area of operations are unique, the baseline augmentation for the modular brigade operating in this phase is as follows:

- Embedded provincial reconstruction team.
- Forward engineering support team-augmenting.
- Explosive ordnance disposal team.
- Sociocultural experts.

- Military police company (provincial police transition team).
- Military police team (district police transition team).
- Three military police majors (district police transition team).
- Two judge advocate general teams.
- Civil affairs company.
- Tactical psychological operations company.
- Public affairs team.
- Contingency contracting team.
- Financial management support team.
- Military transition team (specific augmentation personnel include two lieutenant colonels, thirty-one majors, and twenty-three master sergeants).

TRANSFORMATION PHASE

4-18. In the transformation phase, the BCT seeks to assist FSF to stabilize the operational environment in a crisis or vulnerable state. The operational environment in this phase is more permissive than the initial response phase; however, military forces will often be required to provide security to some actors. Activities in this phase normally include a broad range of post-conflict reconstruction, stabilization, and capacity-building efforts, which the embedded provincial reconstruction team is essential for long-term success. Objectives in this phase include continuing efforts to improve the security situation, reducing the threat to the populace, building host-nation capacity across the stability sectors, and facilitating the comprehensive approach to assist FSF.

4-19. The transformation phase represents a broad range of SFA activities that the BCT could perform to support FSF. The initial response phase differs from the transformation phase in the foreign security force's capability to provide for a safe and secure environment. More specifically, FSF have a level of proficiency so they no longer need a permanent U.S. and foreign security forces relationship for tactical operations. However, they may still need full-time advisors and support, sustainment, and medical assistance. The embedded provincial reconstruction team will continue to play a vital role in assisting governance and development efforts throughout this phase.

4-20. Figure 4-4 illustrates the BCT conducting SFA in the transformation phase. In this example, the BCT's span of control expands from assisting one foreign division to assisting two foreign divisions. The figure also depicts how a BCT's operational area has similarly expanded. SFA activities in this phase seek to establish conditions so the host nation's security sector can provide a secure environment with its own security forces.

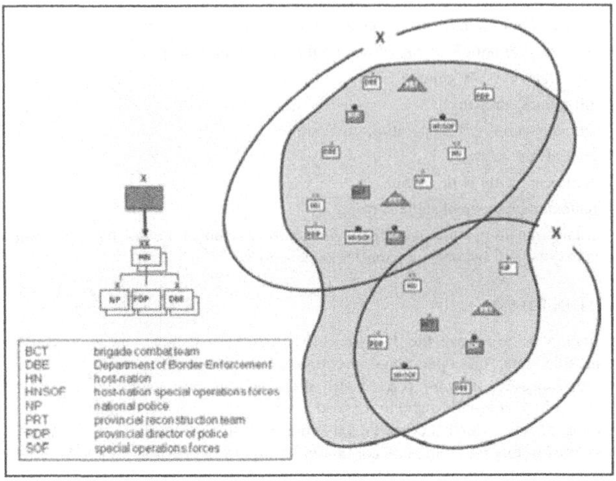

Figure 4-4. Example of transformation phase span of control and area of operations

4-21. The BCT requires augmentation for the transformation phase, like the initial response phase. It often requires similar augmentation due to its greater span of control and area or operations. Column three of table 4-1 summarizes augmentation during the transformation phase with two foreign divisions. The baseline augmentation for the brigade in this phase is as follows:

- Embedded provincial reconstruction team.
- Forward engineering support team-augmenting.
- Explosive ordnance disposal team.
- Sociocultural experts.
- Military police company (provincial police transition team).
- Two military police teams (district police transition team).
- Six military police majors (district police transition team).
- Four judge advocate general teams.
- Civil affairs company.
- Tactical psychological operations company.
- Public affairs team.
- Contingency contracting team.
- Financial management support team.
- Military transition team (specific augmentation personnel include two lieutenant colonels and two master sergeants).

4-22. As conditions of the operational environment improve, the BCT may require a transition to assist three foreign divisions, rather than two divisions as depicted in figure 4-4. This represents a significant expansion of the span of control and area of operations, with a corresponding expansion of required augmentation. Figure 4-5 depicts the BCT supporting three foreign divisions.

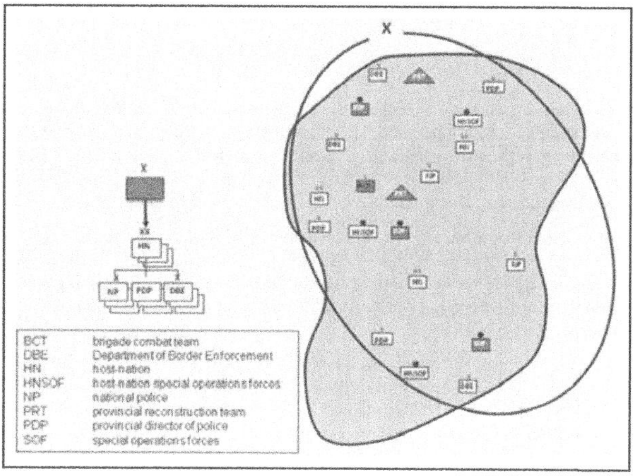

Figure 4-5. Example of a three-division span of control and area of operations

4-23. As in other phases, the BCT requires augmentation to support three divisions. Column four of table 4-1 summarizes augmentation during the transformation phase with three foreign divisions. The baseline augmentation for the brigade in this phase is as follows:

- Embedded provincial reconstruction team.
- Forward engineering support team-augmenting.
- Explosive ordnance disposal team.
- Sociocultural experts.
- Military police company (provincial police transition team).
- Three military police teams (district police transition team).
- Nine military police majors (district police transition team).
- Six judge advocate general teams.
- Civil affairs company.
- Tactical psychological operations company.
- Public affairs team.
- Contingency contracting team.
- Financial management support team.
- Military transition team (specific augmentation personnel include four lieutenant colonels and four master sergeants).

Fostering Sustainability Phase

4-24. In this phase, the focus of SFA continues to shift toward assisting institutions required to sustain FSF. This phase encompasses long-term efforts to assist FSF. In this phase, FSF are capable and viable. They conduct independent operations and can provide a safe, secure internal environment. While the brigade may be initially required during this phase, it is normally withdrawn during this phase. The determination to withdraw the brigade is based on the policy and conditions of the operational environment. Special operations forces, provincial reconstruction teams, and other forces may remain to support a theater security cooperation plan. Thus, remaining SFA missions are normally special operations forces missions, not a BCT mission (special operations forces will normally be involved in all phases of SFA).

TRANSITIONS

4-25. SFA can require a vast array of transitions. These transitions are conducted according to the tactical, operational, and strategic conditions identified during planning (which must support the internal defense and development strategy and using an overall comprehensive approach). Transitions can occur simultaneously or sequentially in different levels or war and in separate echelons. This includes potentially having tactical transitions for different units within one brigade's area of operations.

4-26. There are several major transitions. First, the supported unit may be the U.S. BCT in the beginning of an initial response phase but the foreign security forces will transition to be the supported unit. At this point in the transformation phase, the area in which the BCT conducts SFA will expand. This expansion can occur multiple times during the transformation phase, which is based on conditions, especially the capability and capacity of FSF.

4-27. Operations never proceed exactly as planned. To facilitate flexibility, commanders incorporate branches and sequels into the overall plan. Visualizing and planning branches and sequels are important because they involve transitions. Unless planned, prepared for, and executed efficiently, transitions can reduce the tempo of the operation, slow its momentum, and surrender the initiative to the adversary.

SUBORDINATE UNITS

4-28. The modular brigade augmented for SFA has subordinate units whose sole focus is working with FSF. These advisor teams may be formed from brigade organic resources, external augmentation, or a combination. These teams optimally are embedded with the counterpart unit, or they may reside on a U.S. camp and commute to FSF they support. The method depends on U.S. policy, direction from higher headquarters, the conditions of the operational environment, and capacity of the foreign security force's camps to accommodate the U.S. forces.

4-29. The military transition team (MTT) is a key subordinate unit for the brigade conducting SFA. The MTT mission is to assist FSF units. Figure 4-6 illustrates the relationship between a brigade conducting SFA and a MTT assisting a foreign brigade. In this example, the brigade has a battalion and a division MTT as subordinate units, although it would have other subordinate units. The brigade MTT is subordinate to the battalion and has operational control of one of the parent U.S. battalion's company teams.

4-30. When the brigade provides organic forces to form the basis for the brigade MTT, the company team is the foundation of a brigade MTT. The company team will be augmented with additional personnel and assets. The U.S. brigade must facilitate the MTT operations, such as force protection, transportation, communications, and reachback.

4-31. As companies do not have organic intelligence analysts, personnel must be used to form company intelligence support teams (COISTs). A COIST, an ad hoc organization, assists the MTT and its FSF. Ideally, the COIST should consist of six individuals, which will allow for 24 hour operations and the depth in the team to both integrate with the staff and to successfully support the company's mission. The team should consist of an officer, a noncommissioned officer, and four Soldiers, although the cell can be minimally manned with four personnel. The company intelligence cell works most effectively if its work complements the battalion S-2's work.

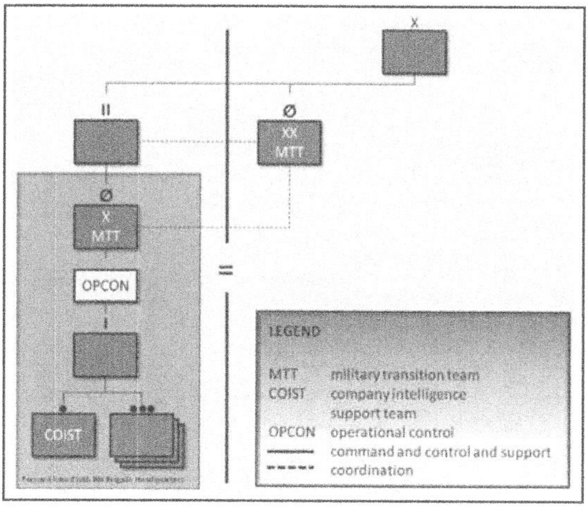

Figure 4-6. Military transition teams

SAMPLE TROOP-TO-TASK MODEL

4-32. The BCT may support assigned to multiple missions, including potentially supporting multiple foreign units in their operational area. Additionally, these FSF organizations may each report through different host-nation government channels and even to different ministries. To synchronize efforts in this case, U.S. forces must achieve unity of command and effort. Similarly, each of the FSF organizational commanders should synchronize their efforts with the host-nation government representatives, as appropriate.

4-33. Figure 4-7 depicts a sample troop-to-task model for a U.S. brigade. It depicts one company team acting as a quick reaction force; thus, the battalion retains the capability to conduct independent full spectrum operations. The figure depicts how U.S. company teams may form the basis of a brigade MTT or a brigade border transition team (and subordinate point of entry transition teams). The figure also depicts how a company team may provide police support. This support is in the form of two platoons supporting police transition teams and a third platoon in a joint security station providing support to a police transition team.

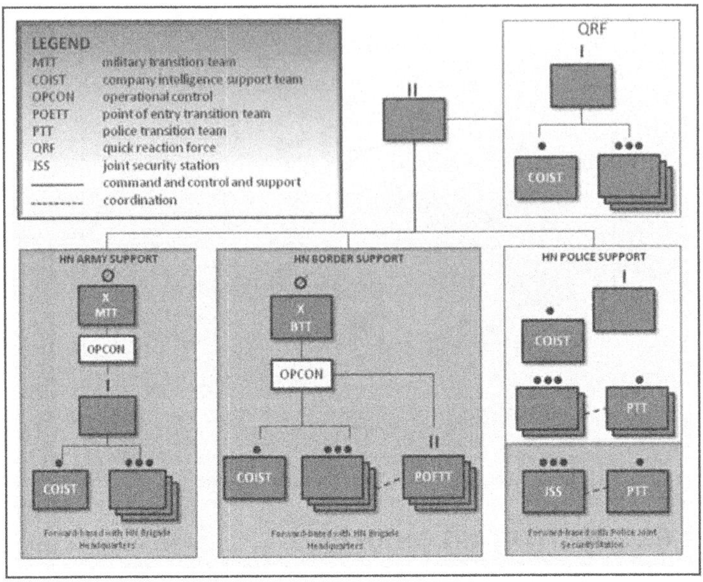

Figure 4-7. Battalions conducting security force assistance

4-34. Subordinate units of BCTs conducting SFA are best located inside a FSF base—not an American base. Co-location facilitates integration with the FSF. It allows the two forces to form mutual understanding and trust. Co-location and the close cooperation it facilitates often improves the population's perception of the legitimacy U.S. and foreign security forces, which can be an essential condition of the overall mission's end state.

4-35. If force protection requires, a U.S. area may be established in the FSF base, although this is not optimal. Key considerations for co-location may include the threat, FSF acceptance, physical space inside the FSF base, sustainment capabilities, medical facilities, and availability of quick reaction forces.

4-36. When U.S. forces are operating out of smaller outposts in an urban environment, the local populace sees the integration and presence of the U.S. and foreign security forces working together. This integration not only enhances overall operational effectiveness and trust, living and working together builds legitimacy of the two forces as well as FSF; it reinforces trust between the FSF and the people they are tasked to protect.

Chapter 5
Unit Operations

Unit operations require continuous activities. Commanders plan for premission, deployment, in-country preparation, employment, redeployment, and postdeployment activities.

PREMISSION

5-1. After receiving a mission, the unit continues detailed preparation activities. The unit obtains supplies, equipment, and training materials and prepares for deployment. Personnel prepare for and rehearse classes to be given in-country. They hone their military skills and conduct extensive area orientation. Paragraphs 5-2 through 5-10 discuss staff actions particular to security force assistance (SFA).

OPERATION PLAN

5-2. The S-3 disseminates the operation plan to unit staff and subordinates. The S-3 ensures predeployment training for Soldiers, to include preparation for training foreign security forces (FSF) and rehearsals for movement. The S-3 reviews the program of instruction for training FSF, to include getting approval from the commander and higher headquarters, if necessary.

CIVIL-MILITARY OPERATIONS

5-3. The civil-military operations section ensures the operation plan minimizes how operations affect the civilian population and addresses ways to mitigate the civilian impact on military operations. Civil-military operations also ensure the operation plan discusses coordination with any augmenting actors. These actors might include civil affairs battalions or brigades, provincial reconstruction teams, or United States Agency for International Development project officers in the operational area. During predeployment training, Soldiers receive training, materials, and briefings on the operational area. This training can cover the history, culture, religion, language, tribal affiliations, local politics, and cultural sensitivities as well as any significant nongovernmental organizations operating in the operational area.

TRAINING

5-4. As discussed in chapter 2, advisors should focus their premission training on the specific requirements of developing FSF. The training emphasizes the host-nation culture and language and provides cultural tips for developing a good rapport with foreign personnel. All premission training for SFA follows the training principles described in FM 7-0.

5-5. Based on the higher commander's training guidance, unit commanders assign missions and approve the draft mission-essential task list that supports SFA. The staff plans, conducts, and evaluates training to support this guidance and the approved mission-essential task list for SFA missions. Commanders prioritize tasks that need training. Since there is never enough time to train in every area, commanders focus on tasks essential for mission accomplishment.

5-6. Once commanders select tasks for training, the staff builds the training schedule and plans on these tasks. The staff provides the training requirements to the commander. After approving the list of tasks to be trained, the commander includes the tasks in the unit training schedule. The staff then coordinates the support and resource requirements with the S-3 and S-4. Finally, the commander ensures standards are enforced during training.

Evaluation

5-7. Evaluations can be either internal or external. Internal evaluations occur at all levels, and they must be inherent in all training. External evaluations are usually more formal and conducted by a headquarters two levels above the unit being evaluated. This subject must be carefully planned and discussed with FSF leaders to account for cultural sensitivities and current capabilities.

5-8. A critical weakness in training is the failure to evaluate each task every time it is executed. Every training exercise provides potential for evaluation feedback. Every evaluation is also a training session. Leaders and trainers must continually evaluate to optimize training. Evaluation must occur as training takes place. Emphasis is on direct, on-the-spot evaluations. However, leaders allow Soldiers to complete the task first. Leaders plan after action reviews at frequent, logical intervals during exercises. This technique allows the correction of shortcomings while they are still fresh in everyone's mind. The after action review eliminates reinforcing bad habits.

Specified Training

5-9. Augmentation elements require area orientation, refresher combat training, field training exercises, and the like. Unit training objectives are for developing capabilities to conduct internal defense and development (IDAD) activities for tactical operations, intelligence operations, psychological operations, populace and resources control operations, and civil affairs and advisory assistance operations in the host-nation language. Units identified for SFA begin intensified training immediately upon deployment notification.

5-10. After deployment to the host nation and before commitment to operations, the unit may receive in-country training at host-nation training centers or at designated training locations. This training helps personnel become psychologically and physically acclimated to the host nation. This training also allows commanders and staff some time to coordinate and plan within their own command and with civilian and military joint and multinational organizations. After commitment, training continues and is stressed between operations, using needed improvements identified in operations as the basis for training.

DEPLOYMENT

5-11. SFA is often conducted in semipermissive environments in which U.S. forces are guests of the host-nation or partner organization. Units move into the operational area by following their deployment standing operating procedures and the operation plan. Paragraphs 5-12 through 5-25 highlight organizations and procedures critical to SFA deployment. This information, however, does not obviate the need for extensive standing operating procedures and an operation plan.

PREDEPLOYMENT SITE SURVEY

5-12. The predeployment site survey unit aims to report accurately to its parent unit the existing FSF mission variables—mission, enemy, terrain and weather, troops and support available, time available, civil considerations (METT-TC). See table 5-1. It also establishes in-country command and control, support, protection, and sustainment relationships for the follow-on unit's mission execution. The site survey coordinates the in-country reception of SFA forces. This coordination wards off most friction during the handoff. This coordination requires all units—outgoing and incoming—to apply a comprehensive approach at all levels of command. Appendix D of FM 3-05.202 discusses site survey procedures in detail.

Table 5-1. Advisor team predeployment site survey checklist

Tasks	Characteristics
Area of operations overview	• Map familiarization. • In sector orientation. • Areas of emphasis.
Graphic control measures	• Unit boundaries. • Routes. • Other United States and multinational support locations. • Maintenance support location. • Medical support locations. • Other points of interest.
United States and multinational unit interviews by specialty staff	• Responsibilities. • Battle rhythm. • Required reports. • Standing operating procedures, continuity book. • Stay-behind equipment.
Briefings on advised unit	• Foreign security forces task organization. • Foreign security forces key leadership. • Intelligence assessment. • Unit operational assessment.
Advised unit introductions	• Among commanders. • Among staffs.

ADVANCE PARTY DEPLOYMENT

5-13. An advance party deploys to the host-nation or partner organization area of operations after the unit completes its mission planning and the proper commander representative declares it deployable. The advance party conducts predeployment activities and coordinates necessary travel documentation and country clearances. The advance party reviews the rules of engagement and any status-of-forces agreements before deploying. If applicable, the proper staff provides an updated threat briefing to the unit. The S-4 section and support company help with out-loading in the unit area. The unit places all supplies and materials on pallets beforehand and prepares the load manifest.

5-14. Upon arrival, the advance party processes through customs and immediately notifies the higher headquarters of its arrival and status. Often the higher headquarters provides point of contacts to the advance party. These contacts help the advance party obtain the proper identification, documentation, and weapons permits.

5-15. The advance party establishes contact with all U.S. and foreign security forces tactical unit commanders and conducts final premission planning activities. The advance party procures secure working, storage, and living areas for the unit and reconnoiters and prepares the training sites. The advance party coordinates for special support equipment to unload heavy supplies and transport the supplies and personnel to the training site.

MAIN BODY DEPLOYMENT

5-16. The unit deploys after completing its planning and the appropriate commander declares the unit ready. The unit follows the same procedures performed by the advance party. Members of the advance party meet the main body on its arrival and brief the main body on any required changes to the operation order.

5-17. After arriving in-country, the unit processes through customs and immediately notifies the higher headquarters of its arrival and status. The unit obtains the proper identification, documentation and weapons permits from U.S. or other sources, if applicable. Often the security assistance organization provides to the unit the necessary point of contacts that expedite this process.

5-18. The unit off-loads equipment and personnel onto vehicles for movement to the training site, or it stores the equipment in a secure area until the processing is completed. FSF secure the equipment but the unit members remember the commander is ultimately accountable for the advisor team's physical and personal security. Supply personnel stay with the equipment and help to guard it.

IN-COUNTRY PREPARATION

5-19. Upon arrival, the commander and S-3 brief the higher headquarters on the planned execution of the mission and reconfirm the required command relationship. Local conditions may require the unit to confirm or establish its in-country and external command and control, support, and sustainment relationships from outside its operational area upon arrival. The unit establishes direct working relationships with its next higher in- or out-of-country supporting element to—

- Determine the limits of the available support and expected reaction time between the initiation of the support request and its fulfillment.
- Confirm or establish communications procedures between the supporting element and the unit, to include alternative and emergency procedures for command and control, all available support, and medical evacuation.

5-20. The unit establishes procedures to promote interagency cooperation and synchronization. The unit—

- Identifies the location of the concerned host-nation, U.S., or other agencies.
- Contacts the concerned agency to establish initial coordination.
- Exchanges information or intelligence.
- Confirms or establishes other coordination protocols as necessary.
- Incorporates the newly established or changed procedures into the plans for mission execution.

5-21. The unit immediately establishes operations security procedures to support its mission execution and identifies rally points incorporated into its defensive, evasion, and personnel recovery plans.

5-22. After receiving a detailed briefing and further guidance from the advance party, unit personnel continue to develop effective rapport with the FSF commander and counterparts. They also assess their working, storage, and living areas for security and verify the location of the training site, communications center, dispensary area, and FSF troop area. With the FSF commander, the unit commander—

- Establishes rapport.
- Conducts introductions in a businesslike, congenial manner.
- Briefs on the unit's mission, its capabilities, and the restrictions and limits imposed on the detachment by the higher U.S. commander.
- Ensures all unit personnel fully support FSF and firmly believes a joint U.S.-FSF effort will succeed. Requests counterpart linkup be made under the mutual supervision of the FSF commander and the unit commander.
- Ensures all current unit plans are tentative and that assistance is needed to finalize them.
- Deduces or solicits the actual estimate of unit capabilities and perceived advisory assistance and material requirements.
- Recommends the most desirable courses of action (COAs) while emphasizing how they satisfy present conditions, achieve the desired training, and meet advisory assistance goals.
- Informs the higher in-country U.S. commander of any significant changes in the unit's plan to assist FSF.

5-23. The S-2 supervises the dissemination of intelligence and other operationally pertinent information within the unit and, as applicable, to higher, lower, or adjacent units or agencies. The S-2 also—

- Monitors the implementation of the intelligence collection plans to include updating the commander's priority intelligence requirements (PIRs), conducting area assessment, and coordinating for additional intelligence support.
- Establishes liaison with FSF intelligence and security agencies (within the guidelines provided by applicable higher authority). Assesses the intelligence threat and resulting security requirements, including coordination with the S-3 on specific security and operations security measures.

5-24. Through the S-2, the commander's PIRs are based on the latest information available and requirements for additional PIRs that arise from modified estimates and plans. The S-2 also—

- Analyzes the foreign unit's status to finalize unit plans for advisory assistance. These plans can include task organization of unit with counterparts, staff functions for planning SFA, and advisory assistance for executing SFA.
- Explains analysis to counterparts and encourages them to help with—and participate in—analyzing, preparing, and briefing the analysis to the foreign unit commander.
- Prepares and briefs the plans for training and advisory assistance.
- Helps the foreign unit inspect the available facilities to identify deficiencies. If the unit finds deficiencies, the S-2 prepares estimates of COAs for the FSF commander to correct them.
- Supervises the preparation of the facilities with their counterparts and informs unit and FSF commanders on the status of the facilities.

5-25. The unit ensures its security is based on the present or anticipated threat. Some recommended actions the unit may take include—

- Hardening its positions based on available means and requirements to maintain low visibility.
- Maintaining unit internal guard system with at least one Soldier who is awake and knows the locations of all other unit personnel. The guard reacts to an emergency by following an internal alert plan and starting defensive actions.
- Maintaining communications with all subordinate unit personnel deployed outside the immediate area controlled by the main body.
- Establishing plans for immediate defensive actions in the event of an attack or a loss of rapport with hostile reaction.
- Discussing visible security measures with foreign counterparts to ensure understanding and to maintain effective rapport. Unit personnel do not divulge sensitive information for the sake of possible rapport benefits.
- Encouraging the foreign unit, through counterparts, to adopt additional security measures identified when analyzing the foreign unit's status and inspecting its facilities.
- Coordinating defensive measures with the foreign unit to develop a mutual defensive plan. Unit personnel obtain from the unit's present reaction and defensive plans for attack. They encourage the foreign unit to conduct mutual full-force rehearsals of defensive plans; if unsuccessful, the unit conducts internal rehearsals of the plans.

EMPLOYMENT

5-26. Once the unit has deployed and arrived in-country, it begins employment. Employment is the conduct of SFA to support the FSF commander. Employment occurs within the advisor team itself. It includes foreign counterparts in the mission planning (preparing the FSF for the mission itself) to increase the capability and capacity of their planning processes, as well as increase the probability of success.

PERSONNEL

5-27. During employment, individual unit personnel perform functions described in paragraphs 5-28 through 5-38.

Commander

5-28. Before the mission, the U.S. commander advises and assists the FSF commander. The latter issues planning guidance for planning the execution of the mission and clarifies commander's intent. The U.S. commander advises and assists the FSF unit commander in the command and control process of tactical operations. By accompanying the FSF unit commander when the mission is received from higher headquarters, the U.S. commander assists any subsequent missions. The U.S. commander monitors how FSF understand the commander's intent and all specified or implied tasks.

5-29. During execution of the mission, the U.S. commander helps the foreign unit commander provide command and control during operations. After monitoring the tactical situation, the U.S. commander recommends changes to the COAs to exploit the situation. After monitoring the flow of information, the U.S. commander recommends improvements to use intelligence collection assets and to keep subordinates reporting required information.

Staff

5-30. Before the mission, the advisor staff advises and assists the foreign counterparts in preparing estimated COAs for essential tasks. The staff helps write tentative plans based on the planning guidance and unit's standing operating procedures. These plans include primary, alternate, contingency, and emergency plans.

5-31. During execution, the staff helps foreign counterparts coordinate the execution of the tasks. They disseminate portions of their plans to other personnel, senior and subordinate staff sections, and supporting elements. They help notify higher, lower, or adjacent staff sections of modified estimates and plans. Finally, the staff—with the S-2 and the S-2 counterpart—helps update the commander's critical information requirements with the latest information and requirements that arise from operations.

Executive Officer

5-32. The executive officer performs the organizational analysis of the unit's coordinating staff sections to ensure efficiency during the planning process according to initial planning guidance. With the foreign counterpart, this officer directs foreign staff sections as they develop estimates and plans. The executive officer monitors the liaison and coordination with FSF higher headquarters, recommending changes to improve efficiency.

Adjutant

5-33. The adjutant provides advice, assists, and makes recommendations to the foreign counterpart for conducting human resources. This includes monitoring the maintenance of foreign unit strength, accountability of casualties, unit morale, and postal activities. This may also include concerns with the foreign pay system, leave procedures, and casualty pay procedures.

Intelligence Officer

5-34. The intelligence officer advises and assists the monitoring of FSF operations security to protect classified and sensitive material and operations and recommends improvements. By helping the foreign counterpart update the situation map, the intelligence officer helps to keep the map current. The intelligence officer recommends improvements to the standing operating procedures of the tactical operations center communications so the intelligence section receives situation reports. This officer helps the counterpart monitor the collection, evaluation, interpretation, and dissemination of information. The intelligence officer assists in the examination of captured insurgent documents and material. This officer helps gather and disseminate intelligence reports from available sources to ensure the exploitation of all assets. The intelligence officer helps the counterpart to brief and debrief patrols operating as a part of reconnaissance and surveillance activities. Finally, the intelligence officer assists, with the advisor operations officer, developing reconnaissance and surveillance plans.

Operations Officer

5-35. The operations officer helps the foreign counterpart to prepare tactical plans using estimates, predictions, and information. This officer monitors command and communications nets, assists in preparing all orders and plans, and helps to supervise the training and preparation for operations. Finally, the operations officer monitors the planning process and makes recommendations for consistency with IDAD goals.

Supply Officer

5-36. The supply officer advises and assists the foreign counterpart in maintaining equipment readiness; monitoring the support provided to the foreign unit, its subunits, and attachments; and recommending improvements. Finally, this officer helps to supervise the use of transportation assets.

Civil-Military Team

5-37. Upon deployment, civil-military teams advise the commander and staff on civil-military considerations and coordinate efforts of any civil affairs units supporting the unit. The civil-military team mentors the counterpart team on the supported foreign element staff on civil-military operations, the legal and moral obligations of military commanders to civilian populations under their control, and the importance of respecting human rights. The civil-military team may introduce the counterpart to relevant nongovernmental organizations, United States Agency for International Development project officers, and provincial reconstruction team staff.

Company-Level Advisors

5-38. Advisors sent to elements below battalion level help foreign counterparts to analyze the mission and commander's intent from higher headquarters. They assist FSF leaders restate the mission, conduct an initial risk assessment, identify a tentative decisive point, and define their own intent. They also assist their foreign counterparts to analyze the mission variables. From these variables, advisors help their foreign counterparts to develop a COA that meets the higher headquarters concept of operations and commander's intent. Finally, they advise and assist in the conduct of operations and the flow of information to the FSF higher commander.

TRAINING

5-39. Training FSF during SFA parallels the training of U.S. forces. Training follows the principles of training described in FM 7-0. This training tailors the objectives to account for cultural differences and focus on the FSF mission and problem set.

Training Assessment

5-40. The commander of the advising unit begins with a training assessment of training plans designed during predeployment and passed on from the preceding U.S. units conducted in coordination with the FSF commander. This assessment is important to both evaluate the foreign security force but also as an exercise to establish and enhance the working relationship between U.S. and foreign security forces. The training assessment should cover all aspects of leadership, training, sustainment, and professionalization. To support an assessment, the advising unit should analyze the following specific foreign unit considerations:

- The unit's mission and mission-essential task list and capability to execute them.
- Staff capabilities.
- Personnel and equipment authorization.
- Physical condition.

- Any past or present foreign influence on training and combat operations.
- Operational deficiencies identified during recent operations or exercises with U.S. personnel.
- Sustainment capabilities, to include sustainment training programs.
- Internal training programs and personnel.
- Training facilities.

5-41. The level of professionalism of FSF, both units and individuals, should also be assessed. Adhering to established rules of engagement, ethics that meet the established laws and regulations of the commanding authority, laws for land warfare, and human rights, promotions based on merit and the security forces' support of civilian leaders and political goals all fall within this assessment.

5-42. The advisor unit, working with FSF leaders, also evaluates current members of the security forces for past military skills and positions. Often military reorganizations arbitrarily shift personnel to fill vacancies outside their knowledge and experience.

Training Plan

5-43. After completing the training assessment, advisors analyze the prepared training plan and determine if changes are necessary. The training plan stresses the deficiencies identified in the training assessment. It also identifies those in the host nation able to help train FSF and to strengthen the legitimacy of the process. Finally, the training plan should consider the unit's eventual self-sustainment. As the foreign security force gains sufficient capacity and capabilities to perform independently, advisors transition from a leading role to a mentoring role.

5-44. The plan for training FSF uses a comprehensive approach, supports the IDAD strategy, and has many echelons. Working with other echelons and actors within the area of operations can provide support and expertise that enhance the training and operations process as well as limit redundancy.

Program of Instruction

5-45. In coordination with the FSF commander and training staff, the advising unit develops a program of instruction. This program incorporates all training objectives that satisfy the training requirements identified during the assessment. The training program must support these requirements. The FSF commander approves the program of instruction prior to execution. When executing the program of instruction, advisors adhere to a training schedule consistent with changes in the mission variables. Advisors ensure through their counterparts and the FSF commander that all personnel receive training. Foreign counterpart trainers rehearse all classes approved on the program of instruction.

Presentation of Instruction

5-46. To present the training material properly, trainers follow the lesson outlines approved in the program of instruction. All training clearly states the task, conditions, and standards desired during each lesson, ensuring the foreign students understand them. Trainers state all warning and safety instructions in the host-nation language. The training to reinforce the concepts includes demonstrations of the execution of each task, stressing the execution as a step-by-step process. Trainers monitor students' progress during instruction and practical exercises, correcting mistakes as they are made.

Training Methods

5-47. An effective method of training used by U.S. forces is the crawl-walk-run method of teaching individual tasks, battle drills, collective tasks, and field exercises. (See FM 7-0.) This method is employed to develop well-trained leaders and units. During all phases, the training must include the mission of the unit in the context of the higher unit's mission to assist with the practical application of the training. Identifying the higher commander's mission and intent, as well as the tasks and purposes of other units in the area, would also add context to the training. This method can also be expanded to include the role of other actors.

5-48. At a minimum, units continue individual training to improve and sustain individual task proficiency while training on collective tasks. Collective training requires interaction among individuals or organizations to perform tasks, actions, and activities that contribute to achieving mission-essential task proficiency. Collective training includes performing collective, individual, and leader tasks associated with each training objective, action, or activity.

Collective Training

5-49. Collective training starts at the squad level. Squad battle drills provide key building blocks that support operations. Trainers can link battle drills and collective tasks through a logical, tactical scenario in a situational training exercise. Although this exercise is mission-oriented, it results in more than mission proficiency. Battle drills and collective tasks support situational training exercises, while these exercises support operations. Advisors must understand the operational environment when training FSF; training incorporates how internal and external threats and civilians affect the environment. Flexibility in using U.S. doctrine in training enhances efforts to make training realistic. Trainers modify U.S. doctrine to fit the FSF level of expertise, their command and control, the tactical situation, and sustainment base. Often the structure and capabilities of FSF differ from that required by U.S. doctrine. When FSF counter an insurgency, these exercises emphasize interplay among psychological and tactical, populace and resources control, intelligence, and civil affairs operations. Similarly, scenarios for each tactical exercise must include interplay among the various possible aspects of IDAD.

Individual Training

5-50. Training individuals of FSF emphasizes physical and mental conditioning, tactical training, basic rifle marksmanship, first aid, combatives, and the operational environment. Individual training also includes general tactics and techniques of security operations and the motivation, operations, and objectives of internal and external threats. Tough and realistic training conditions troops to mentally and physically withstand the strain of continuous offensive operations.

5-51. Foreign personnel cross-train on all types of weapons, communications and other equipment, and skills particular to their unit. Personnel losses must never cause weapons, communications equipment, or essential skills to be lost due to a lack of fully trained replacement personnel.

Small-Unit Leader Training

5-52. SFA frequently entail rapidly changing circumstances; thus, FSF small-unit leaders must be able to plan and execute operations with little guidance. Therefore, trainers stress small-unit leadership training which can occur concurrently with individual training. Tools the trainer uses to train leaders are manuals, previously established training, tactical exercises without troops, battle simulations, and unit missions. This training develops aggressiveness, tactical proficiency, and initiative. Small-unit leader training should include combined arms technical training such as procedures for forward observer and close air support. Leadership training must also include land navigation in difficult terrain and under conditions of limited visibility. Mission readiness and the health and welfare of subordinates are also continuous parts of training.

After Action Review

5-53. In any training, the after action review provides the critical link between training and evaluation. It is a professional discussion that includes the training participants and focuses directly on the training goals. An after action review occurs after all collective foreign unit training. Effective after action reviews review training goals with the concerned FSF unit commander. Advisor personnel ask leading questions, surface important tactical lessons, explore alternative COAs, assist the retention teaching points, and keep the after action review positive.

5-54. The advisor encourages the FSF unit commander to review the training with the entire unit or key subordinate leaders. At this review, the FSF commander stresses how to strengthen the chain of command. During the after action review, advisors avoid criticizing or embarrassing the FSF unit commander or subordinates.

5-55. If possible, after action reviews occur during the field portion of the training when the unit assembles at logical stopping points. These reviews provide feedback from instructors and leaders that increases and reinforces learning. This feedback provides a richer database for key points. During the after action review, the senior evaluator draws information from unit leaders to form possible alternative COAs.

Assessment

5-56. In all operations, including SFA, success is generally measured for three periods: short-, mid-, and long-term. Success in SFA is normally defined within the context of these three periods. In the short-term period, FSF make steady progress in fighting threats, meeting political milestones, building democratic institutions, and standing up security forces. In the mid-term period, FSF lead fighting threats and provide security, have a functioning government, and work towards achieving economic potential. In the long-term period, FSF are peaceful, united, stable, and secure; integrated into the international community; and a full partner in international security concerns.

Measures of Effectiveness

5-57. A *measure of effectiveness* is a criterion used to assess changes in system behavior, capability, or operational environment that is tied to measuring the attainment of an end state, achievement of an objective, or creation of an effect (JP 3-0). A measure of effectiveness provides a benchmark against which the commander assesses progress toward accomplishing the mission.

5-58. The following are tools to assess the effects of operations in meeting expectations defined for the short, mid, and long term. Although all SFA is not focused on a host nation, these tools are based on the four IDAD pillars:

- **Balanced development** attempts to achieve national goals through political, social, and economic programs.
- **Security** includes all activities implemented to protect the populace from the threat and to provide a safe environment for national development.
- **Neutralization** is a political concept that encompasses the physical and psychological separation of the threatening elements from the population.
- **Mobilization** provides organized manpower and resources and includes all activities to motivate and organize popular support of the government.

Measures of Performance

5-59. A *measure of performance* is a criterion used to assess friendly actions that is tied to measuring task accomplishment (JP 3-0). This quantitative analysis determines whether the task or action was performed to standard. The standard may be derived from a procedure or may be time-sensitive. Examples of measures of performance are listed in the appropriate Army tasks in FM 7-15.

5-60. The advisor helps the foreign counterpart to monitor the current situation for unanticipated successes, failures, or enemy actions. As the commanders assess the progress of foreign security operations, they look for opportunities, threats, and acceptable progress. Throughout the operation, the commanders visualize, describe, and direct changes to the operation, feeding the assessments of progress or regression back into the planning process.

REDEPLOYMENT

5-61. Redeployment is the return of forces and materiel to the home or mobilization station. Before redeployment, a unit conducting SFA inventories all supplies and equipment (jointly, if possible) to be passed on to the foreign unit or the relieving U.S. or multinational unit. These actions ensure all items are accounted for and the custody chain for property and equipment is unbroken. The outgoing unit also—

- Prepares a final evaluation report showing its estimate of the foreign unit's capabilities and giving an opinion of its future employment.
- Discusses the foreign unit's performance with its commander. The unit submits a copy of the foreign unit's final evaluation to the next higher FSF commander.

- Passes custody of training schedules, lesson plans, foreign operational records, and the foreign unit's final evaluation to the foreign unit and relieving U.S. or coalition unit.
- Presents debriefings, after action reviews, and the foreign unit's final evaluation to the higher in-country U.S. commander.

5-62. The outgoing unit commander supervises the redeployment. This commander approves prepared redeployment plans or, as necessary, directs changes to the redeployment plans. The commander and operations officer assess and select alternatives or changes to the present redeployment COAs based on the intelligence officer's intelligence estimate. They develop and disseminate a fragmentary order for selected alternatives or changes.

5-63. Unit personnel prepare all accompanying supplies and equipment for shipment. They brief ground support personnel on equipment handling procedures, routes, convoy procedures, and actions to take if a terrorist or insurgent incident occurs. They load personnel and equipment for movement to the departure airfield or airport following the load plan in the unit's operation order. They maintain accountability for all their personnel, supplies, and equipment.

Mission Handoff Procedures

5-64. During long-term SFA, conditions determine the rotation of in-theater units. Time is not the only governing factor. Changes in the operational environment may require reshaping force packages as situations change. In addition, internal administrative concerns might prompt or support a commander's decision to rotate units. Regardless, mission handoff is necessary and defined as the process of passing an ongoing mission from one unit to another with no discernible loss of continuity.

5-65. The overall authority for the handoff and the subsequent transfer of authority lies with the commander ordering the change. The authority for determining the handoff process lies with the incoming commander assuming responsibility for the mission. This changeover process may affect conditions under which the mission will continue.

Redeployment Considerations

5-66. Although intended for a direct handoff between U.S. units, commanders must make specific considerations when making a handoff to a multinational force. The considerations include mission, enemy, terrain and weather, troops and support available, time available, and civil considerations. For units relieved of a function by a government agency, procedures typically entail longer handoff times and more complex coordination. However, the other areas of consideration still apply and may in fact be a greater issue for an agency. Outgoing units that have past, present, or future projects planned with agencies prepare to transfer these projects to responsible agents in the incoming unit.

Mission

5-67. The outgoing unit must prepare a detailed study of their unit's mission statement and detail the current and implied mission tasks. The mission may also require an incoming unit with additional skill sets, such as advanced special operations, near real-time connectivity, civil affairs functional specialists, or complex media production ability. Knowing the mission, commander's concept of the mission, commander's critical information requirement, PIRs, and information requirements helps the incoming unit understand the mission. After a complete in-depth study of the operational area, the outgoing unit should help the incoming staff complete the handoff in a manner that allows for continued, uninterrupted mission accomplishment. The changeover must not allow the adversary to gain any operational advantages.

Enemy

5-68. The outgoing unit must provide the latest available intelligence on all threats that affect the mission. This intelligence includes comprehensive data on terrorists and terrorist-related incidents over the previous several months. In addition to the normal intelligence provided to the incoming unit on a regular basis, the situation may call for a liaison from the outgoing unit. Operations security is critical to prevent the enemy from discovering the impending handoff and then exploiting the fluidity of the change and the concentration of U.S. or multinational forces.

5-69. The in-country unit provides continuous information updates to the incoming unit. PIRs and information requirements were established for the original mission along with operational, strategic, and tactical information. The incoming unit must become familiar with the ongoing and upcoming mission PIRs and information requirements along with their links to planned nonlethal effects.

Troops and Support Available

5-70. To the incoming unit, learning about the friendly forces is as important as knowing the enemy situation. The unit must be familiarized with the command and control structure with which it will work daily. The incoming unit must know all friendly units in adjacent operational environments. It also must be aware of joint and multinational units and the capabilities of their mission support base. The incoming unit must also be aware of other operations, units, and their capabilities.

5-71. Mission handoff is the time for the outgoing unit leadership to introduce the incoming unit leadership to leaders of FSF. This practice helps prevent breaks in intelligence networks already developed by the outgoing unit and reduces the time the incoming unit needs to build trust. The outgoing unit supports the incoming unit's plans to prepare for the transition in counterpart relations. However, potential or anticipated friction between a foreign security force and the incoming unit may cause the relief to occur more slowly. Therefore, incoming and outgoing units need time for overlap to allow for in-country, face-to-face contact with their counterparts before the mission handoff. If possible, outgoing unit members should provide biographical data on their counterparts to include photographs. This information allows unit members to become familiar with their counterparts before deployment and to determine which advisor techniques may need more emphasis. Execution of the mission must continue within the capabilities of the incoming unit, the foreign unit, and the available supporting assets.

Time

5-72. The depth and dispersion of units and the number of operations conducted determines the time required to exchange units. Ideally, an overlap occurs to allow the incoming unit to become familiar with the operational environment and to establish rapport with their foreign counterparts. However, the handoff must occur as quickly as possible. The longer it takes, the more vulnerable become personnel in the operational environment. However, both outgoing and incoming units should not sacrifice continued and uninterrupted execution of ongoing operations for speed. The incoming unit needs enough time to observe training techniques and procedures and to conduct debriefing on lessons learned. Outgoing unit personnel need sufficient time to put incoming unit personnel in contact with agency counterparts.

Civil Considerations

5-73. The outgoing unit provides an in-depth area study, giving close attention to local problems. Often sister units can provide general demographic data that can be expanded upon for unit-specific needs. Popular support for U.S. activities occurring in the operational environment may directly influence changes in the mission statement. The outgoing unit provides this critical information and describes all completed and progressing civic action projects. The incoming unit considers how the host-nation government functions and international civilian or government agencies are involved in or influence the situation in its operational environment.

POSTDEPLOYMENT

5-74. The unit commander debriefs to provide an overview of the mission and all relevant information subsets. Topics can range from military geography, political parties, and military forces to insurgents, security forces, and ongoing agency operations. Concerns specific to a foreign security force and relevant in one SFA operation may not be relevant in another operation. The commander should codify and debrief any information subset that may affect mission success.

Postmission Debriefing

5-75. Redeployment is not the end of the mission. Upon arrival at the redeployment location, the unit undergoes an extensive debriefing. The intelligence staff section officer typically organizes and conducts the debriefing. The intelligence officer coordinates with higher-level intelligence organizations to take part in the returning unit's debriefing, particularly if other organizations tasked the unit to obtain information. All deployed personnel, to include attachments, must be available for the debriefing.

Documentation

5-76. After the debriefing, the unit commander and staff prepare two documents—an after action review and a report of lessons learned.

5-77. The after action review states the who, what, when, where, and how of the operation. It is a permanent record of the major activities of the unit from receipt of mission to debriefing. As such, it is an extremely important template on which past missions may be compared and future missions planned. Within 48 hours of being debriefed, the unit normally submits an after action review through command channels to the higher command. The intelligence and operations officers at each echelon keep copies of unit after action reviews.

5-78. Shortly after completion of the after action review, or simultaneously with its submission, the unit submits a report of lessons learned. This report documents the commander and staff's reflection of the operation and recommendations for the future. Units often prepare the lessons according to the six warfighting functions: movement and maneuver, intelligence, fire support, sustainment, command and control, and protection. This method addresses what worked and did not work during the operation, why it did or did not work, and what changes or substitutions are needed for existing tactics, techniques, and procedures in the unit.

5-79. As a minimum, the lessons learned report is forwarded to appropriate Service- and joint-level lessons learned agencies to allow integration into subsequent unit training and leader education. These agencies might include the Center for Army Lessons Learned and Joint Center for Operational Analysis.

Chapter 6
Sustainment

A brigade combat team may be required to execute operations across the spectrum of conflict. Hence, providing security force assistance (SFA) may require the sustainment staff to analyze the mission to determine support requirements and available capabilities. Since brigade combat teams conducting SFA may be responsible for a large operational area, an accurate and thorough assessment of U.S. assets as well as the foreign security force's assets is critical to mission success. The sustainment planner understands that though the mission can change, methods by which logistics are distributed within a theater of operations do not. Army sustainment enables the commander to execute a mission and sustain the force through full spectrum operations. Sustainment assets are allocated based on those requirements; no standard arrangement fits all situations.

As discussed in chapter 2, one of the principles of SFA is to ensure long-term sustainment. This principle dictates that all policies, strategies, and programs must be durable, consistent, and sustainable by foreign security forces (FSF) and their government once U.S. forces have completed assistance. This can be accomplished by developing the foreign security force's capability and capacity for asset production, management, performance, and maintenance.

SUSTAINMENT IN FOREIGN SECURITY FORCES CAPACITY

6-1. Sustainment is the provision of the logistics, personnel services, and medical support necessary to maintain operations until mission accomplishment. The endurance of SFA forces is primarily a function of their sustainment. Sustainment determines the depth to which the forces can conduct decisive operations, allowing the commander to seize, retain, and exploit the initiative. Endurance is the ability to employ combat power anywhere for protracted periods. Endurance stems from the ability to generate, protect, and sustain a force. It involves anticipating requirements and making the most effective, efficient use of available resources. Sustainment also enables strategic and operational reach. SFA forces require strategic and operational reach to deploy and immediately conduct operations anywhere with little or no advanced notice. FM 4-0 discusses sustainment in detail.

CORE LOGISTIC CAPABILITIES

6-2. Similar to any operation, the core logistic capabilities in SFA provide a framework to facilitate integrated decisionmaking, synchronize and allocate resources, and optimize logistic processes. The challenges associated with support cut across all the core logistic capabilities: supply, maintenance, deployment and distribution, engineering, medical support, logistic services, and operational contract support.

Supply

6-3. Supply operations include identifying requirements; selecting supply sources; scheduling deliveries; receiving, verifying, and transferring products; and authorizing supplier payments. To provide responsive supply operations in SFA to both the advising organization and FSF, logistic planning requires a collaborative environment that includes operations planning, maintenance operations, and the distribution system.

Maintenance

6-4. In SFA, maintenance operations aim to reduce repair cycle times by replacing components, modules, and assemblies as far forward as possible as well as maximizing reliance on parts distribution, visibility, and replacement. Advisors and their foreign counterparts require maintenance capacity for their freedom of action. Force movement and maneuver depends on sufficient readiness levels of transportation equipment and weapons systems. In SFA, maintenance encompasses both intermediate and organizational levels and consists of shop maintenance supporting the supply chain, transportation assets, and weapon systems.

Deployment and Distribution

6-5. The deployment and distribution capability moves forces and logistic support on time by meeting the required delivery date and providing time-definite delivery. By sharing critical information, units create unity of effort among diverse distribution organizations. This unity of effort enhances foreign support efforts and provides end-to-end support to satisfy deployment execution and sustainment operations for advisors and host-nation forces.

Engineering

6-6. Engineering provides capabilities to maximize freedom of action. These include capabilities and activities that support the maneuver of land combat forces; modify, maintain, or protect the physical environment; and contribute to a clear understanding of the physical environment by providing geospatial information and services to commanders and staffs.

Medical Support

6-7. A component of Army health systems oversee operational management of the health service support and force health protection missions for training, predeployment, deployment, and postdeployment operations. Army health system includes all mission support services performed, provided, or arranged by Army Medical Department to support health service support and force health protection mission requirements for civilian and military joint and multinational forces. For more information on Army health systems, see FM 4-02.

Health Service Support

6-8. All support and services performed, provided, and arranged by Army Medical Department promote, improve, conserve, or restore the mental and physical well-being of Soldiers and personnel in other Services, agencies, and organizations as directed. This includes casualty care, medical evacuation, and medical logistics. The health service support mission is a function of the sustainment warfighting function.

Force Health Protection

6-9. Force health protection measures promote, improve, or conserve the mental and physical well-being of Soldiers. These measures enable a healthy and fit force, prevent injury and illness, and protect the force from health hazards. They include some prevention aspects of Army Medical Department functions, combat and operational stress control, dental services, and laboratory services. The force health protection mission is a function of the protection warfighting function.

Logistic Services

6-10. Logistic services comprise the support capabilities that collectively enable the United States to rapidly provide global sustainment for U.S. forces. Logistic services include many disparate activities that are highly scalable capabilities. Included in this area are food, water and ice, base camp, and hygiene services.

Contract Support

6-11. Operational contract support provides acquisition functions to obtain materials, services, and construction of facilities for U.S. and host-nation forces in the local operating areas. The Department of Defense increasingly relies on contractors to perform many functions and tasks and then provides technical oversight. The expeditionary nature of SFA creates an even greater need for this capability.

PRINCIPLES OF SUSTAINMENT

6-12. The principles of sustainment are critical to maintaining combat power, maintaining strategic and operational reach, and providing Army forces with endurance. The principles of sustainment are anticipation, responsiveness, simplicity, economy, survivability, continuity, improvisation, and integration. While these principles are independent, they are also interrelated when used in planning and executing operations. For more information on sustainment principles, see FM 4-0.

MISSIONS

6-13. All brigade support battalions have the same general capability to manage distribution and sustainment operations. The brigade support battalion provides support within an assigned operational area. It is force tailored and task-organized and uses subordinate companies, platoons, and teams to perform specific functions. In the sustainment role, the brigade support battalion is primarily concerned with the continuous management and distribution of stocks and allocation of maintenance in the operational area. Such assistance provides operational reach to maneuver commanders.

6-14. The brigade support battalion contains the following sustainment capabilities:

- Provides supplies, field services, field maintenance, recovery, and field feeding for itself and its assigned subordinates.
- Plans and conducts base and base cluster self-defense; defends against level I threats, assists in destruction of level II threats, and escapes or evades against level III threats.
- Has ability to operate as part of an Army or joint force.
- Coordinates host-nation support.
- Has ability to deploy an advance party to support early entry operations.
- Provides opening, distribution, and sustainment management information and advice to commanders and staff within its operational area.
- Exercises technical supervision over operations for all assigned units.
- Provides limited materiel management for internal stocks—classes I, II, III, IV, V, VII, and IX and maintenance management of internal assets.
- Provides a liaison team to augment other headquarters as necessary.
- Manages and maintains the its property records.
- Maintains data in support of Army equipment status reporting database and Army equipment status reporting system.
- Appoints contracting officer representatives to oversee contractor performance, certify receipt of services, and act as liaisons between the requiring activity and the contracting officer.

6-15. A brigade combat team often has attached companies such as an engineering company, a military intelligence company, an antitank company, or a network support company. Similar to the brigade headquarters and headquarters company, these attached companies are supported by the brigade support battalion regardless of where they are located on the battlefield. If one or part of these companies is task-organized to a maneuver battalion, it remains under the support of the brigade support battalion.

Critical Considerations in Building Host-Nation Sustainment Capacity

6-16. Effective SFA planning begins with accurately assessing and understanding U.S. policy objectives, determining the end state, and identifying resources available. Some possible logistic planning considerations—

- Have a holistic perspective.
- Address the causes, not the symptoms.
- Understand the desired end state.
- Understand the level of FSF senior leader support.

6-17. Logistic planning during SFA uses a holistic perspective. Operational environments can range from peacetime cooperation to counterinsurgency, stabilization to reconstruction. SFA forces determine the foreign security force's logistic and reconstruction requirements based on their internal defense and development strategy; they develop a sustainment end state to support the plan. Rebuilding the host nation's infrastructure and industrial capacity not only provides long-term job opportunities, but also builds trust and legitimacy in FSF, the host-nation government, and partner organizations.

6-18. Logistic planning during SFA addresses the causes, not the symptoms. Operational logistic support problems are the symptoms, not the causes. They result from problems that exist in the supporting systems or in the society's industrial capacity and economic strength. Teaching unit-level inventory management alone will not result in any sustainment improvement.

6-19. Logistic planning during SFA understands the desired end state. Depending on the U.S. strategic interest in a host nation and its security forces, levels of engagement can vary significantly to achieve the desired end state. Sustainment capacity-building activities can be force tailored and scaled to assist in any of the sustainment capability areas.

6-20. Logistic planning during SFA understands the level of FSF senior leader support. These leaders develop and set the policies for resource allocation. They support the various development projects and skills enhancement assistance activities. These senior-level advisors must fully understand the problems identified and an assistance strategy developed by SFA planners. From the beginning, SFA planners work with their counterparts to decide the plans and strategy to promote the vision and ensure continued effort.

6-21. SFA depends on capable logistic support for maneuver and sustainment. Without an improvement in sustainment capacity, FSF cannot significantly improve regardless of how much effort is devoted to advising and training. When planning for sustainment capacity building, planners thoroughly understand the logistic systems to identify causes that lead to functional deficiencies. They must have a holistic perspective and understand the various goals and resources required to formulate an operational approach that effectively and efficiently achieves U.S. objectives.

ASSET MANAGEMENT AND PERFORMANCE

6-22. One tool at the disposal of the commander and planners during an evaluation of foreign infrastructure and industrial capacity is called an asset management plan. This plan, commonly employed in the industrial community, manages an organization's infrastructure and assets to deliver a standard of service. An asset management plan provides information to FSF so they can analyze the capacity, condition, and current and future requirements of their facilities and infrastructure. Physical development should include facilities, transportation networks, area development, utilities, and other infrastructure components.

6-23. A description of the perceived problems of current foreign facilities, or problems created by missing assets, is also important. The description includes how the advisor intends to address deficiencies in foreign facilities and infrastructure. Each organization needs to determine each facility's function and value to their organizational mission by analyzing the current asset performance, costs, benefits, and improvements.

Current Asset Performance

6-24. Asset performance is the most difficult portion of the asset management plan process to capture while assessing conditions of facilities and infrastructure. While historical records may exist, combat operations and possible instability of the host-nation government likely render these records irrelevant. Remaining structures, when measured against U.S. construction standards (designed to protect occupants from man-made or natural disasters), often are deficient, even if untouched by hostilities. Most existing facilities require significant modification to be considered for future use. However, FSF may use them in the interim until new facilities are completed.

6-25. Asset performance may also use gap analysis. Gap analysis defines the present state, the desired end state, and the gap between them. Once a gap has been identified, the commander and staff can develop courses of action to rectify or bridge the gap. Gap analysis alone is not adequate for all problem sets, as goals may evolve and emerge during the course of problem solving. Most problems have many alternative solutions.

Costs

6-26. Establishing the short, medium, and long-term costs for facilities and infrastructure, from construction to control, is a critical step in prioritizing which facilities get built or refurbished first, and which get support at all. Units develop a forward-looking cost-profile for operating, maintaining, refurbishing, and replacing assets. This profile explains to civilian and military officials the implications and continued costs to maintain assets. SFA requires a plan for U.S. forces to hand off assets to FSF.

6-27. In constructing these models, the ideal cost-profile extends to cover the life of the longest-lived asset in the system. By doing so, the profile estimates the whole-life cost as well as determines the average annual costs. Furthermore, this cost evaluation may need to be revisited and updated annually, forming the basis for an annual bid for funding.

Benefits

6-28. Assets should provide some measure of benefit to U.S. and foreign security forces that can be measured or explained during and after SFA. This usually involves translating asset costs and projected performance into a monetary figure. Other benefits may be social or environmental, which may be difficult to quantify in monetary terms. Regardless, some attempt to record all the relevant benefits is important; the more quantitative, the better.

6-29. A benefit evaluation is vital to demonstrating that the ongoing cost of an asset system is reasonable given the scale of the benefits. The expenditure incurred by asset management needs to be justified in some way.

Improvements

6-30. How assets of FSF perform may be improved by acquisition, enhancement, or some other means. An improvement may address the potential reduction of performance and its effect on the rest of the asset system. The improvements represent a potential change to the performance and are typically managed as a project. Such management comes with an appraisal of the additional expenditure, comparing different options and selecting a preferred option based on the commander's guidance and host-nation requirements.

OPERATIONAL CONTRACT SUPPORT

6-31. Contracting is a key source of support for deployed forces across the spectrum of conflict. Because of the importance and unique challenges of operational contract support, the commander and staff need to fully understand their role in planning for and managing contracted support in the operational area. This includes understanding the types of contracted support, contract management, and contract close out.

TYPES OF CONTRACT SUPPORT

6-32. Current doctrine describes three broad types of contracted support: theater of operations support, external support, and systems support.

Theater of Operations Support

6-33. Contracts for theater of operations support provide support to deployed forces under prearranged contracts or contracts awarded from the mission area. They do so by contracting officers under the command and control of the combat sustainment support battalion. Contractors for theater of operations support acquire goods, services, and minor construction support, usually from local commercial sources, to meet the immediate needs of operational commanders. Theater of operations support contracts are the type of contract typically associated with contingency contracting. Sustainment brigades often are the requiring activity for theater of operations support contract actions.

External Support

6-34. Contracts for external support provide a variety of support to deployed forces. External support contracts may be prearranged contracts or contracts awarded during the contingency. These contracts support the mission and may include a mix of personnel including U.S. citizens, third-country nationals, and host-nation subcontractor employees. The largest and most commonly used external support contract is the logistics civilian augmentation program. This Army program is commonly used to provide life support, transportation support, and other support functions to deployed Army and joint forces.

System Support

6-35. System support contracts are prearranged and include, but are not limited to, newly fielded weapons systems, command and control infrastructure, and communications equipment. System support contractors, made up mostly of U.S. citizens, provide support in garrison and may deploy with the force to both training and real-world operations. The unit does not normally have a significant role to play in planning for or coordinating system support contracts other than coordinating and executing support of system support contract-related personnel.

CONTRACT MANAGEMENT

6-36. Contract management requires nominating and tracking contracting officer representatives for every service contract. It will also nominate a receiving official for all supply contracts. To ensure contractors provide the service or item per the contract, they need a quality contracting officer representative and official support. SFA units must identify and coordinate funding requirements in advance with the supporting resource management staff element to secure the correct amount and type of contract funding associated with the mission.

CONTRACT CLOSE OUT

6-37. In long-term operations like SFA, the advising unit ensures direct coordination and transfer of information related to operational contract support before mission handoff. Additionally, when the advance party arrives in the operational area, a designated Soldier actively seeks out current information on local contract support capabilities, policies, and procedures. This individual coordinates the formal handover of existing contract management responsibilities from the redeploying unit. This Soldier must know when recurring service contracts end since it takes 30 to 60 days to obtain funding approval. If the unit waits until the contract is about to expire before requesting additional funds, the sustainment brigade could lose the contracted service until funds are available.

EXTERNAL AGENCIES SUPPORT MANAGEMENT

6-38. Some external agencies help support the management of FSF. External support comes from interagencies, the host nation, and multinational organizations.

INTERAGENCY SUPPORT

6-39. The Department of Defense performs both supported and supporting roles with other agencies. It often is the lead agency for logistic activities in SFA. When tasked to provide military support to civil agencies, Army forces perform a supporting role. Sustainment forces may be tasked to support the U.S. ambassador that, as the President's representative, serves as senior U.S. Government officer in the country, or may employ resources of other U.S. Government agencies or private firms. Whether supported or supporting, close coordination is the key to efficient and effective interagency operations. In the absence of an American Embassy, or during operations in austere environments or remote locations, the advising unit may be the only organization able to provide supplies, services (including force protection), and life support. More typically, during the early stages of the advising unit deployment, the unit may find the American Embassy to be the primary source of supplies, services, and life support.

6-40. Nongovernmental organizations do not operate in either the military or governmental hierarchy. Their relationship with an advisor organization is neither supported nor supporting. An associate or partnership relationship exists between military forces and engaged nongovernmental organizations, ideally focused on unity of effort. If formed, the focal point where U.S. forces provide coordinated support to nongovernmental organizations would be the civil-military operations center of a joint task force headquarters.

HOST-NATION SUPPORT

6-41. Host-nation support and local procurement may provide sustainment, operational support, and tactical support. Host-nation support agreements fulfilling the command requirements for support need to be prenegotiated. Such support arrangements must be integrated into the distribution plan and coordinated with joint, allied, and multinational partners to prevent competition for resources and ensure high priority requirements are met. Host-nation support may include functional or area support and may use host-nation facilities, government agencies, civilians, or military units. Preestablished arrangements for host-nation support can reduce the requirement for early deployment of U.S. assets and can offset requirements for early strategic lift by reducing requirements for moving resources to the theater of operations.

MULTINATIONAL SUPPORT

6-42. Multinational support may consist of support provided from one multinational partner to another. One or more of the following organizational and management options facilitates multinational support:

- National support elements provide national support.
- Individual acquisition and cross-servicing agreements provide limited support.
- A lead nation provides specific support to other contributing nation forces.
- A role-specialist nation provides a specific common supply item or service.
- A multinational integrated sustainment unit provides limited common supply and support.
- A multinational joint sustainment center manages common-user logistic support.

6-43. In all cases, the multinational force commander directs specific multinational support within the applicable laws and regulations of the host nation. When operating within a formal alliance, the advising unit executes support following applicable standardization agreements.

FACILITATING THE REDEPLOYMENT PROCESS

6-44. Sustainment assets facilitate the redeployment and the retrograde process through its command and control of the distribution system. Many of the procedures are used to deploy forces, draw pre-positioned stocks, conduct reception staging and onward movement, and distribute supplies within the theater of operations apply to the redeployment process. Two factors in particular complicate redeployment operations: supporting operations and meeting requirements.

6-45. The same elements that operate and manage the theater of operations distribution system during deployment and sustaining operations perform similar roles during redeployment. When redeployment,

deployment, and sustaining operations occur simultaneously, the theater sustainment command may find it necessary to rebalance its forces or change the missions of subordinate units to effectively support ongoing operations and redeployment.

6-46. Requirements vary widely depending on the nature and scale of redeployment operations, infrastructure, and other factors. For example, redeployment operations could range from personnel only to entire units and their equipment. Depending on the political and military strategy, unit rotations may occur while decisive operations continue unabated or during operational transitions. Key considerations include, but are not limited to, several characteristics. These characteristics include the size of the force redeploying and deploying; infrastructure requirements and limitations; security requirements; traffic circulation; staging areas; distribution system capacity; quantity of supplies and materials to be redistributed; the amount and kinds of materials that require disposal agricultural inspections; and establishing accountability of retrograde cargo. The challenge is effective coordination and synchronization, vertically and horizontally, to ensure responsive simultaneous support to not only ongoing distributed operations, but also redeployment.

Chapter 7
The Advisor

Advisors are the most prominent group of U.S. personnel that serve with foreign security forces (FSF). They live, work, and fight with their FSF. The relationship between advisors and FSF is vital. Advisors are not liaison officers, nor do they command FSF units.

ROLES OF THE ADVISOR

7-1. The military advisor has three roles involving different responsibilities. First and foremost, advisors are members of a U.S. military organization with a well-defined chain of command and familiar responsibilities. Second, advisors embed themselves with their counterparts. Third, advisors are interpreters and communicators between U.S. forces and their foreign counterparts.

7-2. As members of military organizations, advisors receive and execute the orders of superiors. These orders may conflict with the orders their counterparts receive. Among other duties, advisors must act unobtrusively, but nonetheless positively, as inspector general—often observing, evaluating, and reporting on the performance of counterparts and their assigned unit.

7-3. Secondly, advisors live, eat, and work with the officers and men of their host units. Often, advisors soon regard themselves as one of them. The sharing of common hardships and dangers forges potent emotional ties. The success and good name of their units become matters of prime and personal importance to the advisor.

7-4. Finally, advisors are interpreters and communicators between U.S. superiors and foreign counterparts. Advisors must introduce and explain one to the other; they help resolve the myriad of problems, misunderstandings, and suspicions which arise in any human organization, particularly when people of starkly different cultures approach difficult tasks together. Advisors with quick and easy access to influential counterparts can sometimes be the best possible means of communicating.

7-5. To be effective, advisors obviously must gain their counterparts' trust and confidence. This relationship, however, is only a prelude to the advisor's major objective: inspiring and influencing a counterpart to effective action. In pursuing this goal—constantly, relentlessly, and forcefully, yet patiently, persuasively, and diplomatically—advisors must recognize conditions which can benefit or handicap their cause.

CONSIDERATIONS OF THE ADVISOR

7-6. An advisor follows ten considerations when serving with FSF.

BY, WITH, AND THROUGH

7-7. All planned operations of a combat advisory mission must be conducted by, with, and through the FSF. Not counting immediate action battle drill responses, the mark of an effective advisory effort is the amount of personal involvement the FSF take in their own operations. Civilians must see that they are secured by their own security forces to promote the legitimacy of the host nation and their capacity and to build trust and confidence. In a word, combat advisors are shadows whose presence is felt at all times.

Empathy Leads to Competency

7-8. Empathy can be defined as identification with and understanding of another's situation, feelings, and motives. This is tough for experienced U.S. leaders and often harder to explain, but it is the key to the success of an advisory mission. Truly understanding other human beings and where they come from allows for honest relationships to develop. In most cultures, the place to begin understanding another person's feelings and experiences is by understanding the other person's narrative. The narrative is a collective group's identity as an interpretation of both ancient history and recent collective experiences. Delving into the narrative, understanding it and how it affects people, is the beginning of empathy.

Success Is Built on Personal Relationships

7-9. Developing a sense of empathy allows advisors to begin nurturing relationships with those they are tasked to advise. No amount of resources and firepower can compensate for the lack of a relationship between advisors and their counterparts. It must be honest, genuine, and heartfelt. Mutual respect, trust, and understanding create success. Both parties rely on each other for mission accomplishment and often for survival. This relationship is likely to be challenged on numerous occasions; only one built on mutual trust can survive and ensure mission success. Furthermore, advisors learn to pick their battles. Camaraderie and rapport with FSF can vanish in an instant if one does not prioritize and define critical lines in the relationship.

Advisors Are Between U.S. and Foreign Security Forces

7-10. Living and fighting with FSF allows for the development of true bonds, empathy, and trust. These desired results have a price to pay. When addressing their own Army, advisors soon realize that they are also not one of them, increasing their level of frustration. Advisors are often alone, navigating between two military systems and two cultures, never quite fitting in with either one.

Advisors Will Never Win, Nor Should They

7-11. Often advisors find it difficult to understand what victory looks like in this environment. Progress tends to occur at a glacial pace and cannot usually be tracked on a day-to-day basis. The advisor attaining a tactical objective does not achieve success; success is achieved by the foreign security force achieving the objective. Advisors find it difficult to satisfy their own units, and they never fully satisfy the demands of their FSF. They are figuratively and literally caught in the middle. Advising is the art of striving to make a win-win situation for all parties. The advisor and foreign security force relationship is one in which the advisor removes the obstacles in the way. Only a plan by FSF will succeed—and it will only succeed if it is, in fact, their plan.

Advisors Are Not Commanders

7-12. Advisors do not command FSF personnel or units. Advisors provide advice, training, and access to coalition resources to FSF. They are not intended to lead FSF in combat; they are responsible for the command and control of their own team of advisors. However, they can and should influence the commander of the foreign unit.

Advisors Are Honest Brokers

7-13. Advisors are advocates for FSF with conventional forces. They transparently assess capabilities and reveal limitations of foreign units to the FSF's and advisors' higher command. They also provide access to U.S. resources—from close air support to medical evacuation—to FSF they advise.

Advisors Live with Shades of Gray

7-14. Black and white is not available to the advisor. Caught between two cultures, systems, and narratives, the advisor works within a gray area. Unlike operations in a conventional unit, many of the checks and balances are not present. Advisors often find themselves isolated with great autonomy and no supervision. These conditions present moral and ethical dilemmas daily. Advisors need to be comfortable in this environment. Often they lose sleep at night and question their actions and those of their FSF. Thus, this mission calls for Soldiers of enormous character, moral courage, and intellect.

Talent Is Everything, But Advisors Must Understand Rank

7-15. Military forces around the world approach rank and prestige in completely different ways. In some cultures, rank equals experience; in others, it is nobility or education. Often many militaries have rewarded loyalty with promotions, which results in rank-heavy armies. Some FSF recognize talent, while others recognize rank. Advisors likely advise counterparts much more senior in rank. They must understand that rank on the uniform is important to many armies.

Make Do

7-16. Advisors will never have everything they believe they need to succeed. However, their creativity often distinguishes between success and failure. Austerity and economy of force can fail to meet the prevalent expectations that FSF have of a U.S. unit in regards to resources. It is up to the advisor team to close the expectation gap by its creativity and imagination. Scrounging, bartering, and negotiating are daily activities of advisors. An enormous amount of energy must be devoted to these activities to achieve mission success and endear them to their counterparts.

PERSONALITY TRAITS OF THE ADVISOR

7-17. Not every Soldier is well suited to perform advisory functions; even those considered to be the best and most experienced have failed at being an advisor. Effective advisors are only the most capable individuals. Advisors are Soldiers known to take the initiative and who set the standards for others; however, they are also patient and personable enough to work effectively with FSF. Recognizing that not all Soldiers are capable of performing as advisors, leaders should immediately remove advisors who do not exhibit these qualities.

7-18. Because advisors operate in very subjective environments, it is difficult to establish objective criteria by which to assess potential advisors. However, research and experience indicate that several personality traits greatly enhance the advisor's ability to adapt and thrive in a foreign culture. These traits include—

- Tolerance for ambiguity.
- Realistic when setting goals and tasks.
- Open-mindedness.
- Ability to withhold judgment.
- Empathy.
- Communicativeness.
- Flexibility.
- Curiosity.
- Warmth in human relations.
- Motivation of self and others.
- Self-reliance.
- Strong sense of self.
- Tolerance for differences.
- Perceptiveness.
- Ability to accept and learn from failure.
- Sense of humor.

7-19. Of the traits listed above, no single trait is paramount. Advisor selection and training programs seek to develop an understanding of the contributions of each of them.

SKILLS OF THE ADVISOR

7-20. Advisor-specific skills fall into two subcategories: enabling skills and developing skills. Enabling skills build on the individual and collective skills of advisors. Developing skills are the main advisor tasks of teaching, coaching, and advising. Advisors use these skills to develop the capabilities of the foreign security force.

ENABLING SKILLS

7-21. Advisors use enabling skills to work individually and in a group. Enabling skills help advisors communicate across cultures, build rapport, influence, and negotiate. Successfully employing these enabling skills sets the conditions for advisors to move forward with their mission.

Individual Skills

7-22. Individual skills help advisors properly perform their mission. Initially advisors should be selected based on subject matter expertise and not just rank. However, tactical proficiency does not necessarily equate to proficiency as an advisor. Modern-day military advisors must be able to teach and advise FSF. To do this, they must be knowledgeable and proficient in procedures and able to impart this knowledge effectively. They may also be able to advise FSF in the areas of intelligence, communications, operations, and logistics. Some advisors may be required to call in U.S. supporting arms, principally artillery and air support, and coordinate both air and ground casualty evacuation. Finally, advisors should be proficient in basic combat skills. These common skills, warrior skills, and survival skills are often required by advisors. Since these skills are critical to an advisor due to the isolated and independent nature of the mission, advisors should refresh them during predeployment training.

Collective Skills

7-23. As a group of advisors form a team, they must practice their warrior tasks, organize to cover the functions associated with any team, and familiarize its members with their duties and responsibilities. As with individual advisor skills, some are refined before deploying and others are learned in country. Figure 7-1 depicts this graphically.

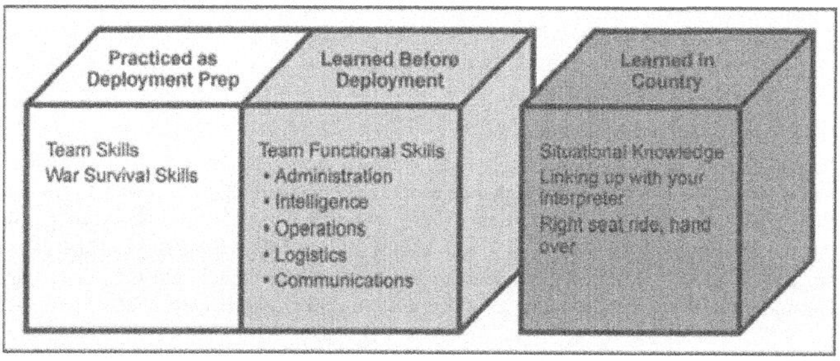

Figure 7-1. Collective advisor skills

DEVELOPING SKILLS

7-24. Developing skills—teaching, coaching, and advising—are required for every advisor. The advisor will make an initial assessment to determine the foreign unit's proficiency to determine the appropriate developing skill with which to start. Figure 7-2 illustrates the appropriate level of advice that corresponds to the foreign unit's capabilities.

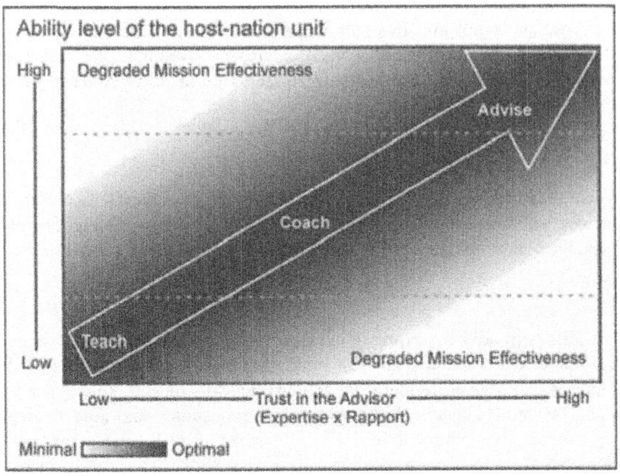

Figure 7-2. Appropriate advising level

7-25. Advisor teams have limited positional power and depend on personal influence to positively affect FSF efforts through teaching, coaching, and advising. Advisor team members often work with people of higher rank or grade than themselves. Advisors should remember the foreign unit's experience and capabilities and carefully choose opportunities to inject or impart knowledge. Foreign units are most receptive to advisor teams that teach unobtrusively. Foreign units most value those who are subtle in their teaching, coaching, and advising. Those advisors who master the ability to create a climate in which FSF personnel feel they are teaching themselves often prove the most effective.

Teaching

7-26. Teaching includes training and education. Methods of teaching can include classroom lectures, seminars, hands-on training, training exercises, and simulations.

Coaching

7-27. Coaching relies on guiding to bring out and enhance capabilities already present. Coaching refers to the function of helping someone through a set of tasks. Those being coached may or may not have appreciated their potential. The coach helps them understand their current level of performance and instructs them how to reach the next level of knowledge and skill. Coaching requires identifying short- and long-term goals and devising a plan to achieve those goals. The coach and the person being coached discuss strengths, weaknesses, and a course of action to sustain or improve the goals.

Advising

7-28. Advising is a combat multiplier that boosts supported unit capability. Advisors have experience in particular areas in which they are advising, but are not required to have similar backgrounds. Advisor relationships are not based on superior to subordinate relationships.

7-29. Advisors provide an expert opinion, advice, and counsel by focusing on both personal development (interpersonal and communication skills) and professional development (technical and tactical knowledge). Advising develops mutual trust and respect.

Advisor Versus Trainer

7-30. Advisors teach, coach, and advise work in a cycle. As FSF master one skill, the advisor can move on to other skills and initiate the process for the new skills. If FSF require additional teaching or coaching, the advisor can take them to that part of the cycle. Eventually, the teaching and coaching should decrease, and most of the advisor's time will be spent as an advisor providing an expert second opinion.

7-31. Ultimately, the question of what is the difference between being a trainer or an advisor will arise. Figure 7-3 shows that every advisor will have to be able to be a trainer, but all trainers are not expected to perform the more sophisticated task of advising. Hence, competent junior officers and enlisted Soldiers can be trainers, but only more experienced individuals can be advisors.

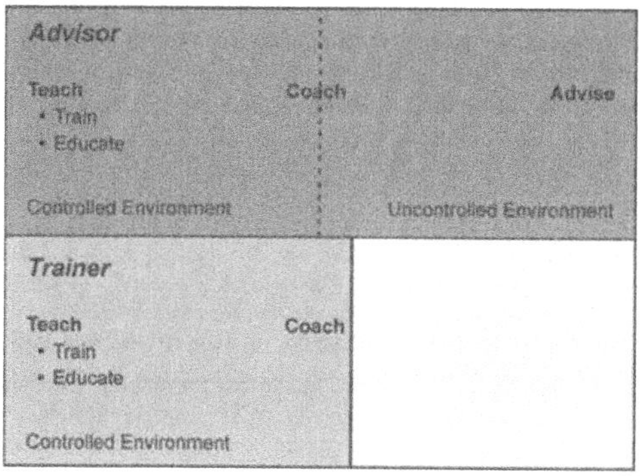

Figure 7-3. Advisor versus trainer

Chapter 8
Culture and Communication

As an advisor, understanding the host-nation population is a crucial element of premission planning and the development of foreign security forces (FSF). Prior knowledge of sociocultural differences aids in building effective relationships and prevents embarrassment, loss of rapport, and compromise of the mission. This chapter discusses three aspects of the sociocultural environment to provide a framework for analysis and comparison of foreign cultures for operational purposes. These aspects are society, culture, and communications.

SOCIETY

8-1. People who have a common culture, share a group identity, occupy a common territory, and are subject to the same political authority can be identified as a society. A society is not easily created or destroyed. Though they share the above factors, societies are not homogeneous nor static. A society usually has a dominant culture but can also have a vast number of secondary cultures, each influencing the other.

8-2. Advisors derive much of their effectiveness from their ability to understand and work with counterparts from a foreign society. Advisors must also consider societies outside their immediate operational environment whose actions, opinions, or political influence can affect the mission.

SOCIAL STRUCTURE

8-3. Society is composed of both social structure and culture. Social structure refers to the relations among groups of persons within a system of groups and is persistent over time. For the purposes of security force assistance, social structure includes groups, institutions, organizations, roles and statuses, and social norms. FM 3-24 contains a more in-depth discussion of social structure.

Groups

8-4. Within societies, groups are composed of two or more people that interact based on mutual interests and interrelated statuses and roles. These groups can be based on race, ethnicity, religion, or kinship. A racial group is defined by its members or by outside groups as different based on physical characteristics or descent. Race is a social category that has no biological basis. An ethnic group is a community whose learned cultural practices, language, history, ancestry, or religion distinguish them from others. Similar to racial groups, ethnic groups can be defined by their members or by outside groups. Religious groups may be subsets of larger ethnic groups and an ethnic group may contain members of different religions. Kinship groups are based on a link by blood relation. It may also be used to indicate a group of persons who live and work together to satisfy basic collective needs and goals. The role of a kinship group is generally seen to care for and educate offspring. This group provides for such needs as food, shelter, and clothing and prepares offspring to be functional members of the society in which they live.

Institutions

8-5. Institutions are the basic building blocks of societies. They are made up of groups that are organized to achieve a common task. Institutions are continuous through many generations and are built to continue to function even as individuals are replaced. An example of an institution from FM 3-24 is a military institution that brings together groups and individuals whose statuses and roles concern defense and security.

Organizations

8-6. Organizations are institutions that have a bounded membership with defined goals and established operations. They include fixed facilities from which they conduct business or operations and a set means of financial or logistic support. The four types of organizations are communicating, religious, economic, and social. Communicating organizations are designed to influence perceptions. Religious organizations are created to regulate societal norms, reaffirm beliefs and values, and provide social support. Economic organizations provide employment and assist in economic development. Social organizations provide support to groups in society, create social networks, and can influence ideologies. Organizations may belong to more than one category and exist to affect the behavior, values, and beliefs of society. As such, advisors must identify influential organizations both inside and outside of the FSF with which they are partnered.

Roles and Statuses

8-7. As members of a group, individuals interact based on assumed or given social positions. These social positions are referred to as statuses. Societies generally associate particular statuses with particular social groups (such as family, lineage, ethnicity, or religion). Each status inherently contains a set of expected behaviors known as a role, which includes how a person of that status should act, feel, and think. A status includes expectations about how others in a society should treat a person of that status.

Social Norms

8-8. Violation of a role derived from a given status results in social disapproval. The standard of conduct for social roles is a social norm. Norms may be either moral (incest prohibition, homicide prohibition) or customary (prayer before a meal, removing shoes before entering a house). Advisors must understand the roles, statuses, and social norms of groups they are advising. This knowledge will clarify the expected behavior of their FSF partners and their role as advisors.

Language

8-9. Language is a learned element of culture that is a system of symbols that people use to communicate with one another. Successful communication within a society requires an understanding of the social setting, appropriate behaviors towards different statuses and roles, and nonverbal cues. The languages used in the advisors' operational environment have a major impact on the advising mission. Languages must be identified to facilitate language training during predeployment and employment, as well as in the requisitioning of translators. Translators are critical for communicating with FSF and interacting with local citizens and leaders.

CULTURE

8-10. Social structure comprises the relationships among groups, institutions, and individuals in a society; in contrast, culture provides meaning to individuals within the society. Social structure can be thought of as a skeleton with culture being the muscle on the bones. The two are mutually dependent and reinforcing. A change in one results in a change in the other.

8-11. Culture influences an individual's range of action and ideas, including what to do and not do, how to do or not do it, and whom to do it with or not to do it with. It is neither a random jumble of customs nor a laundry list of accepted practices and taboos that an advisor should know. It influences judgments about what is right and wrong and what is important and unimportant. Cultural rules are flexible in practice; the customs and actions that an individual practices are based largely on that particular environment.

Cultural Characteristics

8-12. Culture is shaped by history, language, religion, and customs. Members of society use these shared beliefs, values, customs, and behaviors to cope with their world and with one another. It is learned behavior that is shared by members within a society. However, behaviors do not always define a culture.

Frequently, there are more telling cultural signs in the meaning attached to a behavior than in the actual behavior. Culture influences its members' identity, values, belief systems, and cultural forms. As such, it is arbitrary; advisors should make no assumptions regarding what a society considers right and wrong, good and bad. Culture is changeable through the social interactions between people and groups, both inside and outside of that society. TC 31-73 discusses cultural characteristics in great detail.

Identity

8-13. Each individual belongs to multiple groups through birth, assimilation, or achievement. Each group to which individuals belong influences their values and beliefs. Individuals consciously or unconsciously rank their identities into primary and secondary identities. Primary identities are those seen as core to their being; frequently these are national, racial, and religious. In contrast, secondary identities are more peripheral and may include such things as an individual's occupation or hobby.

Values

8-14. Values guide how individuals interact with the world around them. As deeply held beliefs about rightness, values shape how individuals think things should be. Values may never be articulated—in fact, they may not even be consciously held. At times, the beliefs of an individual may even contradict one another. Advisors should have some understanding of a counterpart's value system. Usually values are nonnegotiable. In working with a counterpart with conflicting values, advisors should persuade the individual that the advocated action or behavior really does not conflict with their values.

Belief Systems

8-15. Beliefs primarily take two forms: values (core) and opinions (peripheral). Values, or core beliefs, are not easily changed. Opinions, or peripheral beliefs, stem from thought and decisions. The totality of the identities, values, and beliefs that an individual holds compose that person's belief system.

Forms of Beliefs

8-16. Values are broad moral conclusions about the way to live life. These core beliefs are unstated, taken for granted, resistant to change, and not consciously considered. They reflect strong personal certainties drawn from faith, childhood environments, and life experiences. They lend shape and order to people's lives, helping them to decide if they settled on the right actions, friends, leaders, religions, and careers.

8-17. Opinions, or peripheral beliefs, are immediate conclusions about some aspect of the environment. They are usually quite specific and are open to debate, consciously considered, and easiest to change.

Types of Belief Systems

8-18. Belief systems involve cultural narratives and myths. The interpretation of these narratives and myths provide people with insight into how they should feel, think, and behave. The most prominent systems of beliefs tend to be those associated with religions and ideologies. However, any system of belief in which the interpretation of narratives impacts people's behavior (such as superstitions) can be a contributing component of a society's culture.

8-19. A belief system acts as a filter for new information. It is the lens through which people perceive the world. Beliefs of one group as to what is rational, normal, or true may appear strange, irrational, or illogical to outsiders. Understanding belief systems of various groups in an operational environment allows advisors to work with their foreign counterparts and influence civilians more effectively.

8-20. Belief systems play a fundamental role in the development of a society's values. Careful analysis of a region's belief systems is crucial to understanding the local value system. These systems influence religious doctrine, narratives, and myths and provide enormous insight into the culture of the community. Ultimately, the values which are most prized within a belief system are religious. These beliefs affect that culture's systems of social organization, learned behaviors, and relationship to the physical environment.

Religion as a Belief System

8-21. At its most basic level, religion is both the individual and communal expression of contact with the sacred. Notions of the sacred can vary widely in societies and may include deities and sacred items. In most cases, religions also involve symbols, rituals, and ethics or dogma. Religion can be both an individual practice and a group phenomenon.

8-22. Most societies have some variety of religion—typically a set of sacred beliefs and rituals that provide a common understanding of moral codes and proper conduct. Expressing a common religious understanding helps people to make sense of their world. Often religion provides explanations for human suffering, natural disasters, broken relationships, inequalities, and death. Some societies also have secular equivalents of religion. For instance, America is often remarked to have a civil or civic religion that can be seen in the images, symbols, ideas, and behavior of the political realm.

8-23. Because religion is such an integral part of culture, when advisors carefully prepare and analyze the mission, they must examine the religions and religious groups in the operational environment. Religious beliefs, leaders, and institutions are central to how many societies view the world. When planning any advising operation, planners consider how religion affects FSF and civilians.

Broad Impact of Belief Systems

8-24. In almost every society, followers of dominant religions have a strong societal impact. This influence may be exerted at varying degrees upon the individual citizen, sections of society, the economy, the military, and the political structure. TC 31-73 discusses these in detail.

The Importance of Understanding Culture

8-25. Advisors derive their effectiveness from their ability to understand and work with foreign counterparts from another culture. They must understand enough of their own culture and their counterpart's culture to accurately convey ideas, concepts, and purpose without causing counterproductive consequences.

8-26. Advisors must be aware of aspects of the local culture and history that influence behavior in their operational environment. Advisor team members must understand the reasons and motivations underlying personal interaction and must practice patience when working with their counterparts. Group norms guide individual behavior, and advisors must understand how individuals in a society tend to interact as members of a group, whether a race, ethnic, or kinship group.

8-27. Cultural understanding is not derived from demographic information provided to the military through country briefs prior to deployment. It is gained from studying, interacting, and understanding the people, religion, history, customs, and social and political structures within an area. For true understanding, it is necessary to live among the people, gradually understanding the subtleties and nuances of their culture.

Cultural Education

8-28. Culture influences a person's perception of the world. People from one society tend to view people from other societies through the prism of their own culture. Often, their perceptions are biased, are inaccurate, or involve assumptions of superiority. For example, Americans view their society and its culture very differently from the rest of the world. Likewise, as foreigners, advisors view local cultures through the prism of their own idealized culture. This creates a large gap between the culture the advisor observes and the host nation's real culture.

8-29. To narrow the gap of cultural understanding an advisor will encounter, a program of cultural learning should be undertaken. The end state of this cultural learning should be to gain the trust of foreign partners and civilians. Advisors will do so by demonstrating their understanding and sensitivity towards the host nation's culture and social structures.

8-30. An awareness of the cultural aspects of the operational environment can significantly enhance the effectiveness of an advisor. An extensive study program is best, but even a concise program is worthwhile if it provides advisors with the necessary cultural information. The best programs involve credible teachers who intimately understand the culture and have personal experience in the society. Training is dynamic and moves beyond the standard lecture and handout format. Small groups or a discussion panel work quite well. Cultural behavior and cultural familiarization handbooks can also be used. Since handbooks vary greatly in accuracy and quality, credible subject matter experts should be consulted to determine which documents to incorporate into the training. Table 8-1 discusses the training recommendations and tools.

Table 8-1. Cultural education recommendation

	Training Recommendations
Cultural Education	• Comparison of cultural values and social structures (United States compared to those of the operational area). • Local customs and traditions (for example, greetings do's and don'ts). • Geopolitical history (pre-colonial to contemporary and the orientation of each faction or party). • The role of religion in daily life.
	Training Tools
Resources	• Guest speakers native to the country of interest (for example, nongovernmental organization staff, foreign students, recent immigrants, or selected refugees). • Others who have worked in or studied the mission area (for example, special forces personnel, diplomats, and scholars). • Cultural familiarization handbooks.
Format	• Combination of briefings, small group discussions, and question and answer periods. • Handouts to augment—not replace—speakers. • Visual media, specifically slides and videos of the mission area.

8-31. Advisors with previous operational experience have addressed the importance of cultural education. Many civilian agencies have developed intercultural effectiveness programs designed for their personnel working overseas. Often these programs can be revised and tailored for military and paramilitary groups. If available, these programs are an excellent resource for predeployment preparation.

Cultural Gap

8-32. Each culture has its own societal rules regarding who a person may speak to, how and when the person may speak, and what topics the person may speak about. Many cultures rely on nonverbal signals to communicate. In such cultures, posture, expression, and actions convey more than spoken or written words. Advisors work to bridge the cultural gap by learning the language as well as nonverbal communication cues. To bridge the gap, advisors develop a sense of cultural awareness, recognize differences, learn to adapt, develop tolerance, and maintain personal contact.

Develop a Sense of Cultural Awareness

8-33. Advisors must recognize they are a product of their own culture. They must learn as much as possible about the culture of the people with whom they need to communicate. When communicating with people across cultures, advisors abandon any sentiments of ethnocentrism—the tendency of individuals to judge all other groups according to their own group's standards, behaviors, and customs. Such notions lead an individual to see other groups as inferior by comparison.

Recognize Differences

8-34. Each culture has its own way of accomplishing required daily tasks. Advisors must understand that each particular society may approach things differently; it does not mean that they are inefficient or less valid. Differences should not be seen as negative. Respect for counterparts must be maintained at all times.

Learn to Adapt

8-35. Advisors must be flexible and ready to adapt and adjust their behavior; however, they must be careful not to overdo their adjustment. Individuals who are overly flexible are often perceived as being insincere. The successful advisor must strive to act in a way that is appropriate to the target culture. Above all, advisors should be themselves and show sincerity.

Develop Tolerance

8-36. Advisors must develop a tolerance for deviations from accepted norms. Events or activities that may seem extraordinary to newcomers may be common practice in the culture. Advisors must be aware that members of the foreign culture may be astounded by that which is commonplace in the United States. Careful observation should be made before judgments are rendered about seemingly peculiar behavior.

Maintain Personal Contact

8-37. Personal contact is the most effective way to bridge organizational barriers. Interorganizational stumbling blocks are very real and the prejudices that arise from them are exacerbated by misunderstanding and ignorance. Often it is too easy to attribute negative attitudes and hostile motives to faceless groups. The advisor must employ superior interpersonal skills and work directly and closely with individual members of other organizations. This type of contact can effectively reinforce commonality and diminish the impact of disparity. These personal relationships are the key to effective interorganizational relationships.

Culture Shock and Adaptation

8-38. The term culture shock describes the anxiety experienced by persons in a completely new environment. It is not the result of a single event. Rather, the condition develops slowly from a series of minor events or conditions. Culture shock expresses the feelings of not having a set direction, not knowing what to do or how to do things in a new environment, and not knowing what is appropriate or inappropriate.

8-39. Culture shock occurs because the mind and body have to go through a period of psychological and physiological adjustment when individuals move from a familiar environment to an unfamiliar one. The cues received by all of the senses suddenly change. During the day the foreigner is bombarded with unfamiliar sights, sounds, smells, tastes, languages, gestures, rules, requirements, interactions, demands, systems, and expectations.

8-40. Some differences that advisors experience between their lives at their home station and their lives when deployed to a foreign location are obvious. These differences include language, climate, religion, food, educational system, and the absence of family and friends. Other differences may not be as obvious. These differences include how people make decisions, spend their leisure time, resolve conflicts, express their emotions, and use their hands, faces, and bodies to express meaning.

Stages of Shock

8-41. Five distinct stages occur in the culture shock process. These include enthusiasm, withdrawal, reemergence, achievement and reentry. Figure 8-1 depicts the stages of shock.

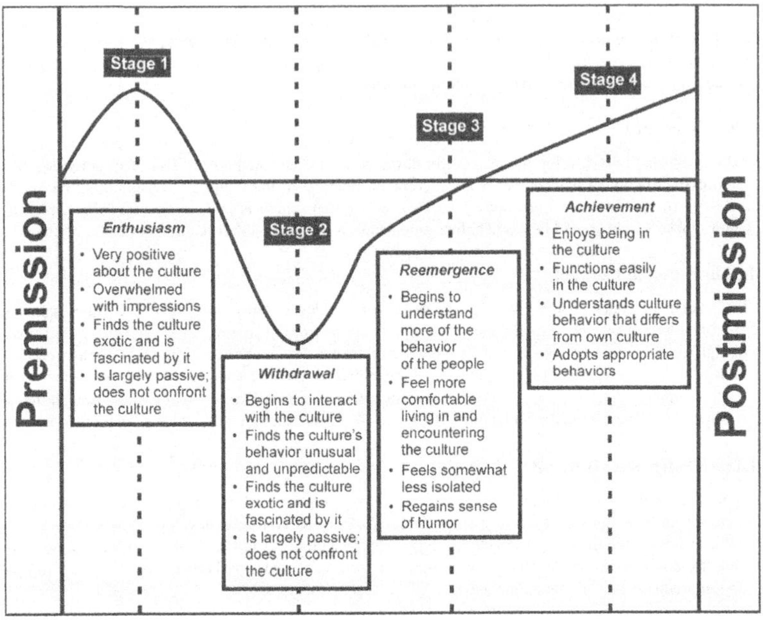

Figure 8-1. Stages of culture shock

Stage 1 – Enthusiasm

8-42. In this stage, a person may feel self-confident and pleasantly challenged. The person may be delighted by all of the new things encountered.

Stage 2 – Withdrawal

8-43. In this stage, a person begins to encounter difficulties and minor crises in daily life. It may be difficult to make oneself understood and foreign counterparts may prove more difficult to work with than anticipated. In this stage, one may have feelings of discontent, impatience, anger, sadness, and even incompetence.

Stage 3 – Reemergence

8-44. This stage is characterized by gaining some understanding of the new culture. A person may not feel as isolated, and a feeling of direction emerges. The individual is more familiar with the environment and is better able to adjust to the circumstances.

Stage 4 – Achievement

8-45. In this stage, the person realizes that the new culture has both good and bad things to offer. This stage is one of integration and the person is increasingly able to function in the new setting. A sense of accomplishment, a reduction of routine annoyances, and a more solid sense of belonging accompany this integration.

Stage 5 – Reentry

8-46. This stage occurs when people return to their own culture. They may find that things are not as they once were. Changes that occurred in their absence—their family, their friends, their communities, and themselves—combine to present a distorted image of home that differs greatly from the one imagined or remembered.

Culture Shock Symptoms

8-47. Being immersed in a society with extreme cultural differences tends to cause feelings of uncertainty and anxiety. The body and mind may react in unusual ways. Some persons may experience more pronounced physical symptoms of stress, such as chronic headaches or upset stomachs. Although uncomfortable, some degree of culture shock is a normal part of the adjustment process. Some common reactions include:

- Feeling irritable with (or even anger toward) one's own group or organization.
- Feeling isolated or alone.
- Tiring easily.
- Changing normal sleep patterns (either too much sleep or not enough).
- Suffering minor (but persistent) body pains, especially in the head, neck, back, and stomach.
- Experiencing feelings of hostility and contempt toward local people.
- Withdrawing from the local population (that is, spending excessive amounts of time alone reading or listening to music).

Overcoming Culture Shock

8-48. An important tool to overcome the obstacles of a new cultural environment is familiarity with the language. An ability to communicate in the new culture, even at the most basic level, reduces the effect and shortens the period needed for adjustment. Other means of overcoming culture shock can include having previous experience in the area (including a regular form of physical activity in personal routines) and acknowledging even slight progress in adjusting to the new culture.

8-49. TC 31-73 discusses culture shock in depth. It identifies the symptoms, cues, and treatment.

CROSS-CULTURAL COMMUNICATIONS

8-50. Communication is the transfer of messages from one person to another. These messages may be passed along verbally, in writing, or by signals (gestures, body language). The sender encodes the message and the receiver decodes it. The type and style of encoding used is based upon the sender's history, beliefs, values, opinions, and belief systems. Communication is determined by culture; as a result it can be either high context or low context. High context communication places an emphasis on clearly stating issues and problems and working towards resolving them in a straightforward manner. Low context communication places a premium on building relationships; as such, problem sets tend to be worked in a less formal matter in societies that utilize this type of communication. Quickly and effectively identifying whether a society uses high or low context communication is a key to effectively conveying material between the advisor and the host-nation population.

8-51. The receiver decodes messages based upon lifestyle, group membership, status and role, worldview, language, and social norms. Communication is a two-way process in which the encoding and decoding methods can affect both sending and receiving. Effective communication occurs when the message is perceived and responded to in the manner the sender intended. Ineffective communication can occur from poorly chosen words, flawed timing, a confused mixture of verbal and nonverbal signals, poor listening skills, and a failure to take culture into account. Communications can be broken into both levels and factors.

LEVELS OF COMMUNICATIONS

8-52. The cross-cultural communications capabilities required by an effective advisor to FSF can be described in three levels. At the lowest level is awareness followed by knowledge. The highest level is reached when these two are combined with well-trained and refined skills. TC 31-73 discusses communications in detail.

Awareness

8-53. Awareness is the basic level of cross-cultural capability. Awareness of cultural differences and their impact is the first prerequisite for successful work with a counterpart. Simply being sensitive to the fact that differences exist and carefully observing actions and reactions can assist the advisor in adjusting behavior and modifying actions to achieve greater influence with the counterpart. Awareness is not region specific and can be instilled in the advisor with relatively little training.

Knowledge

8-54. Knowledge of the details and nuances of a specific culture is the next level of cross-cultural capability. Advisors attain this level through academic study and immersion. Such knowledge is inherently area-specific and rarely transfers from one target area (or culture) to another. Developing in-depth regional knowledge necessary for effective cross-cultural communications requires extensive and time-consuming training. Appropriate personnel assignment policies must support this training.

Skills

8-55. Skills fundamental to effective cross-cultural communications, when combined with awareness and knowledge, form the highest level of cross-cultural capability. These skills include professional competence, language, nonverbal communication, negotiation, interpersonal skills, observation, problem solving, leadership, instruction, and fitness. Although some individuals show greater natural talent for these skills than others, all advisors require continual training in each to achieve and retain their full potential as advisors to foreign counterparts. These skills are discussed in more detail in TC 31-73.

FACTORS AFFECTING COMMUNICATION ACROSS CULTURES

8-56. There are four aspects that can impact communication across cultures. These factors are level of direct communication, perception of time, perception of the individual versus the group, and show of emotion.

Level of Direct Communication

8-57. Cultures fall along a spectrum of formality, with some cultures being direct and explicit in their communication and others less direct and more ambiguous. Cultures that rely on direct communication place more emphasis on verbal communications that are precise and explicit. Context and nonverbal cues are secondary to verbal communication. Cultures that rely on indirect communication rely less on verbal communication and more on context and nonverbal cues. Their verbal communication can seem vague and noncommittal to those from a culture that is more direct.

Perception of Time

8-58. Cultures can differ greatly in their perception of time. This difference divides into two cultural concepts of time: monochronic-time and polychronic-time. In monochronic-time cultures, members emphasize schedules, precise reckoning of time, and promptness. In such cultures, the schedule takes precedence over the interpersonal relation. In polychronic-time cultures, time is viewed as fluid. Members of polychronic-time societies do not observe strict schedules—agendas are subordinate to interpersonal relations.

Perception of the Individual Versus the Group

8-59. Cultures may be individualistic or collectivist in their orientation. An individualist culture is one in which the ties between individuals are loose—where people are expected to take care of themselves and their immediate families. In a collectivist culture, people are raised from birth into strong, cohesive groups. These groups offer a lifetime of protection in exchange for unquestioning loyalty. Advisors must understand from which culture their counterparts come to identify motivations accurately and communicate effectively.

Show of Emotion

8-60. Each culture has its own system of expressing emotion. Some cultures tend to express their emotions and show their feelings openly. Other cultures tend to repress those feelings more. Rather than showing their feelings openly, members of these cultures keep their emotions controlled and subdued. Those from the more expressive culture may view people from the repressive culture as cold or unfeeling. Similarly, those from the more repressive culture may view their more expressive colleagues as eccentric. To communicate effectively, advisors must be able to understand their foreign counterparts' emotional communication and display their own emotions carefully.

Chapter 9
Working with Counterparts

Rapport is a relationship marked by cooperation, conformity, harmony, or affinity. When people discuss good rapport, they describe a relationship founded on mutual trust, understanding, and respect. Relationships characterized by personal dislike, animosity, and other forms of friction often lack rapport. For the advisor, rapport describes the degree of effectiveness.

ESTABLISHMENT OF RAPPORT

9-1. Since advisors are in a unique military position, they establish rapport with their foreign counterparts. That position is one in which the advisor has no positional authority over the actions of their foreign counterparts. This lack of authority means that the doctrinal view of leadership is modified to emphasize interpersonal relationships and deemphasize authoritarian roles. Advisors instead use their people skills to influence the actions and decisions of their counterparts. Advisors use their people skills to building rapport. They have a genuine interest in other people, smile, remember and use people's names, encourage others to talk about themselves, listen to others, discuss what the other person is interested in, and make the other person feel important.

9-2. Advisors obtain certain knowledge before establishing effective rapport. First, they study FM 6-22 to gain basic leadership knowledge for understanding human nature and motivation. Advisors then incorporate information specific to the culture and society of their potential counterparts. This information may take the form of thorough area studies, operational area studies, and other research materials. Knowledge includes the components and techniques of rapport.

COMPONENTS OF RAPPORT

9-3. Three components compose rapport: understanding, respect, and trust. All the components of rapport are two-way streets; to have counterparts share about their culture, language, and experiences, advisors have to be willing and able to share also. Building this rapport may require the advisor to establish a personal level of understanding with their counterparts. Understanding is the first component of rapport. It begins before deployment and may include foreign cultural studies, language training, and equipment and doctrine familiarization. Once in country, advisors continue to broaden their understanding by observing and asking questions.

9-4. Respect is the next component of rapport. It is a reciprocal commodity. The foreign counterpart should grow to respect advisors, what they know, and how they perform. Advisors should look for those things that they respect in their counterparts. This may prove challenging. The counterpart may not fit the mold of U.S. or other coalition officers. Lacking formal training or education, the foreign counterpart may even be illiterate. The uniform standards of the host nation often may differ from the advisors' standards. Start with the basics and expand over time. Additional factors may affect the counterpart's willingness and motivation. For example, by accepting the duties, the counterparts (and their family) may be in mortal danger. The counterpart may have proven to be a fearless warrior, even without formal training. In short, traits deserving respect will exist even before the relationship matures. Mutual respect grows through shared experiences and shared dangers. Advisors should live, eat, and if authorized, fight with their counterparts. By sharing their hardships and dangers, advisors build respect.

9-5. Trust is the final critical component for building rapport. Trust grows gradually from understanding and respect. Building trust needs to begin on day one, but it will not mature until later in the relationship.

Advisors should begin by showing they are reliable and should do everything they say they will do. They should be in the right place at the right time. Two things can enhance this growth. First, the advisor should start out with confidence-building missions. Early success builds self-confidence and trust. Second, advisors should not promise any support they might not be able to deliver.

Techniques of Rapport

9-6. Foreign security forces (FSF) respect advisors who actually show them the right way to do a task rather than just talk. Experience has shown that foreign units will not hesitate to go into danger if their advisors are right there with them. To further contribute to building rapport, advisors should maintain enthusiasm and a positive attitude; learn the language or how to use an interpreter; respect a counterpart's rank, age, status, and experience; develop negotiation skills; and improve interpersonal skills.

9-7. Other rapport-building techniques can include being confident, competent, and capable. Advisors never expect or demand their counterparts to do something that they are unable or unwilling to do. They demonstrate everything. Advisors learn their counterparts' names and spend time with them at meals and during holidays; they prepare mentally to interact with counterparts at all times. Advisors avoid creating an American-only enclave; they avoid giving the impression of favoring any one group. They frequently assess their counterparts' perceptions. Advisors recognize threats to discipline and enforce the chain of command.

ACCOMPLISHMENT OF RAPPORT

9-8. Advisors measure rapport by how well they can motivate their counterparts to act. The basic techniques of motivation (in the absence of authority) are advising, setting the example, seeking compromise, and coercing. TC 31-73 discusses rapport in detail.

Advising

9-9. Advising the counterpart to select a particular course of action is only effective if the counterpart trusts the advisor is professionally competent. If the counterpart does not perceive the proposed solution as realistic, the advisor's competence is questioned. Advisors carefully explain to their counterparts that recommended courses of action are realistic and will be effective.

Setting the Example

9-10. Setting the example for the counterpart must be an ongoing effort for the advisor to avoid the appearance of a "do as I say, not as I do" attitude. In setting the example, advisors explain that what they are doing is the most effective action for the situation. This is particularly true when the counterpart really does not understand the behavior (or purpose).

Seeking Compromise

9-11. When seeking compromise with the foreign counterpart, advisors may create a situation in which the counterpart has a personal interest in successful execution. In some cultures, seeking a compromise may allow the counterpart to save face. Furthermore, in certain situations the counterpart—because of practical experience—may have a better solution to the problem at hand.

9-12. When seeking a compromise in certain cultures, the perceived competence of the advisors may suffer. This may be mitigated somewhat by approaching the compromise as two professionals (the advisor and the counterpart) reaching a mutual conclusion. To reach an effective compromise, advisors may have to conduct negotiations.

> ***Note***: Two areas of concern must never be compromised for the sake of maintaining rapport—operations security and human rights.

COERCING

9-13. Coercion is the least desirable method of motivation because it can cause irreparable damage to the relationship. Advisors should use coercion only in extreme circumstances. Advisors must avoid unintentionally forcing their counterparts into action. Instead, they establish and maintain good rapport by conveying they are sincerely interested in their counterparts, their nation, and its cause. Advisors will not belittle their counterparts' efforts. Advisors must demonstrate that they have come to help because they believe the counterparts' goals are just, fair, and deserving of success.

POLICIES AND AGREEMENTS

9-14. Advisors must remember their primary aim is to forward U.S. policy. Therefore, they establish relationships with foreign personnel, other U.S. forces, and other government agencies. Advisors promote U.S. policy through agreements, command relationships, and operations. TC 31-73 discusses policy in greater detail.

UNITED STATES AND HOST-NATION POLICIES

9-15. Advisors not only support U.S. national policy in the areas where they operate, they forward the policies of the host-nation government (unless otherwise directed by higher authority). Advisors maintain communications with higher headquarters to ensure activities align with U.S. objectives.

9-16. Host-nation policies may dictate practices and procedures that appear inefficient or uneconomical. Advisors should avoid criticizing or condemning such practices and procedures until they thoroughly understand them. Advisors should make private recommendations critical or contrary to host-nation policy.

AGREEMENTS

9-17. The status of advisors is normally specified by a status-of-forces agreement or other arrangement between the U.S. and host-nation governments. These agreements vary widely; they may offer little political protection or may provide for full diplomatic immunity. If no such agreement exists, advisors are subject to local laws and customs. Even if a status-of-forces agreement or other agreement exists that provides immunity, advisors are still expected to observe local laws.

COMMAND RELATIONSHIPS

9-18. Advisors must use the U.S. chain of command to obtain and disseminate guidance and assistance. They carefully distinguish between the U.S. and foreign security forces' chain of command. In particular, advisors prevent their counterparts from attempting to control subordinates through the U.S. chain of command. Advisors provide recommendations—not orders—to their counterparts. Only the counterpart should issue orders to subordinates.

9-19. Nongovernmental organizations and other government agencies may be co-located in the operational area. Advisors integrate the efforts of all organizations into their mission planning and should seek opportunities for integrated efforts with their counterparts. This type of unified action at the tactical level can be achieved through the planning process to integrate nonlethal actions. Holistically, nonlethal actions include development and diplomatic efforts.

CONDUCT OF OPERATIONS

9-20. Advisors can expect to participate in stability operations. These operations use military and paramilitary forces, as part of the interdepartmental team, to support internal defense and development operations. U.S. assistance may include advice on military organization, training, operations, doctrine, and materiel. In addition, the United States may provide and control U.S. logistics and sustainment for FSF. The objective of this assistance is to increase the capability of foreign organizations to perform their missions and operate efficiently in the given operational environment.

LEVELS OF ADVISORS

9-21. Organizations and individuals possessing greater skills and resources assist by imparting their knowledge through assistance efforts. The success of assistance depends largely on effective interaction between advisors and their foreign counterparts. Advisors may operate in the national, operational, and tactical levels. The national level is considered the largest national subdivision. This may be a nation's combined military headquarters, Service components, national-level government officials, or a combination. Operational-level advisors work at a level below the national-level advisor. This may be with military corps or division officers, with regional government officials, or a combination. Depending upon the size of the nation and its military, operational-level advisors may be assigned as low as brigades. Tactical-level advisors normally advise military units at or below the brigade level and provide counsel and assistance across the spectrum of conflict. They advise and assist counterparts in developing unit combat effectiveness and serve as liaisons between FSF and U.S. combat, support, sustainment and protection forces. Chapter 3 of TC 31-73 discusses these levels in detail.

CONSIDERATIONS OF RAPPORT

9-22. When building and maintaining rapport, advisors consider intelligence, human rights, role shock, ambiguity, personal relationships, bureaucracy, responsiveness, and diversity. These considerations are discussed in greater detail in TC 31-73.

INTELLIGENCE

9-23. To promote national policy and attain overall objectives, advisors may need to plan and conduct intelligence training. The advisor should assist the counterpart in developing a local intelligence collection program, training intelligence personnel in their respective specialties, and utilizing trained intelligence personnel properly.

9-24. The advisor should be familiar with area studies and area assessments. Pertinent documents should be compared to detect trends or changes. The advisor also should evaluate the foreign intelligence staff, its standing operating procedures, its chain of command, intelligence projects, and reference material available from other intelligence agencies. Advisors should prepare and maintain a list of essential elements of information.

HUMAN RIGHTS

9-25. Advisors stress the consequences of mistreating suspects, prisoners, or other persons taken into custody. These persons must be treated per Article 3 of the Third Geneva Convention. Article 3 requires care for the sick and wounded, protection of prisoners and detainees of all types from abuse or other harm. Murder, mutilation, and torture are expressly forbidden, as is humiliating or degrading treatment. Advisors must never be active participants in the conduct of such punishment. Furthermore, they must explain to their counterparts that they are obliged to report any atrocities of which they have knowledge.

ROLE SHOCK

9-26. Role shock results from the discrepancy between the roles individuals expect to play and the roles they actually do play. Role shock may also be a product of the tension created by individuals trying to do jobs themselves versus advising others how to do the jobs. Role shock tends to increase until about the anticipated midpoint of the tour and seldom disappears completely until redeployment. Symptoms of role shock resemble those produced by other stressful life events.

9-27. Many advisors become entangled in complexities of bureaucracy—both foreign and American—and become increasingly frustrated when seemingly familiar things fail to respond in expected ways. This continuing frustration can be a key contributor to role shock.

9-28. Although most advisors undergo some degree of role shock, over time they learn to take the personal, living, and social conditions in stride. More challenging are the problems that arise in connection with their jobs—their professional roles, relationships, personal achievement, self-development, self-determination, and self-image. Often advisors find their duties and activities during deployments differ somewhat from those expected. The actual duties and responsibilities are greater in scope, involve technical work outside of their assigned specialties, and require honed administrative (rather than military) skills.

AMBIGUITY

9-29. Most advisory assignments are ambiguous. What an advisor is expected to achieve is not always made clear. Furthermore, assignments may conflict with what they have trained to do, what others expect of them, or what they want or expect to do.

9-30. U.S. and foreign security forces may spend months—even years—negotiating a project. Still, the overall objectives or definition of the mission to be accomplished might be expressed in general terms. Various interpretations of what is to be accomplished arise. Even more views surface regarding the most appropriate means of reaching those goals.

RELATIONSHIPS

9-31. Ideally, foreign counterparts provide a working context for advisors and help them to understand and adjust to the new culture. Historically, those with good working relationships with their counterparts make better adjustments than those who have no counterparts or bad relations with their counterpart. Good working relationships may be difficult to establish and maintain. Most advisors approach their assignments as teachers and mentors. Foreign counterparts may resist such a relationship, particularly if they were expecting someone to take over and do all or part of the job. In addition, many counterparts want their advisors to share in at least some of the blame if something goes wrong.

9-32. The United States has little control over the availability or capability of counterparts. The assigned individual may be technically incompetent, lack professional commitment, have other interests, or be unavailable when needed. In some cases, numerous counterparts are assigned and rotate in their workings with the advisor. A single counterpart may not be assigned until the advisor has been on the scene for some time. In some scenarios, a specific counterpart is never assigned.

BUREAUCRACY

9-33. Although some Americans might disagree, the United States maintains a system that is—by world standards—relatively permissive and nonbureaucratic. As such, advisors may become aggravated when confronted with or obstructed by host-nation bureaucracy, red tape, administrative centralization, and a cumbersome decisionmaking process. Cultures that emphasize consensus in the decisionmaking process often prove quite frustrating for advisors. To be successful, advisors strive to understand the bureaucracy in which they operate.

9-34. Advisors should recognize the value of being or knowing the middleman. This role is sometimes one of the key contributions an outsider can make. Middlemen can move in, around, and out of organizations somewhat independently of the local hierarchy, protocol, and customs. They can link persons and offices that might otherwise have difficulty communicating. Many advisors dislike this role and refuse to practice it; however, the middleman can be a significant catalyst in any advisory operation.

RESPONSIVENESS

9-35. As frustrating as foreign bureaucracies may be, advisors often get even more irritated with the responsiveness of U.S. forces. Advisors frequently fail to account for distances, lead times, diplomatic issues, and other challenges when conducting security force assistance. Moreover, few Soldiers in the unit conducting security force assistance have had substantive experience in working in an interagency environment. These issues are further complicated by the usual tensions between headquarters and field staffs and between administrators and technical specialists. Expectations, values, and styles of behavior vary from group to group.

9-36. Deploying personnel may not be familiar with the tasks of development and the vast range of problems typical of underdeveloped areas. They come in eager to contribute their knowledge and skill, and anxious to make maximum progress in the limited span of their assignment. In so doing, they may fail to recognize that the factors that frustrate them—such as the lack of professional skills and facilities—are actually their reason for being there.

DIVERSITY

9-37. Organizations attract individuals who differ in abilities, knowledge, skills, work ethics, social behaviors, values, and attitudes. Advisors must be able to recognize personality types and develop techniques to cooperate effectively with each type.

RECOMMENDED PRACTICES IN BUILDING RAPPORT

9-38. Advisors may adopt many practices to help work with counterparts. Practices that help advisors build rapport are discussed in paragraphs 9-39 through 9-51.

ENSURE THAT THE ADVISOR PRESENCE IS UNDERSTOOD

9-39. Advisors entering an operational area work under the sponsorship of a military commander, mayor, village head, or some other recognized local leader. Prior to this, the advisor's arrival may need to be coordinated through the district, sector, or other appropriate local office. Clearances from distant national, state, province, or sector governments cannot compensate for a clear, local explanation of why an American advisor is in the area. This particularly applies to small, isolated communities where it is unusual for a stranger to appear for even an hour without being acknowledged and accepted by local leaders. Without explanations from locally respected persons, civilians often arrive at their own explanations, often to the detriment of the advisor effort.

FIND A BASIS FOR COMMON INTEREST WITH THE LOCAL PEOPLE

9-40. If advisors show appreciation for civilians as individuals, cultural gaps and language barriers can be overcome and common ground can usually be found. Advisors should listen and show interest when the locals speak. Initial conversations usually center on universal matters, such as food, shelter, clothing, health, and education. Over time, discussions may naturally move to the matter the advisor wants them to consider. Advisors are better received if they know something about earlier contributions in such matters as agriculture, folk art, religion, and architecture. Naturally, advisors prove more effective and appreciated if they can speak the language.

TRY TO UNDERSTAND WHY THINGS ARE DONE THE WAY THEY ARE

9-41. Although some local practices may seem strange, they generally have good reasons behind them. Advisors can discover these reasons with careful observation and a creative imagination. Food habits, family traditions, folk cures, and festive celebrations almost always have a great deal of human experience at their root. The advisor also needs to be aware that many villages contain rival subgroups and factions. This tension needs to be accounted for when working with the local people.

START WITH WHERE THE PEOPLE ARE AND WHAT THEY WANT

9-42. Advisors find out what the local people really want and work with them to achieve this aim. The need felt by the local people often is the best starting point. Once advisors address this aim, people become more appreciative and cooperative—they begin to raise their expectations and become interested in working for other improvements. To address the initial desires of the people, the advisor may need to call in personnel with the specialized skills needed for the particular project. Although this can result in a delay, it will help the advisor to achieve better cooperation.

WORK WITHIN THE LOCAL, CULTURAL FRAMEWORK

9-43. Advisors understand basic cultural matters such as ethnic backgrounds of the people, family relationships, leadership patterns, and value systems. They also need some knowledge of local services such as health, education, communications, and transportation.

HELP PEOPLE BELIEVE THEY CAN IMPROVE THEIR SITUATION

9-44. Most traditional peoples of South America, Asia, and Africa live in a largely static environment. Through their experience with change, they are more fearful of losing status than they are hopeful of bettering their condition. Therefore, changes suggested by advisors are often viewed with fear. Concrete local projects that yield easily observed benefits help convince villagers that they can improve their situation and make them more willing to cooperate in other projects.

BE CONTENT WITH SMALL BEGINNINGS

9-45. Change tends to come slowly in areas where there have been few in recent times. Advisors remember that knowledge—technical or otherwise—is cumulative. Once a small beginning has been made, greater activity and additional changes soon follow. It is easier to achieve momentum than to maintain it. Regardless, advisors start within as promising a framework as possible and with the support needed to sustain the momentum achieved.

UTILIZE LOCAL ORGANIZATIONS AND RECOGNIZE THEIR LEADERS

9-46. The surest way for an activity to be continued after the advisor leaves is for it to have been launched and carried forward within the local organizational and leadership framework. People respond best when their local organizations are recognized as important and useful. Advisors consult and encourage recognized local leaders to contribute as they can. A well-conceived technical activity reflects credit on the local leaders associated with it. Advisors give attention not only to officials and family heads of local groups, but also to the quiet, behind-the-scenes leaders.

HELP THE GOVERNMENT GET ORGANIZED TO SERVE THE PEOPLE

9-47. For advisors to be most effective, they must understand the organization of the local government and how their activities fit into the overall scheme. Usually a set of agreements exists between various local agencies and the national government (usually through some sort of interministerial council). This agreement provides for a coordinated effort in servicing the varied needs of civilians. The advisor may need to work with appropriate agencies to assist in getting such agreements made. If such agreements already exist, advisors should be careful to recognize and strengthen them.

TRAIN AND USE SUBPROFESSIONAL, MULTIPURPOSE LOCAL WORKERS

9-48. Selected young people in the villages can be trained and used as subprofessional, multipurpose village workers. Otherwise, advisors' influence is restricted to where they are standing and the immediate vicinity. Furthermore, advisors may spend so much time establishing and maintaining enough rapport with the villagers that they may become incapable of rendering any real service at all. Often the gap between civilians and the advisor is formidable because of their educational and cultural differences. Often the advisor works with villagers who are poor, illiterate, and devoid of outside contacts. Volunteer or paid local workers have proven quite helpful as they serve as a liaison between villagers and advisors.

EXPECT SLOW PROGRESS

9-49. As civilians begin to see successes from their joint efforts and begin to have new hope, they naturally want a larger hand in their own matters. The advisor may sometimes feel they want to assume more responsibility than they can carry. These evidences of growing pains should be appreciated, for they are a necessary part of becoming able to assume responsibility. The advisor needs to adjust to the growing desires of the people to help themselves.

TRANSFER CONTROLS CONSTRUCTIVELY

9-50. Advisors need to help civilians see how they can build the new—what they want—upon the foundations of the old—what they already have. From the beginning of a project, the advisor's team needs to talk with local leaders. Together they envision how to train local personnel, gather financial support, and transfer responsibilities before transferring control of activities. If operating responsibility is transferred too early, some breakage—usually of material things—will likely occur. If the team keeps control too long, the civilians who wanted to take over may grow disillusioned with them for not relinquishing control when they thought it should have been. Working with local leaders, the team determines when to phase out each technical activity.

EXPECT LITTLE GRATITUDE FROM THOSE HELPED

9-51. People who benefit from assistance sometimes feel defensive. In accepting assistance they are, in a sense, admitting their own insufficiency. The self-esteem of a person, community, or a nation as a whole is delicate. The advisor's team should not, therefore, expect thanks. Rather, the team should approach the people in a spirit of fraternity and humility, taking satisfaction in their forward progress. The team should do its job the best it can and accept work well done as is its own reward.

DYNAMICS OF ADVISING FOREIGN SECURITY FORCES

9-52. Advisors work in a complex environment that is heavily influenced by the unified effort of multiple organizations, relationships, compartmentalization, and organizational cultures. Paragraphs 9-53 through 9-57 discusses the probable dynamics that an advisor must take into account.

ADVISORS AND INTERDEPENDENT OPERATIONS

9-53. Advisors commonly operate in complex environments with interdependent operations. Interdependent operations require near-seamless integration of agencies representing the instruments of national power—diplomatic, informational, military, and economic. This same degree of integration must be extended to include external organizations and those agencies representing the instruments of power in partner states. The negotiation and cross-cultural skills that enhance their performance as advisors are also well suited to the complex interrelationships that characterize such operations.

ORGANIZATIONAL RELATIONSHIPS

9-54. Often organizational relationships can often be misleading and must be clarified. The actual interrelationships among and within organizations seldom follow a line-and-block diagram. Instead, they are heavily influenced by circumstances, personalities, perceptions, and resources. All relationships and lines of authority are subject to negotiation. Advisors understand that clarifying the roles, functions, and responsibilities is a continuous process and that each statement or action sets a precedent for future interrelationships.

9-55. The advisor must become hypersensitive to the most minor organizational disparities. The apparent similarities—the common language, sociological backgrounds, and citizenship—can be disarming and lead the advisor to overlook very real differences. If ignored, these differences can quickly impede synchronization.

COMPARTMENTALIZATION

9-56. Advisors balance security and synchronization. Overcompartmentalizing must be avoided. Often distrust can be prevented simply by not being overly secretive. Although advisors must conduct a careful, continual risk assessment to avoid compromise, the sharing of critical information is necessary for harmonious operations.

Organizational Culture

9-57. Advisors should approach external organizations as they would foreign counterparts—as unique institutions with distinctive cultures. Organizational differences, biases, and approaches must be researched and—whenever possible—accommodated. Each Service in the Department of Defense has unique perspectives that lead to significant differences in objectives and approaches to various problems. Profound differences in perspective may even be found between the different branches of the Army. If not recognized and accommodated, these characteristics may create distrust, disharmony, and disruption during critical operations.

Chapter 10

Cross-Cultural Influencing and Negotiating

Leveraging foreign counterpart cooperation can prove a difficult and critical task. Although coercion may prompt surrogate forces to perform a specific action, it can never lead to established conduct and legitimacy. Lasting advisors can achieve long-term success only by patiently applying their influence.

Advisors analyze their counterparts and determine their motivations and needs. Basic needs, culture, and individual personalities influence people. Advisors determine what these three things are so the appropriate influence techniques can be applied. Then advisors negotiate. Negotiation is fundamental to problem solving using the study, development, and honing of negotiation skills critical to the advisor.

INFLUENCING

10-1. Advisors need to gain influence with their counterparts, leaders, civilians, coalition members, and others beyond their chain of command. Advisors are self-aware, nonjudgmental, and capable. It is not necessary to accept others' beliefs or values, but advisors need to understand others' ways of thinking and operating. Knowing their own culture and understanding why they believe and value what they do helps advisors find a common cause. A common cause unites a diverse set of decisionmakers with whom advisors interact and try to influence.

10-2. Once advisors gain a minimum amount of influence, they can make things happen through others. They can extend their influence beyond their chain of command. Influence is simply getting others to do what you want.

10-3. Advisors identify the types and principles of influencing. They determine the best techniques with which to influence and review vital considerations. Using measures of effectiveness, they check that their influences have the desired results.

TYPES

10-4. Commitment, compliance, and resistance clarify how advisors understand the impact of their attempts to influence their counterparts. These three types of influencing are determined by how committed the foreign security forces (FSF) are to the task.

Commitment

10-5. Commitment is a willing dedication to a requirement or a cause because one wants to. Commitment-focused influence generally produces longer lasting and broader effects. Whereas compliance only changes immediate behavior, commitment reaches deeper—changing attitudes, beliefs, and behavior. For example, when advisors build desire and responsibility for a task among FSF, they often demonstrate more initiative, personal involvement, and creativity.

Compliance

10-6. Compliance is conforming to a requirement or demand because they feel like they have to avoid punishment or to gain a desired reward. Compliance-focused influence is based primarily on the leader's authority. Giving a direct order to a follower is one approach to obtain compliance during a task.

Compliance is appropriate for short-term, immediate requirements and for situations where little risk can be tolerated. Compliance techniques are also appropriate for leaders to use with others who are relatively unfamiliar with their tasks or unwilling or unable to commit fully to the request. If something needs to be done with little time for delay, and there is not a great need for the target of influence to understand why the request is made, then compliance is an acceptable approach. Compliance-focused influence is not particularly effective when a leader's greatest aim is to create initiative and high esteem with a counterpart.

Resistance

10-7. Resistance is a reaction against change. Resistance generally takes one of four forms—reactance, distrust, scrutiny, and inertia. It is manifested behaviorally (I won't do it), affectively (I don't like it), and cognitively (I don't believe it). Recognizing and lowering potential sources of resistance are key components to gaining compliance and commitment.

MOTIVATIONS

10-8. Different motivations influence behavior. One type of motivation is the desire for something. Another is the fear of something undesirable. As an agent of influence, advisors determine how to increase approach motivations and how to decrease avoidance motivations. For instance, recruiting for a special mission can use special badges as incentive (approach). Advisors can assure Soldiers that the mission is relatively safe by providing extra training and better equipment (lowering risks or avoidance).

PRINCIPLES

10-9. Six influencing principles are relevant to virtually any culture and any group. Understanding and applying these principles can increase an advisor's effectiveness in persuading a group or individual:

- Principle of scarcity.
- Principle of authority.
- Principle of social proof.
- Principle of liking.
- Principle of reciprocity.
- Principle of consistency.

10-10. With the principle of scarcity, people value more of what they can have less. They typically associate greater value with things that are rare, dwindling in availability, or difficult to acquire.

10-11. In the principle of authority, people are more easily persuaded by individuals perceived to be legitimate authorities or experts. They defer to experts who provide shortcuts to decisions requiring specialized information.

10-12. With the principle of social proof, people often look to the behavior of those around them for direction about what choices to make. This action is heightened when those around them are similar in terms of age, education, social standing, and experience.

10-13. In the principle of liking, people prefer to say yes and to comply with the requests of those they like.

10-14. In the principle of reciprocity, if someone grants favors, invited or uninvited, an overpowering need to repay that favor immediately blooms within the recipient of that favor. This human trait transcends all cultures and races.

10-15. With the principle of consistency, the desire for consistency motivates behavior. The drive to look and be consistent is a highly potent weapon of social influence, often causing people to act in ways clearly contrary to their own best interests. People do not like to appear inconsistent to others. It is human nature that people strive to feel good about themselves, which includes behaving per their important values and beliefs. When the behavior is consistent with whom people are and what they value, they feel good. People align with their clear commitments.

TECHNIQUES

10-16. The art of influencing involves knowing what techniques to use based on the situation. Advisors have to know the limits to which can use their influence; their influence is finite. They need to know when to use their influencing skills to achieve the highest payoffs. Their counterparts will only accept so much influencing if they don't completely trust their advisors. Table 10-1 lists positive and negative techniques delineated in FM 6-22. Table 10-2 lists resistance-based influence techniques. These are used when the counterpart resists the advisor's influences.

Table 10-1. Influencing and rapport techniques

Technique	*Description*	*Characteristics*	*Outcomes*
Pressure	Explicit demands, persistent reminders, sometimes appropriate in high stakes, urgent situations	Frequent checking Nagging	Gains compliance Frequent use may cause resentment, undermine morale, and harm relationship
Legitimate Request	Source of authority is used	Source credibility is important	Gains compliance
Exchange	Quid pro quo Reward offered for task completion; punishment threatened for incomplete task	Offer of reward or punishment not tied directly to getting job done, just something that counterpart wants or doesn't want	Gains compliance
Personal Appeal	Request based on friendship or loyalty	Agent of influence highlights talents of counterpart that make target the right person for the job (this is why you are the right person for the job, will you do it for me?)	Gains commitment Builds relationship
Collaboration	Works with target of influence by providing resources of some type, or by removing barriers to task completion	Offer of assistance is directly tied to request	Gains commitment Builds relationship
Rational Persuasion	Using facts, evidence, or logic in a persuasive message	Relies on agent being viewed as expert	Gains commitment Builds relationship
Apprising	Explaining the consequences of a potential action or decision	Decision is good for unit, career, or mission accomplishment	Gains commitment Builds relationship
Inspiration	Arousing strong emotion to build conviction or commitment	Appealing to values of the counterpart	Gains commitment Builds relationship
Participation	Requests for counterpart to be a part of planning, active participation in development of solution	Promotes commitment Possibly more feasible and culturally appropriate course of action	Gains commitment Builds relationship
Relationship Building	Build positive rapport and trust	Provide the foundation for using influence or inspiring	Gains commitment

Table 10-2. Resistance-based techniques

Technique	Description	Characteristics	Outcomes
Sidestep Resistance	Agent defines own role as consultant or advisor, but not commander Pushes the consequence of choice into future Depersonalize task to gain commitment to general idea Focus on end state instead of expended effort	Does not sell, just advises to prevent immediate resistance Puts less focus on immediate impact, more focus on long-term effect Stating "leaders should" instead of "you should" Stress mission accomplishment	Gains compliance Potentially gains commitment if paired with other influence techniques like rational persuasion
Address Resistance Directly	Address reluctance by lowering costs associated with tasks	Provide some of own resources through collaboration or counterargue concerns	Builds commitment
Address Resistance Indirectly	Build confidence, esteem, and self-efficacy for the task	Acknowledge difficulties while showing support for counterpart's abilities	Builds commitment
Distract Resistance	Interfere with the development of counterarguments	Distracting attention	Gains compliance
Disrupt Resistance	Disrupt complacency to bring attention to message	Shake up to motivate	Gains compliance
Consume Resistance	Resistance is energy	Wear them down prior to opportunities to resist. Persist with requests until they concede	Gains compliance
Use Resistance to Promote Change	Reverse psychology	Issuing a challenge	Gains compliance; risky when you don't know target of influence really well

10-17. To succeed with any influence technique, the influence attempt should be perceived as authentic and sincere. Positive influence comes from leaders who do what is right for Army forces, the mission, the team, and each individual Soldier. Negative influence—real and perceived—emanates from leaders who focus more on personal gain and lack self-awareness. False perception may trigger unintended side effects such as resentment of the leader and the deterioration of unit cohesion. With each and every influence attempt, Army leaders should ensure that the influence objective aligns with Army Values, ethics, the Uniform Code of Military Justice, the Warrior Ethos, and the Civilian Creed.

Considerations when Choosing Techniques for Influencing

10-18. Advisors clearly understand what their counterparts value and fear. This information allows advisors to quickly relate the objective of the influence to something that their counterparts treasure. Also, a strong sense of self supports the advisor's recognition of which techniques they can successfully execute. Someone with an intimidating presence may achieve compliance with pressure more quickly than someone without that presence. Understanding the context of the request is important. For instance, using participation in the midst of flying bullets would not make sense. However, a long-term project that impacts many people might be a worthy situation for participation. Advisors also consider—

- The motivation of their counterparts.
- The relationship they have with their counterparts.
- What they do or not have to offer.
- An understanding of the second- and third-order effects their recommendations will have.
- What has worked in the past.
- Possession of an appropriate power based for use of a technique in particular context.
- Consistency of technique with prevailing social norms and role expectations.
- Level of resistance encountered or anticipated.
- Costs of using a technique in relation to likely benefits.

10-19. The cost of a technique refers to the amount of effort, favor, or resources that accompany the technique. A personal appeal that relies on friendship is quite effective; however, counterparts might expect the favor to be returned. If resources or time are plentiful, then the exchange might not seem expensive, but if they are not plentiful, then advisors must choose a different technique. Advisors balance the costs and benefits of each technique.

Moderating Factors

10-20. Sometimes moderating factors diminish the impact of any influence technique. These factors may include cultural misunderstanding, level of trust, or constraints on the counterpart like threats or corruption. Advisors analyze and learn about these factors. Advisors must be flexible when using influencing techniques. When one technique doesn't work, switching strategies or adding techniques may be appropriate.

CONSIDERATIONS

10-21. Seven considerations need to be addressed to influence behavior. These include determining the goal; who to influence; motives; beliefs, values, and attitudes; cultural beliefs, values, and attitudes; and susceptibility. These techniques are designed to help advisors achieve their goals. Just knowing that the counterpart wants something is not enough. Whether it is time off, new weapons, or money, advisors have a limited supply of stuff they can promise; they can only bend so much on rules, regulations, and laws. It is at this point that the advisor enters the realm of negotiations.

Determine the Goal

10-22. Advisors influence others to achieve some purpose. To be successful at exerting influence, advisors have an end state or goal in mind. Sometimes the goal will be very specific. Many goals are less distinct and measurable but are still valid and meaningful.

Determine Who to Influence

10-23. Advisors develop a mental list of those whom they might need to influence. Getting a FSF platoon to take action may be as simple as influencing the commander, but the advisor may have to influence the soldiers and commander of the platoon. When practical, advisors consider many variations of whom to influence. They look for circumstances in the immediate environment that significantly affect their counterparts. Individual or group characteristics to consider can include—

- Gender.
- Religion.
- Age.
- Socioeconomic status.
- Ethnicity.
- Political affiliation.

- Level of education (important in determining how to access and persuade).
- Occupation.

10-24. The challenge is to determine which characteristics impact emotional or behavioral responses and under what circumstances. Ethnicity may significantly affect behavior and emotions regarding voting and politics but may have almost no effect on the decision to enlist in the military. Individual and group characteristics may provide additional insight on motivations.

Determine and Prioritze Motives

10-25. Motives are anything that is required or desired (needs and wants). Motives are the reasons behind the decisions people make. Motives come from an inner desire to meet a need or want. Advisors use the knowledge of what motivates others to influence them. Knowing one's counterpart and others who may be influenced gives advisors insight into influencing their behavior. The desires to fulfill, alleviate, or eliminate a need or want provides the motivation to change behavior.

Basic Motives

10-26. Basic motives involve physiological needs—food, water, and air—and safety needs—security and shelter. When people feel hunger, they are motivated to eat; when people feel pain, they are motivated to reduce the source of the pain. Such basic motives are extremely powerful in driving behavior and overwhelm psychological needs and wants.

Social Motives

10-27. Social motives associate access to other tangibles (money, goods, education, infrastructure, and healthcare) or more complex psychological motives with basic motives. For example, people learn to want money because it can be exchanged for food and other desired goods. Power, achievement, reassurance, escape, justice, and acceptance illustrate types of learned social motives. Examples include:

- Wanting better educational opportunities for their children.
- Wanting better paying jobs.
- Wanting their interests represented by the government.
- Wanting revenge for perceived wrongs.
- Wanting self rule.

Prioritize Motives

10-28. Advisors prioritize motives by immediacy of the need or want. Critical motives are immediate needs. Short-term motives are currently being satisfied or active efforts are being made to satisfy them in the near future. Long-term motives are desired but are not immediately important; satisfaction may be delayed until some point in the future.

Determine Beliefs, Values, and Attitudes

10-29. Although often difficult to derive, beliefs, values, and attitudes can prove quite effective in persuasion. Beliefs are what is thought to be true. People do not believe in something if they think it is not true. People vary from being gullible to skeptical. Values prioritize what one believes. Values can be ethical, doctrinal, social, aesthetic, or economic in nature. Attitudes reflect whether one likes or dislikes something. Likes and dislikes can often be traced back to a person's beliefs and values. This includes loves, hates, frustrations, and fears. Frustrations occur when a want is not met. By examining causes of frustration, the advisor may find another motive to address.

Compare Beliefs, Values, and Attitudes

10-30. The predominant cultural beliefs, values, and attitudes; traditions; and norms may differ greatly from the group the advisor is trying to influence. Many Americans act differently in different cultures. This change in behavior relates directly to their predominant culture.

Determine Susceptibility

10-31. Susceptibility is the likelihood that the individual or group will be open to persuasion. Determining susceptibility aids in prioritizing efforts. If susceptibility is moderate to high, then advisors may not need to focus as much effort as they would if the susceptibility is low. Advisors determine susceptibility by assessing the strength of the underlying source of the motivation and the perception of the desired corresponding behavior. The strength of the motivation depends on the perceived risk and the perceived reward, while the perception of the behavior directly links directly to how consistent the behavior is with the individual or groups' values and beliefs.

Perceived Risks and Rewards

10-32. Risks may include the threat of physical harm, financial loss, or disapproval of peers. Rewards may include financial gain, political power, approval of peers, or increased self-esteem. Risks and rewards can vary from immediate- to long-term. Immediate risks and rewards are stronger than long-term risks and rewards. For example, few people will walk a tightrope (immediate risk), but many people smoke (long-term risk). The higher the perceived rewards and the lower the perceived risks, the more likely the behavior will occur.

Consistency with Values and Beliefs

10-33. Advisors determine if the desired behavior is consistent with values and beliefs. Generally, behavior is consistent with the relative moral or ethical, religious, political, and cultural beliefs, values, and norms. Behavior that is inconsistent or incongruent with values and beliefs often decreases over time. Alternately, if behavior continues to be incongruent with values and beliefs, then individuals are likely to justify this behavior by modifying their values and beliefs. The more consistent a desired behavior change is with the individual's values and beliefs, the more likely the behavior will occur.

10-34. The lower the susceptibility, the less likely influencing behavior will be successful, which does not mean that the person cannot still be influenced. The lower the susceptibility, the more conditions will need to be modified for behavior change to occur.

APPLICATION

10-35. Influencing is a dynamic process. Advisors make a plan before attempting to influence. If one technique fails, they have to be able to shift to other techniques. Advisors do not need to sit down and write a plan before attempting to influence people. However, even with apparently positive results, advisors review their tactics and techniques.

CHECKS

10-36. Advisors check measures of effectiveness, unintended consequences, and second- and third-order effects. Advisors assess if their influencing is achieving their immediate measures of effectiveness. However, with every action some unintended consequences may occur or the activity may not prepare counterparts for more complicated tasks later on.

CROSS-CULTURAL NEGOTIATING

10-37. The principal form of negotiation is likely to be cross-cultural negotiation. Cross-cultural negotiation brings with it a series of unique challenges and techniques. Culture fundamentally affects language and behavior. It also significantly impacts the way people handle conflict.

10-38. Negotiations among members of the same culture can be stressful; negotiating with members of other cultures can be exceedingly difficult. Logically, working with other cultures is a basic skill for the Soldier and an absolute requirement when acting in an advisory role. Understanding the components, perspectives, roles, and outcomes will help minimize the difficulty of cross-cultural negotiation. Chapter 4 of TC 31-73 discusses cross-cultural negotiations.

INFLUENCES ON NEGOTIATIONS

10-39. Cultural differences may lead to a conflict between what the parties expect and what their families and communities expect. This has an obvious impact on negotiating behavior. Negotiations are influenced by perceptions, individualism, punctuality and pace, relationship building, and language.

Perceptions

10-40. Most Americans have certain preconceived ideas about people from other cultures. These perceptions may not be factually based, but they exist and influence negotiations. Similarly, foreign negotiators have certain perceptions about Americans. Again, these perceptions may not be based on fact, but they exist. Advisors need to understand how other cultures perceive Americans to appropriately adjust their negotiating style. Successful advisors find ways to capitalize on the positive preconceptions of their counterparts and find ways to neutralize the negative preconceptions.

Individualism

10-41. The United States encourages a largely individualistic culture. As such, the typical American negotiator prefers to enter into negotiations alone or, if required, as part of a very small negotiating team (two or three people, at most). This behavior is largely a by-product of U.S. culture, which focuses on the individual, performance, initiative, and accomplishments. Most other cultures of the world place less emphasis on the individual and more on the group.

Punctuality and Pace

10-42. Americans place a great emphasis on punctuality, viewing it as an indication of an individual's basic ability and commitment. However, other cultures emphasize punctuality differently. Some emphasize punctuality more than Americans. In other cultures, time is considered relative or may not be important at all.

Relationship Building

10-43. Americans are very competitive during negotiations and are inclined to stress short-term results. Interpersonal relationships are of minimal importance to American negotiators. Building long-term relationships only occurs after the successful completion of negotiations.

Language

10-44. U.S. negotiators must be considerate of others' with respect to language. Even if the counterpart is fluent in English, they may expect negotiators to speak their language or use an interpreter. Members of some cultures do not like to speak English as they are very proud of their language. When U.S. negotiator counterparts do not speak English, they will need an interpreter. Counterparts may also expect American negotiators to present detailed written proposals in the local language. Many cultures appreciate foreigners who take the trouble to learn a few simple courtesy words. However, some prefer that foreigners who are not fluent refrain from using more than a few, basic courtesy words.

ELEMENTS FOR CONDUCTING CROSS-CULTURAL NEGOTIATIONS

10-45. The basic elements common to all negotiations must be applied in ways that allow for the differing cultures of the participants. Elements that must be considered include opening strategies, directness, strengthening behaviors, movement, power, face-saving, formal agreements, and mediation.

COMPONENTS OF NEGOTIATING

10-46. Two opposite types of negotiating exist: distributive and integrative. They reflect a philosophy towards negotiating in general. Distributive negotiating is win-lose while integrative is win-win. Distributive negotiating involves dividing a fixed amount of a thing to the parties involved. This negotiation normally involves people who have never had a relationship before. Integrative negotiating involves joining two parties to achieve something together. Successful negotiations can involve a mix of both strategies.

10-47. When using distributive negotiating, advisors know the point at which counterparts will walk away. Advisors push them almost to that point by manipulation and withholding information. The distributive negotiator believes—

- The disputants are adversaries and the goal is victory.
- Fixed resources are divided so one side receives more than the other after making concessions.
- One person's interests or position opposes the other's interests or position.
- The dominant concern is usually maximizing one's own interests.

10-48. Integrative strategies used by advisors include cooperation, sharing information, and mutual problem solving. The integrative negotiator believes—

- The disputants are joint problem solvers seeking a wise decision.
- An understanding of needs will allow resources to be divided and both sides can win.
- The relationship between the parties is what is critical.
- The dominant concern is to maximize joint outcomes.

Perspectives Affecting Negotiations

10-49. Four perspectives impact negotiations: time, environment, self, and nonverbal communication. Advisors consider how each party views these conditions when negotiating.

Time

10-50. Some cultures and individuals see time as a micro view. It is all about here and now. Other cultures and individuals see time in a macro view. Now is a point that bridges the past with the future. For example, imagine a person with the first scheduled event has a deadline to make an agreement now. The next scheduled event is 1500 so that person has two hours to come to an agreement. The second person might agree that they can come to an agreement now; but to the second person, now could be within the next month or two.

Environment

10-51. Some cultures believe they can impact, if not control, their environment. If something should go wrong, somebody is to blame. Every accident is preventable. Other cultures believe the environment impacts the person. Accidents just happen and no one could have done anything about it. One side may be quite ready to commit to something in the future *knowing* that nothing stands in the way, while the other side may be reluctant to talk in certainties of the future *knowing* that factors beyond their control may prevent them from fulfilling their commitments.

Self

10-52. In some cultures, when a person makes a mistake, blame can be isolated to that person. In other cultures, the individual is seen as an element of the group. So a mistake is made not by the individual but rather the whole group. In negotiations, one might accept a personal slight to get the goal achieved, whereas the counterpart represents the group and cannot allow shame to fall on them even though one might accept it personally.

Nonverbal Communication

10-53. Since advisors will often have to work through interpreters, nonverbal communication becomes more important. Some cultures respect people who mean just what they say and maintain their composure. Other cultures respect the eloquence in which something is said or the emotion used to convey it. Where words may be translated improperly, the advisor or counterpart can hear tone and see emotion.

ROLES OF NEGOTIATION

10-54. When entering a negotiation, advisors must know what their role. The three major roles associated with negotiation are negotiator, mediator, and arbitrator. Each role has its own tactics or techniques and separate functions.

Negotiator

10-55. Advisors most often find themselves in the role of a negotiator. A negotiation is the process by which parties exchange commitments or promises to resolve disagreements and reach a settlement. Advisors prepare for each negotiation so they will not be at a disadvantage during the negotiation. To prepare for negotiations advisors examine their options, discover what motivates the parties, list all positive scenarios, and dispute resolutions.

10-56. Advisors examine their options. They question their goals, the importance of the goals, and the amount of leverage they have. They determine who is in power and what to do if no agreement is reached. Advisors uncover what motivates their counterparts. They delve into their counterparts' underlying needs.

10-57. By listing all the scenarios that satisfy their needs, advisors can maximize the gain for all parties. Advisors can add features to the scenarios that increase the opportunities for all parties to be satisfied. When advisors dispute resolutions, they often find some form of compromise. One of the best outcomes of compromise is when all parties leave the bargaining table satisfied but no single party feels like they have won or lost the argument. A skilled negotiator should strive to achieve a win-win result.

10-58. The negotiator represents one side and has a vested interest in the outcome of the negotiation. It is best to avoid situations where there is a conflict of interests. By blindly pushing the coalition agenda, the advisor could lose the trust of the counterpart. Considerations for a negotiator should include:

- If the foreign security forces ask the advisor to negotiate for them, they expect the advisor, as their negotiator, to have their best interests in mind. The coalition may try to use the advisor to push their agenda and influence rather than negotiate.
- If the foreign counterparts take illegal action, advisors cannot use commitment techniques. They need to use some form of compliant technique.
- If the foreign counterpart is not violating any laws, the advisor should focus on commitment techniques. Success will not be achieved all at once.

Mediator

10-59. Advisors may find themselves in the role of a mediator. The mediator has three general roles to perform: facilitator, formulator, and manipulator. Table 10-3 identifies the roles and tactics advisors use.

Table 10-3. Mediator roles and tactics

Roles	Tactics
Facilitator	• Make contact. • Arrange for interaction. • Clarify the situation. • Supply missing information. • Transmit messages between parties. • Ensure all interests are discussed.
Formulator	• Control the pace, formality, and physical environment of the meetings. • Highlight common interests. • Help parties save face. • Keep process focused on the issues. • Make substitutive proposals. • Suggest concessions parties could make.
Manipulator	• Keep parties at the table. • Change parties expectations. • Highlight costs of nonagreement. • Reward concessions. • Promise resources. • Threaten the withdrawal of resources. • Offer to verify compliance. • Threaten to withdraw mediation.

Arbitrator

10-60. The foreign security force may try to put the advisor in the position of an arbitrator. An arbitration is the process by which the parties to a dispute submit their differences to the judgment of an impartial person or group selected by mutual consent or statutory provision. They want the advisor to decide between two positions. This is a dangerous position for the advisor. When a win-win solution can be found, the advisor can improve rapport with both parties, but when it is a win-lose situation, the advisor could alienate one or both groups.

Expected Outcomes of Negotiation

10-61. The outcome of negotiations sometimes can determine the outcome. Advisors consider the desired outcome to determine which confrontation preference to use. Negotiation can focuses on physical outcomes of relationships. How advisors combine their preferences to their focus affects their perceptions of the outcome.

Confrontation Preferences

10-62. Negotiators will have one of three preferences for confrontation: avoidance, collaboration, or competition. Those that prefer avoidance will consider accommodation as a viable means to resolve an issue. Negotiators that prefer collaboration consider compromise as the preferred outcome. Those negotiators who use competition want to win. Intentionally losing in a negotiation to avoid confrontation seems counterintuitive; however, it tends to protect the current relationship. Collaboration requires a mutual respect of the existing relationship, while competition is generally indifferent. It would be fairly easy to take advantage of a negotiator who prefers to avoid confrontation. At the same time, expecting to be able to compromise with a highly competitive negotiator may become a point of frustration. For the competitor, winning may be more important than the deal itself.

Negotiation Focus

10-63. Defining whether the negotiation is a win-win situation is not as simple as which side gained the most from the negotiation. Advisors often focus on physical outcomes whereas diplomats focus on relationships. Every negotiation varies on which factor is more important. Just as car dealers may accept less profit on individual services, they gain overall if the customer returns.

Perception of the Outcome

10-64. Comparing confrontation preferences leads to nine potential combinations. If advisors can determine a participant's preference—avoidance, collaboration, and competition—and focus, they can determine how the participant might feel based on the outcome of the negotiation. If one participant is a collaborator and one is a competitor, then a win-win situation is possible, but neither side will be completely satisfied with the outcome. The competitors will not have gotten a clear win, and the collaborators will not have gotten what they considered fair. Subsequently, they will both have some minor negative feelings toward the other.

Appendix A

Army Special Operations Forces Imperatives

This appendix outlines the imperatives for Army special operations forces conducting security force assistance and foreign internal defense operations. Table A-1 lists these imperatives.

Table A-1. Army special operations forces imperatives

Imperatives	*General Descriptors*	*Specific Actions*
Understand the Operational Environment	• Knowing the U.S. command relationships (military and interagency) • Knowing the population (who, what, when, where, why, and how) • Knowing the enemy and other antigovernment forces (crime, militias, gangs) • Knowing the higher commander's intent • Knowing that the press is a part of the operational environment (news reports of events reach the U.S. population quickly)	• Define relationship as partnership or augmentation • Conduct analysis using ASCOPE (areas, structures, capabilities, organizations, people, events) • Analyze the enemy order of battle • Protect the population
Recognize Political Implications	• Identifying the political (and legal) restrictions and implications of actions by U.S. and foreign security forces • Identifying political implications of establishing and enforcing rules of engagement policy • Identifying short and long term political implications of security force assistance • Understanding the importance of establishing human rights (critical for tactical and strategic success) • Acknowledging the effects of public opinion on the mission locally, in the United States, and internationally • Understanding that insurgents conduct operations to have political effects among the local populace, external actors, and the U.S. and multinational civilian public	• Include rules of engagement in the planning and briefs of operations • Include local government leadership in operations • Reinforce human rights briefings and training to all foreign security forces • Prepare all U.S. and foreign security forces to talk with the press • Prepare information engagements for all operations • Anticipate and neutralize the methods enemy forces pass propaganda

Imperatives	*General Descriptors*	*Specific Actions*
Facilitate Interagency Activities	• Conducting security force assistance as both a military and an interagency effort; taking advantage of the greater impact of collective efforts • Coordinating the scope and limitations of each agency's influence and programs • (as opposed to a purely military program) • Obtaining the active cooperation of the foreign security forces • Including military and civilian agencies working together for unity of effort	• Include interagency members in planning and training as appropriate • Provide assistance to support limited agency resources, allowing agencies to prioritize their primary roles • Conduct regular coordination meetings with civil agencies to synchronize security efforts • Include host-nation agencies whenever including U.S. agencies
Engage the Threat Discriminately	• Avoiding the use of overwhelming firepower—to prevent alienating the population and turning them away from the host-nation government • Giving training and advice to foreign security forces based on the mission and threat • Targeting insurgents, criminal activities (lawlessness), or subversion—to avoid alienating the populace • Minimizing the use of force in populace and resources control measures—weighed for potential gain versus potential cost • Ensuring that rules of engagement and escalation of force are balanced between force protection and safeguarding civilians	• Take civilians into account when considering weapons systems and techniques for operations • Include civilians in risk assessment during mission planning • Train and rehearse rules of engagement and escalation of force to prepare Soldiers to react in accordance to policy • Publicize reasons for use of populace and resource control (for example, curfews protect civilians); avoid enemy propaganda or rumors dictating the psychological response of the populace • Update rules of engagement and escalation of force periodically • Rules of engagement or escalation of force reflects engaging targets that are visually verified and not simply suppressive fire in populated areas
Consider Long-Term Effects	• Building up the foreign security forces and alleviating the root causes of the current situation over time—not just treating the symptoms • Solving problems in a broader political, military, social and economic context • Making policies, plans, and operations consistent with national and theater priorities and the objectives they support	• Understand local grievances and underlying issues • Interact with all aspects of the population a government to identify root causes • Ensure the commander's intent remains focused on higher goals and policy related to addressing root causes, and protecting the local population • Monitor and update rules of engagement as local government and police gain control of rule of law. Conduct operation that adhere to local laws as possible • Understand legal and political constraints (rules of engagement) to avoid strategic failure while achieving tactical success

Imperatives	*General Descriptors*	*Specific Actions*
Ensure Legitimacy of Special Operations	• Reinforcing and enhancing the legitimacy and credibility of the foreign security forces and government • Gaining trust and credibility to ensure advisors and U.S. forces obtain legitimacy with the local population • Obtaining legitimacy in theater, in the United States, and internationally to ensure ongoing support	• Conduct joint and multinational patrols and operations. • Ensure information engagement is common in all operations and contributes to maintaining credibility • Engage the media with the host nation to inform as well as to justify actions • Focus on what the population considers the root causes, not the U.S. perspective • Develop long-term solutions to restore or improve basic services with host-nation leadership
Anticipate and Control Psychological Effects	• Considering the psychological effects of all operations (such as populace and resources control measures, combat operations, and civic action) • Avoiding negative psychological effects that can negate tactical victories • Appreciating the power of people's perceptions versus reality	• All operations are developed, analyzed and executed with information engagement • Continue to gauge the psychological effects with the local population, local government, and security forces as well as the enemy (propaganda and interrogation)
Apply Capabilities Indirectly	• Teaching host-nation forces to plan instead of giving them tasks • Supporting U.S. efforts enhancing the legitimacy and credibility of the host-nation government and foreign security forces • Balancing control and pressure on foreign security forces to ensure they conduct successful tactical operations and accomplish missions • Fulfilling primary role to advise, train, and help foreign security forces take the lead and stabilize the host nation • Influencing a foreign security force without commanding it; not accepting a lack of effort or a high level of corruption	• Conduct training of foreign security forces only to teach basics • Transition to mentor or advisor relationship during planning and operations to encourage the host-nation to take the lead • Use discretion when advising counterparts as not to embarrass foreign security forces leadership • Use host-nation methods whenever possible

Imperatives	*General Descriptors*	*Specific Actions*
Develop Multiple Options	• Considering the complex and irregular environment • Planning to use the full range of capabilities • Anticipating that an operational environment may dictate a change of rules of engagement or type of operations • Ensuring foreign security forces are capable of shifting from one option to another before and during mission execution	• Continue to refine and train on battle drills and contingencies • Develop an understanding of skills and abilities of the members of the unit and community • Elicit the input of foreign security forces on how to address situations to widen the options beyond conventional techniques • Include government agencies, special operations forces, nongovernmental organizations, local police, and the host-nation government to develop additional methods or options for solving problems
Ensure Long-Term Sustainment	• Avoiding advising and training foreign security forces in techniques and procedures beyond their capabilities to sustain • Modifying tactics, techniques, and procedures; training; operations; and sustainment to fit the culture, education, and technological capability of the foreign security forces • Developing host-nation sustainment before handing over economic, social, political, military, and security initiatives and projects • Avoiding host-nation dependency on U.S.-funded programs	• Understand and utilize the equipment and manpower present in the foreign security forces unit's inventory • Assess the capabilities of foreign security forces for leadership, training, sustainment (logistics, administration, and equipment), and ongoing development of a professional security force • Recognize the programs that are durable, consistent, and sustainable by the host nation • Do not begin programs beyond the economic or technological capacity of the host nation
Provide Sufficient Intelligence	• Ensuring detailed, near real-time, all-source tactical intelligence products support all activities and programs • Ensuring detailed and comprehensive intelligence on all aspects of the operational environment and its internal dynamics support operations • Ensuring human intelligence, urban recon, and internal security elements assess internal threats, warn the government, and take action to penetrate and neutralize instability • Considering aspects of the society not directly related to the tactical combat situation when assessing the enemy threat • Using information from intelligence assets to advise, train, and help foreign security forces counterparts and ease interagency efforts	• Ensure host nation forces are conducting operations using accurate, real-time intelligence • Establish priority of effort when identifying intelligence requirements • Assist foreign security forces to prioritize and focus counterintelligence on the commander's critical information requirement to receive timely and specific intelligence • Assess other organizations that impede the rule of law and government functions (such as gangs, criminals, smugglers, militias, mercenaries, drug lords, and narcoterrorists)

Imperatives	General Descriptors	Specific Actions
Balance Security and Synchronization	• Maintaining unity of effort and avoiding compartmentalizing key security forces from planning and intelligence contingencies even when security concerns exist • Avoiding either excessive or insufficient security • Anticipating reasons foreign security forces may not provide intelligence, such as lack of internal security or lack of U.S. intelligence support	• Encourage combined planning, preparation, and operations with foreign security forces' military, police, and governmental units • Resolve conflicting demands on mission planning and execution • Employ measures to mitigate risk of operations security instead of dismissing the host-nation government, police, or military needed for the overall effort • Strive to gain the trust of the local population to gain the most human intelligence and support from the foreign security forces

Appendix B
Legal Considerations

GENERAL LEGAL BASIS

B-1. Law and policy govern the actions of U.S. forces in all military operations, including security force assistance. For U.S. forces to conduct operations, a legal basis must exist. This legal basis profoundly influences many aspects of an operation. It affects the rules of engagement, how U.S. forces organize and train foreign forces, the authority to spend funds to benefit the host nation, and the authority of U.S. forces to detain and interrogate. Under the Constitution of the United States, the President is the Commander in Chief of U.S. forces. Therefore, orders issued by the President or the Secretary of Defense to a combatant commander provide the starting point in determining the legal basis. Laws are legislation passed by Congress and signed into law by the President and treaties to which the United States is party. Policies are executive orders, departmental directives and regulations, and other authoritative statements issued by government officials. This appendix summarizes some of the laws and policies that bear upon U.S. military operations in support of security force assistance. This summary does not replace a consultation with the unit's supporting staff judge advocate.

LEGAL AUTHORITY FOR SECURITY FORCE ASSISTANCE

B-2. If the Secretary of State requests and the Secretary of Defense approves, U.S. forces can participate in security force assistance. The request and approval can go through standing statutory authorities in Title 22, U.S. Code. Title 22 contains the Foreign Assistance Act of 1961, the Arms Export Control Act of 1976, and other laws. It authorizes security assistance, developmental assistance, security force assistance and other forms of bilateral aid.

B-3. The request and approval for security force assistance might also occur under various provisions in Title 10, U.S. Code. Title 10 authorizes certain types of military-to-military contacts, exchanges, exercises, and limited forms of humanitarian and civic assistance in coordination with the U.S. ambassador to the host nation. In such situations, U.S. forces may be granted status as administrative and technical personnel based on a status-of-forces agreement or an exchange of letters with the host nation. This cooperation and assistance is limited to liaison, contacts, training, equipping, and providing defense articles and services. It does not include direct involvement in operations. Assistance to police by U.S. forces is permitted but, generally, Department of Defense (DOD) does not serve as the lead government department. Without receiving a deployment or execution order from the President or Secretary of Defense, U.S. forces may be authorized to make only limited contributions during operations that involve security assistance.

EXISTING UNITED STATES LAW

B-4. The Constitution divides the power to wage war between the executive and legislative branches of government. Under Article I, Congress holds the power to declare war; to raise and support armies; to provide and maintain a navy; and to make all laws necessary and proper for carrying out those responsibilities. Balancing that legislative empowerment, Article II vests the executive power in the President and makes the President the Commander in Chief of the Armed Forces. This bifurcation of the war powers created an area in which the coordinate political branches of government exercise concurrent authority over decisions relating to the use of Armed Forces overseas as an instrument of U.S. foreign policy.

B-5. The Supremacy Clause of the Constitution (Article VI) states, in part, that all treaties made by the United States are the supreme law of the land. Therefore, ratified treaties, such as the United Nations (U.N.) Charter (see paragraphs B-7 through B-8) and the Geneva Conventions, create legal obligations on U.S. forces regarding their ability to perform various types of missions and functions.

B-6. The War Powers Resolution of 1973 ensures the collective judgment of both the executive and legislative branches to deploy U.S. forces by requiring consultation of and reports to Congress, in any of the following circumstances:

- Introduction of troops into actual hostilities.
- Introduction of troops, equipped for combat, into a foreign country.
- Greatly enlarging the number of troops, equipped for combat, in a foreign country.

B-7. The President is required to make such reports within 48 hours of the triggering event, detailing—

- The circumstances necessitating introduction or enlargement of troops.
- The constitutional or legislative authority upon which he bases his action.
- The estimated scope and duration of the deployment or combat action.

B-8. Since the War Powers Resolution was passed over the veto of President Nixon, no President has either conceded its constitutionality or complied fully with its mandates.

International Law and Treaties

B-9. The U.N. Charter became effective on 24 October 1945 after being ratified by the United States and a majority of other signatories. The U.N. Charter mandates that all member states resolve their international disputes peacefully. It requires that member states refrain in their international relations from the threat or use of force. The U.N. Charter also provides that all nations have the right to use self-defense to combat acts of aggression against them until such time as the Security Council takes action.

B-10. The U.N. Charter provides the essential framework of authority for the use of force, effectively defining the foundations for a modern *jus ad bellum.* Inherent in its principles are the requirements for necessity, proportionality, and timeliness. Necessity involves considering—

- The exhaustion or ineffectiveness of peaceful means of resolution.
- The nature of coercion applied by the aggressor State.
- The objectives of each party.
- The likelihood of effective community intervention.

B-11. Proportionality refers to limiting force in magnitude, scope, and duration to that which is reasonably necessary to counter a threat or attack. Timeliness refers to situations in which the delay of a response to an attack or the threat of attack attenuates the immediacy of the threat and the necessity to use force in self-defense.

B-12. U.S. forces obey the law of war. The law of war is a body of international treaties and customs, recognized by the United States as binding. It regulates the conduct of hostilities and protects noncombatants. The main law-of-war protections come from The Hague and Geneva Conventions. They apply at the tactical and operational levels and are summarized in ten rules shown in table B-1.

Table B-1. Summary of the law-of-war rules

• Fight only enemy combatants.
• Do not harm enemies who surrender; disarm them and turn them over to the chain of command.
• Do not kill or torture detainees.
• Collect and care for the wounded, whether friend or foe.
• Do not attack medical personnel, facilities, or equipment.
• Destroy no more than the mission requires.
• Treat all civilians humanely.
• Do not steal; respect private property and possessions.
• Do one's best to prevent violations of the law of war.
• Report all violations of the law of war to superiors.

B-13. During security force assistance operations, commanders must be aware of Common Article 3 of the Geneva Conventions and the status of insurgents under the laws of the host nation. One article contained in all four of the Geneva Conventions—Common Article 3—is specifically intended to apply to internal armed conflicts:

> *In the case of armed conflict not of an international character occurring in the territory of one of the High Contracting Parties, each Party to the conflict shall be bound to apply, as a minimum, the following provisions:*
>
> *(1) Persons taking no active part in the hostilities, including members of armed forces who have laid down their arms and those placed "hors de combat" by sickness, wounds, detention, or any other cause, shall in all circumstances be treated humanely, without any adverse distinction founded on race, colour, religion or faith, sex, birth or wealth, or any other similar criteria. To this end, the following acts are and shall remain prohibited at any time and in any place whatsoever with respect to the above-mentioned persons: (a) Violence to life and person, in particular murder of all kinds, mutilation, cruel treatment and torture; (b) Taking of hostages; (c) Outrages upon personal dignity, in particular humiliating and degrading treatment; (d) The passing of sentences and the carrying out of executions without previous judgment pronounced by a regularly constituted court, affording all the judicial guarantees which are recognized as indispensable by civilized peoples.*
>
> *(2) The wounded and sick shall be collected and cared for. An impartial humanitarian body, such as the International Committee of the Red Cross, may offer its services to the Parties to the conflict. The Parties to the conflict should further endeavor to bring into force, by means of special agreements, all or part of the other provisions of the present Convention. The application of the preceding provisions shall not affect the legal status of the Parties to the Conflict.*

HOST-NATION LAW AND STATUS-OF-FORCES AGREEMENTS

B-14. After considering the type of baseline protections represented by fundamental human rights law, the military leader must be advised in regard to the other bodies of law that leader should integrate into planning and execution. This includes consideration of host-nation law. Because of the nature of most noninternational armed conflict missions, commanders and staffs must understand the technical and pragmatic significance of host-nation law within the area of operations. Although in theory, understanding the application of host-nation law during military operations is perhaps the simplest component, in practice it is perhaps the most difficult.

B-15. Judge advocates must recognize the difference between understanding the technical applicability of host-nation law and that law's application to control the conduct of U.S. forces during operations. In short, the significance of host-nation law declines in proportion to the movement of the operation toward violent conflict. Judge advocates should understand that U.S. forces enter other nations with a legal status that exists anywhere along a notional legal spectrum.

B-16. Judge advocates and planners must be familiar with any status-of-forces agreements or status-of-mission agreements that may be applicable. In any given country, there may be agreements short of status-of-forces agreements, such as diplomatic notes, on point. Relevant international documents affecting military operations may be difficult to locate. Several sources are available for locating applicable international agreements governing the status of U.S. forces or affecting military operations. Department of State (DOS) publications, such as *Treaties in Force,* contain unclassified international agreements. Both the relevant geographic combatant command's legal office and the defense attaché or military assistance group at the embassy normally have access to host-nation or international agreements impacting a military operation.

B-17. Status-of-forces and other forms of agreements frequently exist. They are essentially contractual agreements or treaties between two or more nations that establish the legal status of military personnel in foreign countries. Topics that are usually covered in a status-of-forces agreement include criminal and civil jurisdiction, taxation, and claims for damages and injuries. In the absence of an agreement or some other arrangement with the host nation, DOD personnel in foreign countries may be subject to its laws. Commanders ensure that all personnel understand status of U.S. forces in the area of operations and are trained accordingly.

LEGAL CONSTRAINTS ON THE SECURITY FORCE ASSISTANCE MISSIONS

B-18. U.S. law and regulation play a key role in establishing the parameters by which military forces may conduct security force assistance missions. These parameters tend to constitute constraints on the activities of military units. They range from the rules of engagement in combat situations to the ability to spend government funds in furtherance of a training or support mission.

GENERAL PROHIBITION ON ASSISTANCE TO POLICE

B-19. Historically, DOD is not the lead government department for assisting foreign governments. DOS is the lead when U.S. forces provide security assistance—military training, equipment, and defense articles and services—to host-nation governments. The Foreign Assistance Act of 1961 specifically prohibits assistance to foreign police forces except within specific exceptions and under a Presidential directive. When providing assistance to training, DOS provides the lead role in police assistance through its Bureau of International Narcotics and Law Enforcement Affairs. The President, however, may delegate this role to other agencies. This was done in 2004, when the President signed a decision directive granting the commander, United States Central Command, authority to train and equip Iraqi police.

TRAINING AND EQUIPPING FOREIGN SECURITY FORCES

B-20. All training and equipping of foreign security forces are specifically authorized. U.S. laws require Congress to authorize expenditures for training and equipping foreign forces. The laws of the United States also require DOS to verify that the host nation receiving the assistance is not in violation of human rights. The Secretary of Defense may authorize deployed U.S. forces to train or advise host-nation security forces as part of the mission. In this case, DOD personnel, operations, and maintenance appropriations provide an incidental benefit to those security forces. Numerous other programs to assist foreign forces are paid for with funds appropriated by Congress for that purpose. Consultation with a staff judge advocate or legal advisor early in the planning ensures the availability of funds for missions to train and equip foreign forces.

RULES OF ENGAGEMENT

B-21. *Rules of engagement* are directives issued by competent military authority that delineate the circumstances and limitations under which United States forces will initiate and/or continue combat engagement with other forces encountered (JP 1-02). Often these directives are specific to the operation. If there are no operation-specific rules of engagement, U.S. forces apply standing rules of engagement. When operating with a multinational force, commanders must coordinate the rules of engagement thoroughly.

FISCAL LAW CONSIDERATIONS

B-22. In security force assistance missions, like all operations, commanders require specific authority to expend funds normally found in the annual DOD appropriations act. As a general rule, operation and maintenance funds may not be used for security force assistance missions. In recent years, Congress appropriated additional funds to commanders for conducting more complex stability operations that were not typically covered by operation and maintenance funds. Examples include the Commander's Emergency Response Program (CERP), the Iraq Relief and Reconstruction Fund, Iraq Freedom Fund, and Commander's Humanitarian Relief and Reconstruction.

B-23. DOS has the primary responsibility, authority, and funding to conduct foreign assistance on behalf of the U.S. government. Foreign assistance encompasses any and all assistance to a foreign nation, including security assistance (assistance to the internal police forces and military forces of the foreign nation), development assistance (assistance to the foreign government in projects that will assist the development of the foreign economy or their political institutions), and humanitarian assistance (direct assistance to the population of a foreign nation). The legal authority for DOS to conduct foreign assistance is found in the Foreign Assistance Act of 1961.

B-24. There are two exceptions to the general rule requiring the use of Title 22 funds for foreign assistance. The first exception is for interoperability, safety, and familiarization training. DOD may fund the training (as opposed to goods and services) of foreign militaries with operation and maintenance dollars only when the purpose of the training is to enhance interoperability, familiarization, and safety training. Operation and maintenance funds may not be used for security force assistance training. This exception applies only to interoperability training. The second exception is congressional appropriation or authorization to conduct security foreign assistance. This exception allows DOD to fund foreign assistance operations if Congress has provided a specific appropriation or authorization to execute the mission.

CONSTRAINTS ON FUNDING

B-25. The Foreign Operations Appropriation Act (an annual DOS appropriations act) was enacted in 1997. Known as the Leahy Amendment, it prohibits the U.S. government from providing funds to a unit of the security forces of a foreign country if DOS has credible evidence that the unit has committed gross violations of human rights. The Leahy Amendment contains additional constraints on government funding of security force assistance missions. The law restricts funding until the Secretary of State determines and reports that the government of such country is taking effective measures to bring the responsible members of the security forces unit to justice.

B-26. Congress specifically appropriates funds for foreign assistance. The United States Agency for International Development (USAID) expends such funds under the legal authorities in Title 22. Provisions of Title 10 may also authorize amounts of money for these purposes. Standing funding authorities are narrowly defined and generally require advance coordination within DOD and DOS.

B-27. Effective foreign forces need training and equipment. U.S. laws require Congress to authorize such expenditures. U.S. laws also require DOS to verify that the host nation receiving the assistance is not in violation of human rights.

FUNDING AUTHORIZATION FOR SECURITY FORCE ASSISTANCE

B-28. All training and equipping of foreign security forces must be specifically authorized. Military and civilian personnel, operations, and maintenance appropriations should typically provide only an incidental benefit to those security forces. All other weapons, training, equipment, logistic support, supplies, and services provided to foreign forces must be paid for with funds appropriated by Congress for that purpose. Examples include the Iraq Security Forces Fund and the Afghan Security Forces Fund. Moreover, the President must give specific authority to DOD for its role in training and equipping foreign security forces. In May of 2004, the President signed National Security Presidential Directive (NSPD) 36 that made the commander, U.S. Central Command, under policy guidance from the chief of mission, responsible for coordinating all U.S. government efforts to organize, train, and equip Iraqi Security Forces, including police. Absent such a directive, DOD lacks authority to take the lead in assisting a host nation to train and equip its security forces.

KEY SECURITY ASSISTANCE AND FOREIGN ASSISTANCE FUNDING PROGRAMS

B-29. Congress has appropriated funds to allow DOS to conduct its foreign assistance mission, including the following activities:

- Foreign Military Financing Program.
- International Military Education and Training Program.
- Economic Support Fund.
- Peacekeeping operations.
- Antiterrorism assistance.
- Global humanitarian demining.

- Refugee assistance.
- Personnel details.

B-30. DOS administers the following programs but no funds are appropriated to sustain them:

- Foreign Military Sales Program.
- Foreign Military Lease Program.
- Economy Act Security Assistance.
- U.S. Government Commodities and Services Program.
- Direct Commercial Sales Program.

B-31. Additional special support created by law for the foreign assistance mission includes—

- Excess defense articles.
- Presidential drawdowns.

B-32. DOS directly, or indirectly through USAID, finances numerous development assistance activities to address the following needs:

- Rule of law.
- Governance.
- Agriculture and nutrition.
- Population control.
- Health.
- Education.
- Energy.
- Environment improvement.

B-33. CERP (see paragraph B-19) is an example of a targeted humanitarian assistance fund program. CERP's primary purpose is to enable military commanders to respond to urgent humanitarian relief and reconstruction requirements within their fields of interest by carrying out programs that will immediately assist the host-nation populace. CERP was originally funded with seized Iraqi assets, but Congress later appropriated U.S. funds for the purpose. CERP is not applicable to missions outside Iraq and Afghanistan. Future missions, though, may have similar funding sources established to facilitate a humanitarian assistance mission. Commanders consult with the servicing judge advocate to determine the availability of funds.

Appendix C

Information Engagement

INFORMATION SUPERIORITY

C-1. Information shapes a commander's operational environment at every level. Twenty-first century conflicts occur in operational environments with instant communication. Because information systems are everywhere and exposure to news and opinion media is pervasive, an individual's words and actions can have immediate operational and strategic implications. Gaining and maintaining the trust of key publics has never been more important. Army forces conduct operations in the public eye and must maintain public confidence by being consistent in word and deed.

C-2. *Information superiority* is the operational advantage derived from the ability to collect, process, and disseminate an uninterrupted flow of information while exploiting or denying an adversary's ability to do the same (JP 3-13). Advisors build partner capacity in this mission area. Advisors ensure that host-nation counterparts integrate information superiority into operations as carefully as fires, maneuver, protection, and sustainment. Army information tasks contribute to information superiority. Information tasks are used to shape the operational environment (see FM 3-0). The information tasks are—

- Information engagement.
- Command and control warfare.
- Information protection.
- Operations security.
- Military deception.

C-3. Advisors prioritize the five information tasks based on the host nation's capabilities and strengths. This appendix discusses some information engagement capabilities that advisors can expect to employ while conducting security force assistance, as well as considerations when working with the media. For a further discussion of the other information tasks, see FM 3-0.

INFORMATION ENGAGEMENT CAPABILITIES

C-4. *Information engagement* is the integrated employment of public affairs to inform U.S. and friendly audiences; psychological operations, combat camera, and U.S. government strategic communications and defense support to public diplomacy, and other means necessary to influence foreign audiences; and leader and Soldier engagements to support both efforts (FM 3-0).

C-5. In general, advisors focus on information engagement because land operations occur among civilian populations. Advisors can have a significant impact by focusing on this task. Advisors work with host-nation counterparts to communicate information to the populace, build their trust and confidence in friendly forces, promote popular support for friendly operations, and positively influence public perceptions and behavior. Advisors integrate information engagement throughout the operations process.

C-6. Advisors ensure host-nation counterparts focus their information engagement activities on achieving desired effects locally. However, land operations always occur in a regional and even global context. Therefore, local information engagement activities must support and complement higher headquarters information engagement activities, host-nation and U.S. government strategic communications guidance (when available), and broader host-nation and U.S. government policy where applicable.

C-7. Three information engagement capabilities deserve particular attention as the advisor can directly influence them. These are public affairs, combat camera, and leader and Soldier engagements.

Public Affairs

C-8. First, public affairs can have a profound and lasting effect on public perceptions of operational success or failure. This effect may have strategic consequences, even when related to tactical operations. Therefore, advisors ensure that public affairs efforts are integrated at tactical, operational, and strategic levels. (See paragraphs C-10 through C-20 for more information to help advisors working directly with the media.)

Combat Camera

C-9. Second, combat camera is used to document a wide range of tactical and operational successes and to counter enemy propaganda. If there is no dedicated host-nation combat-camera capability, the advisor recommends that a host-nation soldier be designated, equipped, and trained for this function. If that is not possible, the advisor should obtain a digital camera and, if necessary, photograph relevant host-nation activities personally. Positive photographs can have a tremendous benefit, and digital photos are easily taken and transmitted.

Leader and Soldier Engagements

C-10. Third, leader and Soldier engagements are a fundamental and highly effective method to influence people. They involve face-to-face interaction by leaders and Soldiers to influence the perceptions of the local populace. Military leaders engage key communicators and civilian leaders (often referred to as key leaders) through carefully planned meetings. Examples of key leaders include local tribal elders, elected or appointed officials, military and police commanders, and nongovernmental organization leaders.

WORKING WITH THE MEDIA

C-11. The appropriate public affairs officer normally manages media contacts. However, this is not always possible. A deployed advisor team seldom has a public affairs officer representative.

C-12. Occasions may arise when the host-nation press wants to question advisors. Even if an advisor team lacks a public affairs officer representative, advisors must avoid refusing to speak to the media. Apparent refusal to speak with accredited members of the media, for whatever reason, may inadvertently create strong negative impressions with strategic implications. Therefore, advisor teams designate a media spokesperson and keep this individual well informed. This individual coordinates media contacts for the advisor team. This individual ensures that official, published public affairs officer guidance and command policies regarding media relations, including anonymity and operations security, are observed. The following paragraphs provide general media relations guidance for advisor teams.

Enhancing Relationships with the Media

C-13. Advisors strive to convey a positive impression and enhance relationships between the media and the military. If deemed acceptable within the public affairs guidance and command policies, the advisor team's designated media spokesperson—

- Maintains a list of reliable reporters and editors that cover operations in the team's operational area and keeps them informed of significant activities.
- Answers media inquiries promptly, accurately, and courteously.
- Finds out reporters' deadlines and uses them to the friendly force's advantage.
- Stresses the human aspects of a story that readers, viewers, or listeners can identify with, including the negative impact of opposing operations on the people.
- Points out how advisors and host-nation counterparts are working to address the needs of the unfortunate.
- Encourages the media to observe constructive work the host nation and advisors are doing and to interview them about their efforts.
- Identifies a location in advance for the press to take photos and videos.

Correcting Public Misconceptions

C-14. If information is withheld, journalists often fall back on speculation. Such speculation is usually inaccurate. However, it is often near enough to the truth that it is easily accepted by large sections of the public and possibly by established governments. In addition, partisan sources may find it advantageous to leak part of a story to the press to build up public support for their position. On occasion, such activities can grow into fully orchestrated press campaigns.

C-15. With modern satellites and communications technology, media are able to distribute reports and photographs extremely quickly. Incidents, sometimes fabricated or slanted toward a partisan viewpoint, may be aired in living rooms across the globe the same day.

C-16. However, advisors and host-nation counterparts frequently can correct public misconceptions and provide accuracy, adequate context, and proper characterization promptly. They can turn a potential liability into a major advantage by remaining prepared to coordinate and communicate with the media. They can take advantage of the media's powerful and far-reaching ability to communicate with critical audiences.

C-17. If the media report inaccurate or damaging information, advisors use caution in deciding on a course of action. Advisors (usually through the team's designated media spokesperson) inform their chain of command. Advisors consider the following questions:

- Is it important enough to correct or is it a detail that—in the long term—is not really important?
- How damaging is the charge, criticism, or error? Will a correction simply give greater visibility to an unfavorable point of view?
- Is a correction worth a restatement of the entire problem to new audiences who did not read or see it the first time?
- Is it possible to target only the audience originally exposed to the story?
- Can significant gain result from pointing out an error?

C-18. When incorrect or misleading information is released to the public through host-nation government or media sources, the advisor team determines a course of action appropriate to the situation. It must include a plan to ensure that relevant host-nation government officials and media recognize the inaccuracy of the information. This may prevent future incidents. Operational and combat experience since 2001 shows that in the struggle for information superiority, friendly forces must be candid and honest. Additionally, they must not allow lies disseminated by enemies to stand unchallenged, particularly in areas of great international interest.

Interviews With Advisor Team Members

C-19. Advisor teams sometimes arrange for media interviews with selected advisor team members. Properly conducted interviews can support the mission and even improve morale. When preparing for an interview, the team's designated spokesperson—

- Provides all team members with a single and simple theme to convey to the press, sometimes in the form of a mission narrative (a succinct account of the mission as a whole, explained consistently with the host nation's cultural view).
- Conducts rehearsals in front of video cameras, using mock interviews to practice communication of the theme.
- Selects interviewees based on their understanding of the theme covered, not their comfort level in front of a reporter. (Shyness is a normal trait that may add a human touch to a sensitive situation.)

C-20. If an advisor team member does not know the answer to a reporter's question, the individual should attempt to find out and get back to the reporter or refer the reporter to an appropriate source. This may establish the team as a helpful information source and develop a relationship for future balanced coverage.

C-21. Communicating with the media is the most difficult aspect of information engagement. Anything advisors say or do publicly can impact the mission. Therefore, advisors ensure consistency with national policy, professionalism, and clear communication. Advisors must know and follow the policies of the geographic combatant commander, the American Embassy, and the appropriate public affairs officer regarding media interviews. Additionally, advisors follow the specific guidelines shown in table C-1.

Table C-1. Specific guidelines for media interviews

- Maintain a friendly, yet professional attitude.
- Remain positive, greet interviewers, and welcome questions. (This is difficult under stressful or tragic conditions; however, a calm, mature appearance earns respect from audiences.)
- Use language that is clear and easy to understand.
- Avoid military jargon or terminology that others can misinterpret.
- Never enter an interview unprepared.
- Prepare notes and study them carefully; use facts to back up statements.
- Ensure the facts supporting their position are up to date and come to mind easily.
- Anticipate possible questions and think about various responses.
- Rehearse before interviews on location if possible and, if time permits, role-play tough questions.
- Learn and use reporters' names.
- Research interviewers' organizations, including—
    - The organizations' views or views they defend.
    - Their underlying motivations.
    - How members of the organizations previously conducted interviews.
- Keep the interests of the local nationals or other beneficiaries of the mission at the forefront.
- Avoid talking from the point of view of U.S. interests.
- Finish answering one question before answering another if two or more questions are asked at the same time.
- Avoid speculation and give only factual information that can be verified.
- Use caution in quoting statistics as such data may be easily disputed or reinterpreted.
- Avoid repeating a question that contains incorrect information or inflammatory language (this may result in a misquotation).
- Strive to tell the truth.
- Avoid saying *no comment*. (It gives the appearance the advisor has something to hide.)
- If unsure of something, admit it rather than mistakenly lie.
- If questioned about another agency's activities, refer to that agency for comment.
- Do not speak for other organizations.
- Avoid exaggeration or claims that cannot be backed up.
- Refuse all off-the-record discussions.
- Expect any statement they make to be quoted publicly.
- Avoid criticizing other groups or organizations, and avoid committing information fratricide.
- Remind reporters that photographing recognizable dead Soldiers, charts, maps, supply depots, or electronic warfare assets is off limits.
- Remind reporters that their personal security is not a primary military concern.
- Remain objective and calm.
- Demonstrate conduct befitting a representative the U.S. government
- Never give personal opinions and beliefs.
- Avoid reacting emotionally when interviewers appear skeptical or hostile.
- Maintain focus on the mission.

Glossary

SECTION I – ACRONYMS AND ABBREVIATIONS

ABCA	American, British, Canadian, Australian, and New Zealand Armies Program
AR	Army regulation
BCT	brigade combat team
CCIR	commander's critical information requirement
CERP	Commander's Emergency Response Program
COA	course of action
COIST	company intelligence support team
DA	Department of the Army
DOD	Department of Defense
DODD	Department of Defense directive
DOS	Department of State
DOTMLPF	doctrine, organization, training, materiel, leadership and education, personnel, facilities
FM	field manual
FSF	foreign security forces
IDAD	internal defense and development
ISR	intelligence, surveillance, reconnaissance
JP	joint publication
METT-TC	mission, enemy, terrain and weather, troops and support available, time available, civil considerations (mission variables)
MTT	military transition team
NSPD	national security Presidential directive
PIR	priority intelligence requirement
S-2	intelligence staff officer
S-3	operations staff officer
S-4	logistics staff officer
SFA	security force assistance
TC	training circular
UN	United Nations
US	United States
USAID	United States Agency for International Development

SECTION II – TERMS

foreign internal defense

(joint) Participation by civilian and military agencies of a government in any of the action programs taken by another government to free and protect its society from subversion, lawlessness, and insurgency. (JP 3-05)

foreign military sales

(joint) That portion of United States security assistance authorized by the Foreign Assistance Act of 1961, as amended, and the Arms Export Control Act of 1976, as amended. This assistance differs from the Military Assistance Program and the International Military Education and Training Program in that the recipient provides reimbursement for defense articles and services transferred. (JP 1-02)

***foreign security forces**

Forces, including but not limited to military, paramilitary, police, and intelligence forces; border police, coast guard, and customs officials; and prison guards and correctional personnel, that provide security for a host nation and its relevant population or support a regional security organization's mission.

host nation

(joint) A nation that receives the forces and/or supplies of allied nations, coalition partners, and/or NATO organizations to be located on, to operate in, or to transit through its territory. (JP 3-57)

humanitarian and civic assistance

(joint) Assistance to the local populace provided by predominantly US forces in conjunction with military operations and exercises. This assistance is specifically authorized by Title 10, United States Code, section 401, and funded under separate authorities. Assistance provided under these provisions is limited to (1) medical, dental, and veterinary care provided in rural areas of a country; (2) construction of rudimentary surface transportation systems; (3) well drilling and construction of basic sanitation facilities; and (4) rudimentary construction and repair of public facilities. Assistance must fulfill unit training requirements that incidentally create humanitarian benefit to the local populace. (JP 3-57)

internal defense and development

(joint) The full range of measures taken by a nation to promote its growth and to protect itself from subversion, lawlessness, and insurgency. It focuses on building viable institutions (political, economic, social, and military) that respond to the needs of society. (JP 3-07.1)

international military education and training

(joint) Formal or informal instruction provided to foreign military students, units, and forces on a nonreimbursable (grant) basis by offices or employees of the United States, contract technicians, and contractors. Instruction may include correspondence courses; technical, educational, or informational publications; and media of all kinds. (JP 1-02)

security assistance

(joint) Group of programs authorized by the Foreign Assistance Act of 1961, as amended, and the Arms Export Control Act of 1976, as amended, or other related statutes by which the United States provides defense articles, military training, and other defense-related services, by grant, loan, credit, or cash sales in furtherance of national policies and objectives. (JP 3-57)

security force assistance

The unified action to generate, employ, and sustain local, host-nation, or regional security forces in support of legitimate authority. (FM 3-07)

security sector reform

The set of policies, plans, programs, and activities that a government undertakes to improve the way it provides safety, security, and justice. (FM 3-07)

References

Field manuals and selected joint publications are listed by new number followed by old number.

REQUIRED PUBLICATIONS

These documents must be available to intended users of this publication.

FM 1-02 (101-5-1). *Operational Terms and Graphics.* 21 September 2004.

JP 1-02. *Department of Defense Dictionary of Military and Associated Terms.* 12 April 2001.

RELATED PUBLICATIONS

These documents contain relevant supplemental information.

JOINT AND DEPARTMENT OF DEFENSE PUBLICATIONS

Most joint publications are available online: <http://www.dtic.mil/doctrine/jpcapstonepubs.htm.>

DOD 5105.38-M. *Security Assistance Management Manual.* 3 October 2003. <http://www.dsca.osd.mil/samm/>

DODD 3000.05. *Military Support for Stability, Security, Transition, and Reconstruction (SSTR) Operations.* 28 November 2005. <http://www.dtic.mil/whs/directives/corres/html/300005.htm>

JP 3-0. *Joint Operations.* 17 September 2006.

JP 3-05. *Doctrine for Joint Special Operations.* 17 December 2003.

JP 3-07.1. *Joint Tactics, Techniques, and Procedures for Foreign Internal Defense (FID).* 30 April 2004.

JP 3-07.4. *Joint Counterdrug Operations.* 13 June 2007.

JP 3-57. *Civil-Military Operations.* 8 July 2008.

ARMY PUBLICATIONS

Most Army doctrinal publications are available online: <http://www.army.mil/usapa/doctrine/Active_FM.html>. Army regulations are produced only in electronic media. Most are available online: < http://www.army.mil/usapa/epubs/index.html>.

AR 11-31. *Army International Security Cooperation Policy.* 24 October 2007.

AR 12-1. *Security Assistance, International Logistics, Training, and Technical Assistance Support Policy and Responsibilities.* 24 January 2000.

FM 3-0. *Operations.* 27 February 2008.

FM 3-05.137. *Army Special Operations Forces Foreign Internal Defense.* 30 June 2008.

FM 3-05.202 (FM 31-20-3). *Special Forces Foreign Internal Defense Operations.* 2 February 2007.

FM 3-07. *Stability Operations.* 6 October 2008.

FM 3-24. *Counterinsurgency.* 15 December 2006.

FM 3-34.400 (FM 5-104). *General Engineering.* 9 December 2008.

FM 4-0 (100-10). *Combat Service Support.* 29 August 2003.

FM 4-02 (FM 8-10). *Force Health Protection in a Global Environment.* 13 February 2003.

FM 6-0. *Mission Command: Command and Control of Army Forces.* 11 August 2003.

FM 6-22 (22-100). *Army Leadership.* 12 October 2006.

FM 7-0. *Training for Full Spectrum Operations.* 12 December 2008.

FM 7-15. *The Army Universal Task List.* 27 February 2009.

TC 31-73. *Special Forces Advisor Guide.* 2 July 2008.

OTHER PUBLICATIONS

National Defense Strategy of 2008. Washington, DC: U.S. Government Printing Office, 2008. <www.defenselink.mil/news/2008 National Defense Strategy.pdf>

National Military Strategy of the United States of America. Washington, DC: U.S. Government Printing Office, 2004. <www.defenselink.mil/news/Mar2005/d20050318nms.pdf>

National Security Strategy of 2006. Washington, DC: U.S. Government Printing Office, 2006. <www.marforres.usmc.mil/docs/nss2006.pdf>

Quadrennial Defense Review of 2006. Washington, DC: U.S. Government Printing Office, 2006. <www.defenselink.mil/qdr/report/Report20060203.pdf>

Armed Forces. Title 10, U.S. Code. <http://www.loc.gov/law/help/guide/federal/uscode.php>

Arms Export Control Act of 1976. <http://www.loc.gov/law/help/guide.php>

The United States Constitution. <http://www.house.gov/house/Constitution/Constitution.html>

Foreign Relations and Intercourse. Title 22, U.S. Code. <http://www.loc.gov/law/help/guide/federal/uscode.php>

Foreign Operations Appropriation Act of 1997. <http://www.loc.gov/law/help/guide.php>

Foreign Assistance Act of 1961. <http://www.loc.gov/law/help/guide.php>

The Geneva Conventions of 1949. <http://www.loc.gov/rr/frd/Military_Law/MLR_site-map.html>

National Security Presidential Directive 36. "United States Government Operations in Iraq." 11 May 2004.

Treaties in Force 2007. <http://www.state.gov/s/l/treaty/treaties/2007>

Charter of the United Nations. <http://www.un.org/aboutun/charter/>

War Powers Act of 1973. <http://www.thecre.com/fedlaw/legal22/warpow.htm.>

OUTSIDE RECOMMENDED READING

von Steuben, Friedrick Wilhelm. *Regulations for the Order and Discipline of the Troops of the United States.* Philadelphia: Styner and Cist, 1779.

PRESCRIBED FORMS

None

REFERENCED FORMS

DA Form 2028. Recommended Changes to Publications and Blank Forms

Index

Entries are by paragraph number.

Entries are by paragraph number.

Entries are by paragraph number.

Entries are by paragraph number.

Made in the USA
Monee, IL
05 November 2026

67980412R00184